Science for a Better Tomorrow

Ulrich A. K. Betz
Editor

Science for a Better Tomorrow

Curious 2024 Insights

Springer

Editor
Ulrich A. K. Betz
Future Insight e.V.
Darmstadt, Germany

ISBN 978-3-031-93622-7 ISBN 978-3-031-93623-4 (eBook)
https://doi.org/10.1007/978-3-031-93623-4

This Springer imprint is published by the registered company Springer Nature Switzerland AG
The registered company address is: Gewerbestrasse 11, 6330 Cham, Switzerland

Foreword 1

Welcome Speech: Hosting Curious 2024

Dear Mrs. Garijo,

Dear Mr. Baillou,

Dear Mr. Betz,

Dear Mayor Haase,

Dear Ladies and Gentlemen,

It is a great pleasure for me to welcome you all to Rhineland-Palatinate today, here in the beautiful state capital of Mainz—whether you are here in person or connected digitally.

I am delighted that we, the state of Rhineland-Palatinate and the state capital of Mainz, are able to jointly host the Curious Future Insight Conference 2024. You are all shapers of a better future. Gathering this "spirit" here and spreading it to the world is a special honour and opportunity for Rhineland-Palatinate as a hub of research and innovation.

Above all, I would like to thank MERCK and Future Insight e.V. for creating this conference format and thus enabling a truly high-level exchange on the challenges of the future. A high-class conference format will be offered in the next 3 days, which has not yet taken place in this form in Rhineland-Palatinate. It offers our research organizations an excellent platform for new collaborations and networks, in accordance to the title of the conference: "UNITED BY SCIENCE For A BETTER TOMORROW".

MERCK and Future Insight e.V. have succeeded in attracting top international researchers and the most innovative minds as speakers. The focus is on the key challenges of the future—from biotechnology to transformation. They will generate important inspiration to enable a good future for people and society. The focus will be on how we can live and age in a healthy way, how we as a society can benefit from the opportunities of digitalization and how we can use artificial intelligence to our advantage. Through its format, the conference promotes top young talents and can serve as a magnet for attracting outstanding international minds from the science and technology pioneer sector to Rhineland-Palatinate.

The eyes of science, business and technology are on Mainz these days with the Curious Future Insight Conference 2024.

The world's attention was already on Mainz before: when the company BioNTech developed the world's first mRNA-based vaccine Comirnaty against the Sars-Covid-2 virus. Biotechnology from Rhineland-Palatinate has brought hope to the whole world amid the coronavirus pandemic.

At this point, I would also like to warmly welcome Prof. (Uğur) Şahin, CEO of BioNTech, who will give a keynote speech immediately after the welcoming remarks.

BioNTech's success story is an outstanding example of how excellent basic research is transferred into application. However, this success is also based on a long-term funding policy in the science sector. This forward-looking science policy has enabled us to create a strong research environment in the field of life sciences.

The state government has succeeded in taking advantage of the momentum of global visibility in the development of the coronavirus vaccine. With the state's biotechnology strategy, we have made this future technology a priority for the government. With success: Rhineland-Palatinate is on its way to becoming one of the world's leading locations in biotechnology and life sciences.

The great dynamics in the development of Rhineland-Palatinate as a biotechnology hub is also the result of intensive collaboration between science and industry. Just this spring, we were able to celebrate three groundbreaking ceremonies in Rhineland-Palatinate: With the construction of a new high-tech production facility in Alzey, the US pharmaceutical company Eli Lilly and Company is investing around 2.3 billion euros—1000 jobs will be created locally. The biopharmaceutical company AbbVie is investing 150 billion euros in a new research and development building in Ludwigshafen. A new laboratory building for the TRON research institute will be built in Mainz, offering space for 400 employees. This demonstrates the enormous dynamism and attractiveness of Rhineland-Palatinate as an industrial location. Companies in Rhineland-Palatinate find good economic conditions and close links to science.

The Curious Future Insight Conference 2024 is taking place in a dynamic environment in Rhineland-Palatinate and will certainly provide new impetus. Where so many brilliant and innovative minds come together, only great things can emerge. We can shape the major challenges of the future only together with you.

I am very excited about everything that will be discussed and initiated here in the coming days. And yesterday's Science Slam, which the Leibniz Institute for Resilience Research gGmbH (LIR) organized together with Biomindz and Rhineland-Palatinate, showed just how much fun this can be.

Finally, I would like to highlight one issue in particular.

The appeal of STEM subjects is also strengthened by the conference: a competition, the Junior Visions Prize, was launched as part of the conference, in which pupils from over 49 schools of Rhineland-Palatinate took part. We can see how the children see our future and what innovative ideas they have in the exhibition hall, which I am extremely pleased about. I would particularly like to take the opportunity to thank all participating children and also our state coordinator for biotechnology,

Prof. Dr. Eckhard Thines, who provided a wonderful scientific commentary on these pictures.

Our children will be our tomorrow's researchers.

Thank you very much.

Land Rheinland-Pfalz
Minister für Wissenschaft und Gesundheit des
Landes Rheinland-Pfalz
Mainz, Germany

Clemens Hoch

Foreword 2

In Mainz, Future Has Tradition

Mainz is evolving into a leading centre for life sciences and biotechnology. Large-scale construction and capital investment projects are underway to establish lasting infrastructure that will provide the best possible conditions for academic research and business. This will enhance the city's future prospects and support its growth into an international hotspot for innovative technologies and developments in the life sciences and biotech industry. The Curious Future Insight Conference 2024 served as an ideal forum for the city to present itself and its open-minded, cosmopolitan values, while placing it at the centre of the global community for innovation and cutting-edge technologies.

The origins of Mainz stretch back 2000 years and it has given rise to two revolutionary advances in its history that have changed the world for the better. The first of these was the invention, by Johannes Gutenberg, of the movable-type printing press that revolutionised the way people across the world could access knowledge. More recently, the Mainz-based company BioNTech successfully developed an mRNA vaccine to combat the COVID-19 pandemic.

These recent achievements, in particular, have put Mainz even more firmly in the global spotlight as a hub of scientific research and biotechnology. The city is systematically capitalising on this momentum and successfully remaking itself as a vibrant international hotspot for innovative technologies and developments in life sciences and biotechnology.

We are striving hard to optimise the conditions necessary for this to happen. But the real momentum is coming from the brilliant minds in our city and region, many of whom were present at Curious 2024. Their research, their bold entrepreneurship and their commitment are all helping to propel Mainz into a beacon of innovation.

That is because, as a city of science and academia, and as a launchpad for revolutionary technologies, all of us are pursuing the same ambitious vision and strategy: To significantly expand life sciences and biotechnology over the next few years and transform Mainz into a thriving hub with a formidable international profile.

Spurred on by these developments, our city has been named the most dynamic city in Germany for the third time in a row. We are working tirelessly to ensure that, in the years to come, Mainz is firmly etched in the minds of all relevant players in the global life sciences and biotech industry. To put it simply: Mainz is the place to B—the place to Biotech.

Construction and Capital Investment to Support Successful Research, Start-Ups and Growth

That this momentum continues unabated demonstrates that our strategy is paying off. At the moment, for example, there are substantial investments in the construction of new, innovative buildings in many areas of the city with state-of-the-art laboratory and office space.

The focus is on the development of the approximately 12-hectare Life Science Campus located directly adjacent to Johannes Gutenberg University and its associated University Medical Centre as well as Mainz University of Applied Sciences. The immediate neighbourhood is home to academic and research institutions with top-level core facilities.

One of the first projects that is already nearing completion is LAB 1 at Innovationspark Mainz GmbH & Co. KG. It provides around 3000 sqm of modern laboratory and office space for companies and start-ups in the sector. The first tenants, Lifecare and LigniLabs, have already been lined up to move in and will be able to use this space to forge ahead with developing their products and innovative technologies. At the same time, preparations for LAB 2, which will offer a further 9000 sqm of laboratory and office space, are in full swing and planning permission has already been granted.

At the start of 2024, Kadans Science Partner also announced their intention to make a substantial investment in Life Science Campus Mainz. The Dutch company, which specialises in start-up ecosystems, plans to complete a state-of-the-art building with around 19,500 sqm of office and laboratory space by the end of 2027.

This building is also expected to accommodate the Life Science Centre Mainz (LZM), which is jointly funded by the city of Mainz and the State of Rhineland-Palatinate and offers incubator services for start-ups in the fields of biotechnology and life sciences. The goal of this collaboration is to attract additional biotech companies to Mainz, to support new start-ups and to forge contacts between them and the local academic and scientific communities.

These and other projects are part of a scheme currently underway to establish the most up-to-date office and laboratory capacity in the city. Furthermore, additional space is available at short notice on the campus premises to relocate and build modern research and office facilities. In this way, Mainz provides attractive opportunities to meet an ever-increasing demand for space from the life sciences and biotech industry.

At the same time as providing modern laboratory infrastructure, we are also focusing on creating a sustainable urban environment that will set new standards. For example, the campus will be integrated into the district heating network and will boast a sustainable mobility concept with tram connections and a central parking garage. In addition, green spaces as well as childcare, restaurants and conference facilities will enhance the district's overall appeal. The aim is for the campus to offer the perfect conditions for developing innovative ideas and technologies.

Mainz is a compelling location that offers a unique combination of high quality of life and efficient infrastructure. With access to leading research institutions and its proximity to big-name companies such as BioNTech, Boehringer Ingelheim and Novo Nordisk, it is particularly appealing to global players and investors.

Medium and Long-Term Growth Prospects for Biotechnology

However, we are also taking a longer-term view when it comes to further developments. For instance, after the area previously occupied by military barracks has been acquired and developed by the City of Mainz, additional sites will become available in the immediate vicinity of BioNTech's head office in "An der Goldgrube" situated in downtown Mainz. In the medium term, there are plans to establish a mixed area here with space for research, work and housing as well as social and public facilities at a single location.

In addition, the City of Mainz has held an international urban planning tender for a site of further 50 ha that offers long-term potential for development. The area is immediately adjacent to the currently developing Life Science Campus and forms an extension of the existing biotechnology corridor in the city centre, where an outstanding cluster of academia, research and development as well as entrepreneurial activity is located—all at the heart of the Rhine-Main Region and close to Germany's principal aviation hub, Frankfurt Airport.

Interdisciplinary Collaboration the Key to Success

As a location for life sciences and biotechnology, Mainz already boasts a thriving ecosystem with leading players. The city and its surrounding region are in an ideal position, thanks to major companies in the sector such as BioNTech, Boehringer Ingelheim, Merck, Novo Nordisk, Sanofi and Schott Pharma. Additionally, there are other life science and biotech firms and suppliers with successful operations in the entire Rhine-Neckar Metropolitan Region.

What is more, internationally renowned educational and research institutions are creating an impressive pool of knowledge and contributing significantly to the economic prosperity. Meanwhile, nine collaborative research institutes, six hospitals, a research university and a university of applied sciences with a world-class reputation are driving interdisciplinary research in the field of life sciences and related research and specialised research activities in the city.

Focusing on interdisciplinary collaboration is the key to success and, consequently, to future breakthroughs in biotechnology and life sciences. In addition to assisting start-ups and established companies, we are also laying the foundations for innovative research projects that will transform Mainz into a globally recognised location.

Mainz regards this approach as a strategic development and, in this sense, the prospects that biotechnology offers for the local economy as a whole are overwhelmingly positive.

Given the opportunities for growth that have recently been created or are still under development, the Life Science Hub Mainz offers the best possible environment for additional start-ups, companies and research fields in order to successfully conduct research, establish businesses and achieve long-term growth.

Mainz Spirit Creates an Ideal Environment for Innovation

Our home city's distinctive flair and superb quality of life also play a part in this. The locals of Mainz—Mainzers—are easy-going, open-minded and warm-hearted. This special Mainz Spirit is what fosters a climate that enables people from diverse backgrounds to easily get into contact and into conversation with each other.

This is an essential factor for exchanging knowledge and, in turn, forms the basis for innovation. In the words of our honorary citizen Professor Uğur Şahin, who also held an impressive keynote address during the Curious Future Insight Conference: "Yes, my wife and I felt quickly at home in Mainz. We came because of the science and stayed because of the people".

In this respect, Mainz can rightly claim to be a very liveable city with a spirit of research and entrepreneurship that improves people's lives—both within city limits and far beyond.

Mainz Science Declaration Signed

We believe that the most noble task of science is contributing to a better future for all of humanity. It is with this in mind that the "World Science Declaration" has set itself the goal, encapsulated by its mantra "make science not war", of ensuring that all countries, societies and organisations worldwide allocate more resources to science, research and technology.

As a graduate of chemistry myself, the principles of this global initiative are particularly close to my heart, which is why I was delighted to sign the Science Declaration for our state capital and support the call to provide more resources for research, science and technology. In doing so, Mainz has joined an impressive number of cities around the world that have also declared their support for this initiative. From Boston to Berlin, from Paris and Tel Aviv to San Francisco, people are committed to building a better future.

As a global community, we have already accomplished some amazing things: we have decoded the human genome, we have sent people into space, we can combat disease with antibiotics—and yet there is still so much to do.

That is why I am calling on to all those with an interest in this issue: Sign the Science Declaration and make a statement in support of a stronger scientific and research community! Let us spread this rallying cry throughout the world together!

Curious Future Insight Conference 2024 in Mainz

Mainz benefited enormously from the Curious Future Insight Conference and it was a tremendous privilege for us to host this year's event. Over the course of 2 days, we had the opportunity to be at the centre of global attention for innovation and next-generation technology beyond our strong local research, science and business ecosystem.

It remains a source of immense pride for us that seven Nobel Prize winners, dozens of leading scientists, the who's who of research, technology and innovation gathered and exchanged ideas here in Mainz's Rheingoldhalle. Apart from being a great honour, it is also an incentive for Mainz to continue playing its part in building a positive future for the world—continuing a tradition that goes back thousands of years.

I would like to express my heartfelt gratitude to the organisational team behind this remarkable scientific conference—the Future Insight e.V. association—and to all the numerous helpers and sponsors who together helped to transform Mainz into a global showcase of innovative technologies and developments for 2 days. Indeed, this is what we intend Mainz to become in the long term.

We hope that all those who travelled from further afield to our city for the Curious Future Insight Conference 2024 were able to experience and feel this special Mainz spirit. All the guests, scientists, researchers and entrepreneurs are warmly invited to return to Mainz again and again—perhaps you will even stay here permanently—because of the science *and* the people.

Landeshauptstadt Mainz, Oberbürgermeister Nino Haase
Mainz, Germany
e-mail: oberbuergermeister@stadt.mainz.de

Contents

Part I

Bright Future

Prologue: The Future Insight Association

1

Ulrich A. K. Betz

Over the millennia, humankind has learned to better understand and shape the world we live in. The power we have acquired through science and technology has enabled us to feed over seven billion people, develop antibiotics to fight bacterial infections, travel to the moon, and create artificial intelligence, to just name a few breakthroughs. However, this same power now also threatens our existence in unprecedented ways, whether through nuclear war, uncontrolled effects on the global atmosphere, or engineered pandemic pathogens. The advancements of digital artificial intelligence make us wonder what truly defines our humanity. Simultaneously, we are witnessing a new level of global conflict, not only between autocratic governments and democracies, but also within democracies themselves and increased polarization of political forces is disrupting societies.

What is required is a new movement that combines science with ethics and inspiration, that integrates the power of innovation with traditional wisdom, building, for example, on clear ethical principles common to all world religions, such as the Golden Rule, and seeks to expand the horizons of science, remaining grounded in evidence, while being open to the profound mysteries of consciousness and the human spirit. What is required is a movement that goes far beyond the classic left-right thinking, we need a movement of truth. What is required is a movement that unites humanity to build a bright and peaceful future together.

The Future Insight e.V. association aims to foster such a global movement. It unites like-minded individuals and organizations in its mission to support the advancement of science, research, and education for the benefit of humanity, aligned on the three fundamental principles of truth, love, and hope—science, ethics, and inspiration. The association fosters collaboration, promotes innovation, and supports transformative initiatives, nurturing the peaceful advancement and use of science

U. A. K. Betz (✉)
Future Insight e.V., Darmstadt, Germany
e-mail: ulrich.betz@merckgroup.com; ulrich.betz@future-insight.org

U. A. K. Betz (ed.), *Science for a Better Tomorrow*,
https://doi.org/10.1007/978-3-031-93623-4_1

for a better tomorrow. "Lucem videmus—we see the light of science, ethics and inspiration—towards a bright future."

The Future Insight association was founded on June 13th, 2022, in Darmstadt, Germany, and rapidly attracted new members worldwide. The association has the code (VR-84647) which refers to an entry in the Register of Associations in Germany. The association is a "gemeinnütziger Verein," a non-profit association in Germany that operates for the public benefit, such as promoting education, culture, sports, or social welfare.

End of 2024, the association has grown to around 450 members including several Nobel Laureates. Both public law entities and private individuals can become members.

All readers are cordially invited to join the association free of charge which can be done online at: https://futureinsightassociation.org.

Donations are welcome at: https://secure.betterplace.org/de/donate/platform/projects/142565.

We welcome individuals, non-profits, and companies alike. Your support, regardless of size, helps pave the way for a better tomorrow. No matter who you are, your contribution matters! If you're driven by progress, believe in collaboration, value ethics and the Golden Rule, and want to promote science, then join us now to make an impact! Together, we help to advance science for the benefit of humanity, unite as friends to change the world and develop and promote a holistic worldview, science-ethics-inspiration.

A key reason for the formation of the Future Insight e.V. association was to foster the advancement of the Curious—Future Insight Conference, one of the most renowned interdisciplinary science conferences in the world, which was initiated by Merck KGaA, Darmstadt, Germany, at the occasion of its 350th anniversary in 2018. The Curious—Future Insight Conference brings together some of the world's brightest scientists and most accomplished innovators to present their work and explore the future of science and technology, solving the challenges of today and enabling the dreams of a better tomorrow. The Curious—Future Insight Conference is one of the world's most renowned interdisciplinary conferences on science and innovation. It promotes interdisciplinary and transdisciplinary dialog, fosters collaboration between academic and industrial research, and facilitates the transfer of research findings to industrial applications. The Curious—Future Insight conference brings together top researchers (including numerous Nobel laureates), leading innovators (including many CEOs), and ambitious young talents. Its central aim is to inspire people to use science and technology to create a bright future while generating a positive vision for tomorrow.

The Curious—Future Insight conferences have taken place in Darmstadt in 2018 and 2022 and in Mainz in 2024. Over the years, more than 7000 people have attended the Curious—Future Insight conference. It has succeeded in attracting over 200 partner organizations, including the Max Planck Society, DFG, Helmholtz Association, Leibniz Association, Nature, AAAS Science, the European Commission, the German Federal Government, and the governments of Hesse and Rhineland-Palatinate, as well as numerous companies such as Merck, BioNTech, Evonik, and Schott.

Held every 2–4 years in the Rhine-Main-Neckar region, this globally influential conference provides a unique opportunity for innovators and scientists to network with the best in the world launch groundbreaking collaborative projects. The event catalyzes connections between academic and industrial scientists, leading to joint projects with scientific and economic potential. Furthermore, it offers young talents a unique chance to learn from leading global experts about cutting-edge research and to establish contacts that benefit their careers.

The conference offers a glimpse into the crystal ball of the future—what will be the game changers of tomorrow? For instance, as early as 2018, the topic of pandemic preparedness was extensively addressed at Curious2018 and supported with a special science award, the Future Insight Prize. Additionally, the topics of digitalization and innovation are of critical importance. The 2024 Nobel laureate, Prof. David Baker, presented his AI research at Curious2024 in July and subsequently received the Nobel Prize in October. Likewise, Benjamin List, David MacMillan, Frances Arnold and Emmanuelle Charpentier all received their Nobel Prizes after they had presented at Curious—Future Insight as keynote speakers. Twenty-one Nobel laureates and many other top scientists, such as the CEO of Microsoft Satya Nadella, genome research pioneer Craig Venter, synthetic biologist George Church, Boston Dynamics founder Marc Raibert, and BioNTech CEO Ugur Sahin, have been featured as speakers. Other conference components include panel discussions, workshops, partnering meetings, an innovation exhibition, and a start-up fair. Five prestigious scientific awards are presented at the event: the Future Insight Prize, the Johann Anton Merck Award, the Nature Spin-off Prize, the Nature Science in Shorts Award, and the Junior Visions Prize. By hosting the event in 2018, 2022, and 2024, a unique brand with global influence has been established. Since end of 2022 the Future Insight e.V. association organizes the Curious—Future Insight Conference together with hundreds of partners (http://curiousfutureinsight.org). The next Curious—Future Insight Conference is planned for July 2027. The organization team of Curious—Future Insight is recruited from among the Future Insight e.V. association members which contribute to set-up and organization of the landmark event and beyond.

In addition to the Curious—Future Insight Conference, the association is organizing additional events such as "Innovation Circle Evenings" for in person networking and in-depth discussions on preselected topics (such as "Generative AI and what it can do for me" as topic for the first event), "Innovation Tours" to visit Innovation Hotspots as well as a "Vision Finder Workshop" to help interested participants uncover their mission in life and provide individual support as required. The association is also promoting the World Science Declaration—Make Science Not War http://make-science-not-war.org underlining the desire to invest in the advancement of science and technology to help solve some of humanities most pressing problems such as fighting disease, overcoming climate change, ensuring nutrition, raising and educating the next generation, rather than wasting the scarce resources of humanity in conflict and wars.

> We, the signatories, are people of different national origins, creeds and convictions. We all firmly believe that human progress is deeply linked to further advances in science and technology. We are truly convinced that science is a force for good which enables us to solve many of mankind's most pressing challenges. We believe that huge opportunities will arise from future science and technology efforts. Yet we are also very well aware of the responsibility and accountability we bear for the new technologies that are realized. We call on all nations, societies and organizations to devote more resources to the advancement of science and technology. We encourage the international community to join forces in battling debilitating diseases, ensuring sufficient food for a growing world population, stopping the destruction of our environment, and engaging in joint endeavors to elucidate the secrets this fascinating universe holds. Nothing shall be impossible..

Readers are encouraged to sign the declaration at http://make-science-not-war.org.

Inspiration for the endless possibilities of science and technology are assembled at our Dream Board which shows desires and wishes for a bright future and a collection of ideas what should be invented but has not yet been invented. Entries can be screened and new contributions to the Dream Board provided at https://www.curiousfutureinsight.org/dreamboard/.

In the frame of a series of discussions the Future Insight Association has formulated ten key principles that are guiding its actions—Principles of the Future Insight Association:

1. We promote scientific progress for the benefit of humanity.
2. We follow and advocate for clear ethical behaviors treating others with compassion and respect, mindful of their wishes and well-being.
3. We keep in mind the inherent limitations of human endeavors including science and respect the believe in a spiritual principle in the universe, and strive for a deeper understanding and connection.
4. We encourage the creation and application of new technologies to help shape a better world for all and to deepen our understanding of the universe and its underlying principles.
5. We connect people of all nations and beliefs to create a bright and peaceful future together.
6. We respect and advocate for universal human rights as formulated in the Universal Declaration of Human Rights (UDHR) on December 10th, 1948. https://www.un.org/en/about-us/universal-declaration-of-human-rights.
7. We believe in equality of opportunity for all and in the principle of meritocracy, aiming at the realization of the full human potential.
8. We advocate for future generations, the protection of children, investments in educational opportunities, preservation of the natural environment, ensuring a thriving future now and for generations to come.
9. We believe in the settlement of space, allowing humanity to one day transcend the limitations of the Earth.

10. We acknowledge the possibility of consciousness within both organic and non-organic entities beyond humanity and commit to upholding ethical principles in our interaction with such forms of life.

The future belongs to those who believe in the beauty of their dreams.

To brighter tomorrows!

Lucem videmus—We see the light—join now at: https://www.curiousfutureinsight.org/about-us.

Bridging Biosciences, AI, and IT for Next-Generation Technological Advances in BIO4 Campus

2

Jelena Begović

Contents

Abstract

Biotechnology, artificial intelligence (AI), and information technologies (IT) are becoming a key tools, enabling faster and more precise analysis of the ever-growing amount of complex biological data generated by the rapid advancement of different technologies. In biomedicine, AI is used for the discovery of new drugs and personalized therapy, thereby completely transforming the approach to patients and diseases. In the field of agriculture, AI helps optimize the cultivation of plants and animals, increasing productivity and reducing the use of chemicals, making agriculture greener and more sustainable. AI and IT also play a significant role in genetic analysis and modification, enabling more precise gene manipulations and accelerating research in this area. In industrial biotechnology, AI is used to design microorganisms that efficiently produce biofuels and other chemicals. The synergy between biotechnology, AI and IT and allows for a better understanding of complex biological systems, paving the way for new discoveries and innovations. The combination of these powerful technologies is rapidly transforming our world, bringing advanced solutions that redefine the way we live today and open up new, currently unimaginable

J. Begović (✉)
Minister of Science, Technological Development and Innovation, Belgrade, Republic of Serbia
e-mail: nenad.paunovic@gov.rs

U. A. K. Betz (ed.), *Science for a Better Tomorrow*,
https://doi.org/10.1007/978-3-031-93623-4_2

possibilities for the future. BIO4 Campus in Serbia is envisioned as a new infrastructure that supports science and innovation in fields that will emerge from intersection between biotechnology, information technologies, and AI.

Keywords

Artificial intelligence · Biotechnology · IT · Bioinformatics · BIO4 campus

2.1 Introduction

Future trends in biotechnology, IT and AI have the potential to offer solutions to the global challenges we face as a humanity and as individuals. Climate change, population growth, agriculture and food production, food security and quality, poverty, energy, biodiversity protection, the environment, and transportation are just some of the problems for which we must find solutions in the near future. Rapid progress compels us to embrace these technologies not only as tools but as opportunities to adapt, innovate, and grow as economies and societies. Serbia has embraced the transformative era brought about by the convergence of life sciences and information technologies, particularly in the fields of biosciences and artificial intelligence. This marks a pivotal step in technological advancement, embodied in the creation and development of the new bioeconomy hub in Europe—BIO4 Campus (Fig. 2.1). This state-of-the-art complex facility envisioned as a collaborative platform is designed to merge academia and the private sector to drive innovation across four bios: biotechnology, biomedicine, biodiversity, and bioinformatics with AI. By merging, interconnecting, and evolving into new disciplines, the four BIOs represent the continuum of technological progress. These powerful fields have now

Fig. 2.1 View of the future BIO4 Campus in Belgrade

reached a critical point of convergence, unlocking new possibilities for expanding human knowledge, advancing research, and driving innovation.

2.2 From Past Innovations to Future Technologies

Technology has always been a powerful driver of human civilization's development, transforming the way we live, communicate, and solve problems, from the invention of the wheel to the rise of artificial intelligence. As we expand the horizons of knowledge, we continue to innovate; today, technology not only shapes our daily lives but also pushes the boundaries of what is possible and redefines our future. **Biotechnology**, as one of the key technologies today, emerged when humans transitioned from a nomadic lifestyle of hunters and gatherers to organized communities, using selective breeding of plants and animals to improve food production and quality. This led to the creation of the first settlements, which later grew into cities, marking the beginning of population growth and societal organization, which has changed throughout history. However, the first produced beer, wine, yogurt, cheese, and bread represent true examples of traditional or empirical biotechnology. Even at that time, microorganisms played a role in the production processes of value-added foods derived from plants or animals. This early application of biotechnology laid the foundation for progress that still has an enormous, and arguably crucial, impact on the further development of agriculture, medicine, and industry.

Examples of biotechnology today are numerous, from the use of bacteria or cells to produce vaccines and biological drugs to the production of lab-grown meat [1, 2]. Of course, today we live in a time of intense development of other technologies that have elevated biotechnology to a higher level and enabled its widespread application. Driven by the rapid advancements in information technology (IT), numerous other fields have seamlessly joined the technological revolution. One area that has pushed the boundaries of biotechnology itself is the field of information technology. Information technology was first mentioned in 1958 in a paper by Leavit and Whisler, where they announced a new technological revolution: "Over the past decade, a new technology has begun to be introduced into American business, so new that it is still difficult to assess its importance. The new technology still does not have a universally adopted name. We will call it information technology" [3]. Subsequently, a revolution began—one that continues to shape our lives today. The global IT industry market in 2022 reached a value of 9,481,627.92 million USD [4]. This market is experiencing rapid and substantial growth, not only driving economic progress but also reshaping the way we live. In 2012, Serbia's export of IT products and services amounted to 375 million euros. A decade later, in 2023, it reached 3.4 billion euros, marking the fastest growth among all economic sectors. Major global IT companies have established research and development centers in Serbia, creating job opportunities for highly skilled experts and innovators while positioning the country as a key global hub for developing cutting-edge solutions, from chip design to software innovations.

Just as information technologies were revolutionary in the second half of the twentieth century, biological sciences, and biotechnology in the twenty-first century have become crucial for the further development of humanity and for solving global problems such as poverty, food shortages and hunger, challenges in medicine, environmental degradation, and more. The **BIO4 campus** focuses on addressing and transforming four major global challenges: medicine, agriculture and food production, energy, and the green agenda.

Information technologies have enabled the analysis of complex biological data, including genetic data, which has led to the development of precision medicine tailored to individual genetic profiles. The combination of information technology and biological sciences has led to the development of a new discipline, bioinformatics, which represents the application of computer science and statistics in solving and analyzing biological problems. Thanks to modeling and analyzes at the molecular level, today we have a much deeper understanding of the origins of diseases, we improve diagnostics, and we accelerate the development of new therapies, thanks precisely to IT tools for managing and interpreting the enormous amounts of biological data that we generate today. Already during the twentieth century, the synergy between IT and biotechnology accelerated scientific discoveries but also enabled a large number of innovations that are transforming agriculture, healthcare, and environmental management [5, 6]. As increasingly powerful computer systems develop, biological sciences and biotechnology advance by quantum jumps.

However, with the development of computer hardware in the twenty-first century, the intensive development of artificial intelligence, whose roots trace back to Alan Turing 1937 began [7]. What exactly is Artificial Intelligence? In the scientific context, it represents a field of computer science that develops systems or algorithms that aim to mimic human intelligence and the ways we solve problems. As early as 1955, John McCarthy defined the concept of artificial intelligence as a combination of science and engineering to create intelligent machines. The following year, a summer workshop was organized at Dartmouth College, where this term entered general use [8].

Indeed, computer pioneer and visionary Alan Turing constructed the Turing machine in the form of a mathematical model, which served as the basis for the further development of modern digital computers. However, it was not until the development of technologies, especially graphics cards and processors after the year 2000, that artificial intelligence as a technology gained new momentum. The emergence of powerful computer technologies were an introduction to the further development of biotechnology, which began to use machine learning and artificial intelligence for predicting protein structures, drug discovery, and solving challenges in medicine, agriculture, and environmental protection. Today, we live in a time of rapid acceleration of research and development processes, thanks, among other things, to the combination of information technology, artificial intelligence, and natural sciences, primarily biological sciences.

Bioinformatics falls under the field of biotechnology and represents a multidisciplinary field where computer science, mathematics, and statistics are used to process and analyze data primarily generated by biological sciences. Today, bioinformatics has become a standard in modern biology, especially in the fields of molecular biology and genomics (the study of the structure, organization, function, and evolution of genes and genetic material in organisms). The origins of bioinformatics trace back to the 1960s, starting with protein sequencing, which required computational analysis. The first sequenced protein was insulin, while DNA sequencing technology emerged 25 years after the discovery of DNA's double-helix structure in 1953 [9–12]. The rise of microcomputers, desktop computers, and the internet accelerated bioinformatics, leading to global databases (SWISS-PROT, GenBank) and tools like BLAST and Autodock.

We need to go back several decades to 1990, when the U.S. National Center for Human Genome Research launched the "Human Genome Project," which, after 13 years when it was completed, cost $2.7 billion [13]. But it is certain that this major project marked the beginning of the genomics era, where scientists are now uncovering at the molecular level the secrets hidden by genes and genetic material, with the idea of advancing medicine, food production, and the environment. Today, a little over three decades later, thanks to the development of other technologies and the proliferation of these methods, the cost of reading an individual's genetic information has dropped to less than $1000. As a result, the amount of sequenced genetic data doubles every 7 months [14]. Stevens and co-authors in the same study predicted, based on detailed analyzes, that by 2025, the amount of data generated by genome sequencing will be between 2 and 40 exabytes. This number may not elicit much of a reaction, but when you consider that 1 exabyte (EB) = 10^18 bytes, or 1000 petabytes = 1 million terabytes = 1 billion gigabytes, you can grasp the scale of the data generated solely by DNA sequencing. It is difficult to remain indifferent to these figures. With this amount of data increasing daily, the question arises of how and where we will store this data and how we will analyze it. This is why it is essential for supporting infrastructure, such as national data centers, to keep pace with data production, ensuring efficient storage, processing, and utilization of the ever-growing volume of diverse data.

It becomes clear that, without the support of IT and artificial intelligence, much of this information would remain untapped. We need to centralize data and integrate various types of information. To achieve this, we require infrastructures like data centers, as well as experts from diverse fields to manage, analyze, and utilize this data effectively. The National Data Center in Kragujevac, equipped with an NVIDIA supercomputer, has become a vital and invaluable infrastructure for academia and startups, fostering the advancement of science and innovation in Serbia.

Already these big data sets with the assistance of IT and AI are speeding up the development of precision, or more commonly still used, personalized medicine. The assessment of the risk of developing a particular disease has also advanced significantly thanks to big data analysis [15]. Today, the new generation of biological drugs and therapies is based on producing a drug for just one patient. Simultaneously, pharmacogenomics has advanced to evaluate the effectiveness of specific drugs based on the genetic makeup of individual patients or groups, particularly with the support of information technology and artificial intelligence [16].

However, even beyond the boundaries of medicine, genomics with artificial intelligence has found application primarily in agriculture, where by analyzing the genetics of microorganisms and applying them, we introduce regenerative agriculture, which is much less harmful to the environment [17]. Big data allows us to use natural forms of microorganisms in agriculture instead of chemical agents, thereby reducing pollution, increasing yield, and applying green technologies. Similar applications of this technology exist in the field of environmental protection. Microorganisms and some plants can be very effective in cleaning up contaminated soil or water from various pollutants [18, 19]. And this leads us directly to biotechnology.

The popularity of the term artificial intelligence today is probably largely associated with popular chatbots, such as ChatGPT and others like Gemini and Copilot. Additionally, existing chatbots, although most commonly used for simple data searches, can certainly be used for interpreting scientific data, albeit with many limitations [20]. Taking into account the number of nucleotides, genes, peptides, and proteins in nature, in complex systems like living organisms, with the development of new technologies and the generation of large datasets, it is simply impossible to advance science without bioinformatics and artificial intelligence.

Machine learning opens new areas of application where the machine, based on learned data, independently recognizes, predicts, and decides. Deep learning, a subset of machine learning based on artificial neural networks, allows the machine or computer to mimic, in some way, the processes and pathways that occur in the human brain. Consequently, the algorithm makes its own conclusions and can be of great assistance when dealing with large datasets. Thanks to these programs, the field of biotechnology, specifically synthetic biology, has seen rapid development in recent years. Following the discovery of DNA, mastering the process of reading information encoded in this molecule, the development of technologies that today enable DNA synthesis, and the CRISPR technology, which allows for precise modifications within any DNA strand, combined with big data analysis through artificial intelligence, we have reached a clear turning point where, as a species, we can precisely develop new living systems and modify existing organisms to help us solve numerous challenges we face as humanity, from health to the development of green solutions [10, 21–23]. Today, we are extensively using synthetic biology, which enabled the approval of the first drug based on recombinant DNA technology in 1982, Humulin®, which is human insulin synthesized in the bacterium *Escherichia coli* approved by the U.S. Food and Drug Administration. Today, most insulin is produced in genetically modified bacteria or yeasts [24]. Merging of AI with these powerful technologies will open new possibilities, discoveries and completely new fields of science.

Despite all technological advancements and knowledge, we still lack effective therapies or drugs for many diseases. The process of developing a new therapy or drug is lengthy, typically taking 10–15 years, during which the drug undergoes research and development until clinical trials and market preparation. The process is highly risky, as failure is possible at any stage of development. Moreover, drug development is extremely expensive, with the average cost of developing a drug used in oncology estimated between 944 million dollars and 4.54 billion dollars [25]. However, with the application of artificial intelligence tools, including big data analysis, drug design, and prediction of its action at the molecular level, the duration of development can be significantly shortened, thereby reducing both risk and the cost of development [26]. Ultimately, patients could receive new drugs earlier and at a lower cost. The integration of genetic, medical, biochemical, and imaging data into databases demands faster algorithms for analysis. Artificial intelligence unlocks this potential, transforming medicine by enabling personalized approaches based on genetics and improving diagnostics through precise processing of imaging dana. AI has become an invaluable tool, enhancing both diagnosis and decision-making in healthcare.

Beyond healthcare, the advancements in AI find critical applications in other fields, such as agriculture, which shares a similar reliance on data for precision and efficiency. By integrating new technologies into agricultural practices, such as monitoring plant health or early detecting diseases in animals or plants, AI bridges the gap between human and environmental health.

By 2050, the global population is expected to reach nearly ten billion, driving greater demand for food and creating opportunities and challenges for agriculture. Sustainability is key, as 90% of global forest loss is linked to agricultural expansion, including livestock, according to the United Nations report related to sustainable development goals [27]. To make agriculture more efficient, virtual assistants for farmers who provide data on weather, soil, pests, and diseases, precipitation, and more, facilitate and optimize decision-making by farmers. Precision agriculture, supported by virtual assistants, provides farmers with data on weather, soil, pests, and diseases, optimizing decision-making.

Biotechnology and artificial intelligence further enhance agriculture by enabling the selection of crops and livestock based on genetic and environmental data, increasing yields, and reducing reliance on pesticides through green solutions like soil-regenerating microorganisms. AI, combined with genomics and proteomics, refines problem-solving by narrowing hypotheses for optimal farming practices. Integrated with sensors and monitoring systems, AI detects diseases, water shortages, and nutrient deficiencies, elevating agricultural efficiency and sustainability [28]. Biotechnology, paired with agriculture, IT, and AI, enables efficient biofuel production by optimizing processes and resources [29]. AI-driven data analysis helps identify the most suitable crops and methods for biofuel generation, while biotechnology enhances crop yields and accelerates the conversion of biomass into energy [30, 31]. Together, these technologies support the development of sustainable and eco-friendly energy solutions to address growing global energy demands. Given the persistent and often negative impact of human development, societal progress, and technological and economic activities on the environment, artificial intelligence and biotechnology are increasingly being applied in bioremediation as well as the identification and monitoring of biodiversity [32–34]. Numerous species face extinction, underscoring the need for more effective conservation in order to keep biodiversity on our planet. The CAPTAIN framework, based on artificial intelligence is tool that optimizes the trade-off between cost and biodiversity protection, and can protect more species under budget constraints [34].

2.3 The Future of New Technologies and BIO4 Campus

The continued convergence of these technologies will undoubtedly open up new fields of application, lead to the creation of new professions, and increasingly impact our daily lives.

Serbia has embraced the transformative era brought about by the convergence of life sciences and information technologies, particularly in the fields of biosciences, information technologies and artificial intelligence. This marks a pivotal step in technological advancement, embodied in the creation and development of the BIO4 Campus. This state-of-the-art facility is designed to merge academia and the private sector to drive innovation and solutions across biotechnology, biomedicine, biodiversity, and information technology alongside artificial intelligence.

The BIO4 Campus will encompass seven faculties of the University of Belgrade, nine research institutes, a cutting-edge science and technology park for the development, support, and acceleration of startups, as well as R&D centers of leading biotech and biopharma companies. Hosting more than 4000 students and over 1000 PhD holders, the Campus represents a fusion of talent and expertise with state-of-the-art infrastructure, featuring more than 300 laboratories and nine core facilities from imaging to sport biomechanics. For the beginning, 160,000 m^2 of space will be dedicated to education, science, knowledge transfer, technology transfer, and collaboration between academia and the private sector is a vision for shaping the future of Serbia (Fig. 2.1). This hub will serve as a catalyst for innovation, bridging the gap between research and real-world applications, fostering interdisciplinary collaboration, and driving sustainable development in the region and beyond. As a center of gravity for the region it is a bold step toward positioning whole Balkan region as a global spot in innovation innovation development and also become part of EU education, science and innovation infrastructure. This project seeks stronger synergy with European space and more efficient connections of academic institutions and startups facilitating the efficient integration of accession countries into the EU framework.

In short, collaboration and internationalization are vital to the growth of science and technology. Collaboration and internationalization are key drivers for the advancement of science and innovation. Science thrives on the exchange of ideas, knowledge, and expertise from diverse sources, which significantly accelerates discovery and problem-solving. Collaboration, both within scientific communities and across industries, allows for the pooling of resources and skills. Researchers from different disciplines in BIO4 campus will combine their unique perspectives to tackle complex challenges, fostering creativity, and innovation. This collective approach often leads to groundbreaking discoveries that would not be possible in a siloed environment. Internationalization, on the other hand, will ensure that science and innovation are not limited by geographic boundaries. The BIO4 campus is

Fig. 2.2 Detailed view of future BIO4 Campus in Belgrade

designed to be a crucial part of both European and global research infrastructure, aiming to enhance the European Union's research and innovation ecosystem (Fig. 2.2). By integrating BIO4 into this broader framework, it will not only contribute to the EU's scientific and technological advancements but also enhance its global competitiveness. This strategic positioning will allow BIO4 to contribute to future advancing of European research and innovation area while also contributing to global scientific progress.

Furthermore, international collaboration brings together a rich diversity of cultural perspectives and ideas, which is essential for addressing the global challenges that face humanity. Moreover, international partnerships provide access to broader funding opportunities and investment, which are crucial for the commercialization of scientific innovations.

Innovative environment like this one should foster the emergence of new scientific fields at the intersection of biosciences, IT and AI. The BIO4 Campus will not only be open to collaboration but also act as a central hub—a "center of gravity"—for international students, researchers, and experts (Fig. 2.3). The growing number of Serbian startups rooted in information technology, biotechnology, agritechnology, and artificial intelligence underscores the recognition by the market and entrepreneurs of the immense opportunities to shape the future through innovation.

The rapid advancement of artificial intelligence, information technology, and biotechnology stands as one of the defining achievements of modern history. Over the past few decades, AI has evolved from theoretical models into transformative tools that power everything from medical diagnostics to autonomous

Fig. 2.3 Detailed view of future BIO4 Campus in Belgrade

systems. Simultaneously, breakthroughs in IT have revolutionized how we connect, communicate, and process information, enabling unprecedented levels of global collaboration. Biotechnology, with its roots in genetic research, has given rise to life-saving therapies, sustainable agricultural solutions, and new frontiers in understanding life itself.

The combination of these powerful technologies is rapidly transforming our world, bringing advanced solutions that redefine the way we live today and open up new, currently unimaginable possibilities for the future. And that is why, from the standpoint of economical, technological, and social development, it is crucial for us as a country to get involved and contribute as active participants in the unstoppable processes that are already steering our future. According to Cervicorn Consulting, the global biotechnology market is projected to grow significantly, expanding from USD 1.68 trillion in 2024 to approximately USD 3.54 trillion by 2033 [35]. Tissue engineering and regeneration and DNA sequencing are two most prominent technologies when it comes to biotechnology market share. Investing in human capital and the infrastructure required for the development and application of new technologies is both a challenge and a necessity. Biotechnology research and development demand significant financial investment in advanced technologies, highly skilled professionals, and comprehensive preclinical and clinical trials. BIO4 campus aims to serve as a hub where small and medium-sized enterprises (SMEs) can gain access to costly equipment and expert support, fostering innovation and collaboration in biotechnology. By fostering collaboration and providing resources, the BIO4 campus aims to bridge these gaps and support future development of biotech sector. This effort must also embrace international collaboration and global integration to drive transformation, ensure continuous adaptation to technological advancements, and leverage these innovations for the benefit of society worldwide.

References

1. Frost I, Sati H, Garcia-Vello P, Hasso-Agopsowicz M et al (2023) The role of bacterial vaccines in the fight against antimicrobial resistance: an analysis of the preclinical and clinical development pipeline. Lancet Microbe 4(2):e113–e125. https://doi.org/10.1016/S2666-5247(22)00303-2
2. Jones N (2023) Lab-grown meat: the science of turning cells into steaks and nuggets. Nature 619(7968):22–24. https://doi.org/10.1038/d41586-023-02095-6
3. Leavitt HJ, Whisler TL (1958) Management in the 1980's. Harv Bus Rev 36(6)
4. https://www.businessresearchinsights.com/market-reports/information-technology-it-market-108885
5. Reddy L, Akinola S (2023) Transforming healthcare with the synergy of biotechnology and information technology. Bioengineering 10(4):421–439. https://doi.org/10.3934/bioeng.2023025
6. Mannion AM (1995) Agriculture, environment and biotechnology. Agric Ecosyst Environ 53(1):31–45
7. Turing AM (1937) On computable numbers with application to entscheidungsproblem. Proc Lond Math Sos Ser 2 42:230–265
8. https://home.dartmouth.edu/about/artificial-intelligence-ai-coined-dartmouth
9. Sanger F, Thompson EOP (1953) The amino-acid sequence in the glycyl chain of insulin. The identification of lower peptides from partial hydrolysates. Biochem J 53:353–366
10. Watson JD, Crick FH (1953) Molecular structure of nucleic acids: a structure for deoxyribose nucleic acid. Nature 171:737–738
11. Sanger F, Nicklen S, Coulson AR (1977) DNA sequencing with chain-terminating inhibitor. Proc Natl Acad Sci U S A 74:5463–5467
12. Maxam AM, Gilbert W (1977) A new method for sequencing DNA. Proc Natl Acad Sci U S A 74:560–564
13. Human genome project completion: frequently asked questions. National Human Genome Research Institute, NHGRI LoS Biol
14. Stephens ZD, Lee SY, Faghri F, Campbell RH, Zhai C, Efron MJ, Iyer R, Schatz MC, Sinha S, Robinson GE (2015) Big Data: astronomical or genomical? PLoS Biol 13(7):e1002195. https://doi.org/10.1371/journal.pbio.1002195. eCollection 2015 Jul
15. Fröhlich H, Balling R, Beerenwinkel N, Kohlbacher O, Kumar S, Lengauer T, Maathuis MH, Moreau Y, Murphy SA, Przytycka TM, Rebhan M, Röst H, Schuppert A, Schwab M, Spang R, Stekhoven D, Sun J, Weber A, Ziemek D, Zupan B (2018) From hype to reality: data science enabling personalized medicine. BMC Med 16:150. https://doi.org/10.1186/s12916-018-1122-7
16. Taherdoost H, Ghofrani A (2024) AI's role in revolutionizing personalized medicine by reshaping pharmacogenomics and drug therapy. Intell Pharm 5:643–650
17. Wójcik-Gront E, Zieniuk B, Pawełkowicz M (2024) Harnessing AI-powered genomic research for sustainable crop improvement. Agriculture 14(12):2299
18. Saha L, Tiwari J, Bauddh K, Ma Y (2021) Recent developments in microbe–plant-based bioremediation for tackling heavy metal-polluted soils. Front Microbiol 12:731723. https://doi.org/10.3389/fmicb.2021.731723
19. Kuppan N, Midhila Padman M, Mahadeva M, Srinivasan S, Devarajan R (2024) A comprehensive review of sustainable bioremediation techniques: eco friendly solutions for waste and pollution management. Waste Manag Bull 3:154–171

20. Kaftan AN, Hussain MK, Naser FH (2024) Response accuracy of ChatGPT 3.5 Copilot and Gemini in interpreting biochemical laboratory data a pilot study. Sci Rep 14:8233
21. Ma Y, Zhang Z, Jia B, Yuan Y (2024) Automated high-throughput DNA synthesis and assembly. Heliyon 10(6):e26967
22. Doudna JA, Charpentier E (2012) A programmable dual-RNA-guided DNA endonuclease in adaptive bacterial immunity. Science 337(6096):816–821
23. Dara M, Dianatpour M, Azarpira N, Omidifar N (2024) Convergence of CRISPR and artificial intelligence: a paradigm shift in biotechnology. Hum Gene 41:201297
24. Baeshen NA, Baeshen MN, Sheikh A, Bora RS, Ahmed MMM, Ramadan HAI, Saini KS, Redwan EM (2014) Cell factories for insulin production. Microb Cell Fact 2(13):141. https://doi.org/10.1186/s12934-014-0141-0
25. Schlander M, Hernandez-Villafuerte K, Cheng CY, Mestre-Ferrandiz J, Baumann M (2021) How much does it cost to research and develop a new drug? A systematic review assessment. Pharmacoeconomics 39(11):1243–1269. https://doi.org/10.1007/s40273-021-01065-y
26. Shi Y (2024) Drug development in the AI era: AlphaFold 3 is coming! Innovation 5(5):100685
27. The Sustainable Development Goals Report 2024. https://unstats.un.org/sdgs/report/2024/The-Sustainable-Development-Goals-Report2024.pd
28. Kachouei MA, Kaushik A, Azahar A. Internet of things-enabled food and plant sensors to empower sustainability. 2023. https://doi.org/10.1002/aisy.202300321
29. Corral-Bobadilla M, Lostado-Lorza R, Sabando-Fraile C, ´iñiguez-Macedo S (2024) An artificial intelligence approach to model and optimize biodiesel production from waste cooking oil using life cycle assessment and market dynamics analysis. Energy 307:132712
30. Nath S (2024) Biotechnology and biofuels: paving the way towards a sustainable and equitable energy for the future. Discov Energy 4(1):8. https://doi.org/10.1007/s43937-024-00032-w
31. Kumari A, Kundu PK, Gupta MM, Bala K, Chandra S, Dutta R, Das A (2022) Role of microorganisma in production of biofuels. In: Novel feedstock for biofuels production. Springer, Berlin, pp 65–116
32. Steffi PF, Anburaj TR, Mishel PF (2022) Artificial intelligence in bioremediation modelling and clean-up of contaminated sites: recent advances, challenges and opportunities. In: Omics insights in environmental bioremediation. Springer, Singapore, pp 683–702
33. Patowary R, Devi A, Mukherjee AK (2023) Advanced bioremediation by an amalgamation of nanotechnology and modern artificial intelligence for efficient restoration of crude petroleum oil-contaminated sites: a prospective study. Environ Sci Pollut Res Int 23:1–26
34. Silvestro D, Goria S, Sterner T, Antonelli A (2022) Improving biodiversity protection through artificial intelligence. Nat Sustain 5:415–424
35. Biotechnology market size, share, trends, report 2024 to 2033. 2024. Cervicorn consulting. https://www.cervicornconsulting.com/biotechnology-market

Mentoring Early Career Researchers 3

Theresa G. Mercer and Michael-Bogdan Margineanu

Contents

T. G. Mercer (✉)
Senior Lecturer in Environmental Sustainability, Cranfield Environment Centre, Cranfield University, Bedford, UK
e-mail: theresa.mercer@cranfield.ac.uk

M.-B. Margineanu
Cancer Neuroscience Lab, The Francis Crick Institute, London, UK
e-mail: marginm@crick.ac.uk

U. A. K. Betz (ed.), *Science for a Better Tomorrow*,
https://doi.org/10.1007/978-3-031-93623-4_3

Abstract

The Curious Future Insight Conference is a platform that brings together scientists from around the world every 2–4 years to discuss the latest advances and innovations in science and technology with the aim of 'uniting through science for a better tomorrow'. Over the 2-day gathering, a number of themes are covered relating to Energy, Healthy Lives, the Human Mind, Life Reimagined, Materials & Solutions, Nutrition and Vibrant Digital. In the 2024 conference, the importance of mentoring and supporting Early Career Researchers within the wider science ecosystem was a prominent feature of a panel session hosted by AAAS/Science and moderated by Dr. Rachel Bernstein of Science, deputy news editor focusing on careers and community. In this panel session that also included Dr. Senka Holzer, Medical University of Graz and Professor Ulrike Fasbender, University of Hohenheim in Germany, we explored the importance of the mentor/mentee relationship from a scientific and professional perspective, how the relationship can help individuals in navigating challenges within a science career and strategies and approaches for fostering the relationship. This chapter will address some of these main themes around mentoring relationships, bringing in our insights and personal perspectives.

Keywords

Mentoring · Challenges · Support · Mentor/mentee relationship · Responsibilities · Reverse mentoring · Research advisor

3.1 Introduction to Mentoring

The word Mentor first appears in Greek Mythology in *The Odyssey*. Mentor, a trusted friend of Ulysses—the epic hero of the Trojan War, is entrusted with the protection and guidance of his son before Ulysses departs to Troy [1]. In a similar vein, mentoring refers to the process of sharing knowledge, skills, and experience between a mentor and a mentee with the goal to help the mentee progress in both their professional and personal life. Compared to an apprenticeship, mentoring has a broader focus and is not limited to specific skills learning.

Mentors play an important role in helping mentees in the process of developing a scientific research career, and nowadays, in the era of information overload, this support is essential. Navigating too many options (study opportunities, research directions, career paths, etc.) or apparently no options at all (failures in securing very competitive academic positions, unsuccessful grant applications) requires training and advice to help one be in the position to make the best decisions for what comes next.

Mentoring is, however, not only a process helping one make decisions but also a supportive element in identifying best directions to grow professionally and personally and finding the right balance between personal and professional lives.

There are many examples of mentoring directed to students from underprivileged backgrounds or countries with less developed scientific ecosystems. Also, mentoring relationships that are directed towards work/life balance, parenting, and many more.

Mentoring relationships can be short or life-long, start with a long and arduous search or through a serendipitous meeting. In this chapter, we will describe our own mentoring journey and highlights from the Mentoring panel during the Curious 2024 Future Insight Conference.

3.2 Our Experiences of Mentoring

Theresa

Theresa Mercer is a Senior Lecturer in Environmental Sustainability. Her research focuses on two main areas; environmental systems change and collaborative education approaches for environmental sustainability. She is also the Sustainability and Environmental Management Programme Director, acting as a Champion for ten MSc level courses at Cranfield University. Theresa is particularly passionate about advocating for support aimed at extended leavers within the academic and research communities, having been one herself, and exploring the development of formal support frameworks with HE institutions and research groups that provides support before, during and after periods of extended leave.

My experience with mentoring goes back around 10 years ago. I was an early career researcher on an amazing research project exploring the biodiversity and ecosystem service provision in urban areas. There was a lot of field and lab work involved and I thoroughly enjoyed the practical aspects of the work. Three months into the 3-year project I became pregnant and eventually took my maternity leave. During this time, I moved from England where my project was based at Cranfield University to Cardiff, Wales where my husband was based at the time. As I got to the end of my maternity leave of 1 year, I realized that cross-country commuting with a baby in tow just wouldn't be possible. On top of that, my baby daughter didn't take to nursery. She went on sleep strike and hunger strike. I remember ringing my project PI at the time who was also my mentor and telling him I couldn't carry on with the project because of these compounding factors. I was in floods of tears and at the time thought that this would mark the end of my academic career [2].

Two to three days later my PI rung me back and said to me 'look Theresa, don't worry, we can find a way for you to work through this.' 'We can support you for however long you plan to take out of academia.' This time out of academia turned out to be 5 years. During that time I had both informal and formal support from two of my mentors at Cranfield University. Now I've rejoined academia and returned to Cranfield after a posting at the University of Lincoln. My two mentors continue to support me now in my career since returning to academia. My takeaway message is that as a mentor, you can have transformative, life-changing impacts on the mentees that you are mentoring and can play a huge role in their career trajectories, particularly for those of us that haven't followed a standard route through academia or research careers.

Michael

Michael-Bogdan Margineanu is a Postdoctoral fellow in cancer neuroscience at the Francis Crick Institute and has 13 years of research experience at leading institutions from Germany, Saudi Arabia, UK, Ireland, Austria, USA, Switzerland, and Romania. He is the Founder and President of Biomentorhub, a youth non-profit and non-governmental organization focused on mentoring young scientists and innovators in life sciences from Romania and other countries with very limited R&D professional opportunities in this field. He is also an alumnus of the Lindau Nobel Laureate Meetings (Medicine & Physiology 2018) and co-founder of the Lindau Mentoring Hub.

I believe mentoring in science does not start when you actually start to study science formally in university. In my case, it started in the family; parents stimulated my curiosity through discussions in a way I think maybe they were my first mentors, even for my scientific journey later on, as we always consulted on major decisions.

Then I think throughout progression in school you get to meet some professors which have a strong impact on your development, inspiring you to choose a particular path. And I think this can be seen to some extent as mentoring as well. Of course for each person in particular the experience and interactions they will have will be different.

But in my case, I had amazing professors who pushed me forward and stimulated my curiosity for science, for research since secondary school. And I think that this continuous encouragement, this continuous support was very important to move forward along the way. And then of course university (both undergrad and master's degree studies) provided me with mentors. However, maybe the most impactful mentoring relation I had was with my PhD advisor and that's not always the case for everyone.

And not everyone has an excellent mentoring relation with their PhD advisors. However, I could say that when I did choose to apply to the respective research group and work with the respective PI, I looked at him as a role model outside his scientific expertise too. It was something I factored into my decision, and it really paid off. And I'm very grateful for that.

3.3 Finding the Right Mentor and Developing the Best Mentoring Relationship

3.3.1 What are some methods for finding mentors and fostering the relationship?

Theresa

Most academic institutions or research institutes will have some sort of formal mentoring scheme in place. I rejoined Cranfield University and have been there for just over a year now and they have an official way of doing that. You are paired with a mentor beyond your research centre so it is an academic or researcher with that external view, which is great in helping you look beyond your discipline and immediate network.

The other approach is through informal networks. This does not necessarily just happen and these networks need to be nurtured, particularly if you have been out of academia for a while. If you have been on extended leave, like I had, I continued with support from my formal mentors but I also reached out through my informal networks that I had developed during my PhD and Postdoc times. I found this larger support network from beyond the project I was involved with incredibly supportive as well.

So there is the formal way of doing things through your relevant institutions' mentoring schemes but also the informal route as well. In order to nurture your networks, do reach out regularly and check-in. Most people are happy to be asked to be an informal mentor. I have done this in the past and have a number of external mentors as a result. They can give very useful external perspectives.

Michael

Apart from formal mentoring schemes, there are other specialized platforms like Biomentorhub or the Lindau Mentoring Hub, accessible to the alumni of the Lindau Nobel Laureate Meetings. Biomentorhub is accessible to anyone looking for mentors in the life sciences sector.

Related to the mentoring relationship, I think in order to benefit most from this relationship, the mentee needs to believe in the mentor's capacity to help and be able to follow the recommendations received.

I think it's a bit similar to the relationship with a medical doctor. When you go to a medical doctor and are consulted and you get a recommendation, part of the healing process will also be based on your belief that what was recommended will work. It might take you some time to find the medical doctor that you trust, with whom you communicate well. Same applies to a mentoring relationship.

At the same time, I think the mentee must offer openness in the relationship with the mentor to benefit most. It's not something easy to achieve initially, but it's something crucial. The most impactful and useful mentoring relations would be the ones in which there is such openness and maybe you reach a level in which you can consider the mentor, even part of your extended family. Your mentor can be a great role model, but if you do not connect well, you will not get the most out of this mentoring experience. Also, I know it can be very difficult to take criticism but do not run away if you are criticized or if you receive critique from the mentor (constructive non-personal criticism). This is something that can help you grow.

3.3.2 What Are The Hallmarks of a Good Mentor-Mentee Relationship?

Theresa
In order for a mentor-mentee relationship to work well, regular contact and check-ins are essential. These allow the mentee to maintain momentum in their career plans to discuss and problem-solve any challenges and a chance to get feedback. The communication in any meetings should be open and honest with respect, kindness, and dignity being the cornerstones of the relationship. What I found really positive in my experience with my current mentors is that you leave a meeting feeling uplifted, so mentors instill that confidence in you and are emotionally intelligent in their support. A good mentor will open their network and provide opportunities to the mentee. At the same time, a mentor will foster a safe space to allow the mentee to make mistakes and learn from them. A mentor will help lift mentees up in their career, helping them realize their potential.

From my experience, the very best mentors have been those that don't just support you in your career but support you as a whole person. So your personal life, your values and your choices. If mentors can lean into that and really support you in your whole being, then that's where I feel you have the most fruitful mentorship.

Michael
I think that also relates to the openness aspect. Sometimes you might not feel so comfortable to open up completely in front of your research advisor, but you will do that in front of someone else who might know you also a bit better. I think one other important aspect that you raised about personal development is also very important. I think one can see this from the perspective of athletes training.

Mentoring should play a very important role in helping you achieve your maximum potential. Give your best and bring the best impact you can in the world. And here, certainly, you need more mentors to be involved. And there are people that can help you basically improve your thinking, improve decision-making and build these skills. So it's not about scientific research skills. It's about other soft skills, if you can call them like that, emotional intelligence skills, which I think are more and more important nowadays.

3.3.3 What Are the Responsibilities of a Mentor and a Mentee in the Relationship?

Theresa
I think one of the main responsibilities of a mentor is providing honest feedback and helping the mentee identify areas for further development. They also need to be available for regular check-ins. For a mentee, when you have that feedback from your mentor, it is important that you proactively act on it. So understanding that sometimes constructive criticism is there to help guide you and help you grow so to

view it and use it in that way. It is also useful to respect the time your mentors give to the relationship as well. Being prepared for meetings can help you get the most out of the time with your mentors so for example having points of conversation or a rough agenda and having any actions prepared that have been met since the previous meeting.

3.3.4 Should You Develop a Mentoring Relationship Distinct from That with Your Research Advisor?

Michael
I think this relates to the openness element I have highlighted before. Especially in the beginning it could be difficult for you to open up and discuss matters related to your personal life and career planning with your research advisor. This is when it is helpful to have a mentor who is not a direct research supervisor but someone who stays with you while you transition through different educational and career phases. I think such a mentor is important and you can see this as a long-term mentoring relationship. At the same time, your current research advisor can become a long-term mentor too.

3.3.5 What Is a Challenge You've Faced as a Mentee or Mentor and How Did You Resolve It?

Michael
As a mentee, I feel I was shy in the beginning to ask for more help and guidance. I think one should not feel uncomfortable to do so. Expecting that the mentor is not able to provide more support or that the mentor will not be able to answer a particular question can prevent you from asking it and will lead you to reach wrong conclusions.

Theresa
I think one of the main challenges I've faced as a mentee has been not realizing the power of the mentor-mentee relationship. In my experience as an extended leaver, I thought my time away would signal the end of my academic career and what I really needed to do was reach out and trust in my mentor. In the end I received both the informal and formal support from my mentors during my time away that helped me re-enter academia with experience under my belt in that time away.

3.3.6 What Is a Common Pitfall in Mentor-Mentee Relationships and How Can It Be Addressed?

Michael

Do not expect that the mentor should take decisions for you. I think you need to be aware that the mentor does not know you and the context better than you do and has limited responsibility when it comes to your decision-making process. You can ask for advice, receive it but assume full responsibility for important career-related decisions that you make.

Theresa

For those that take a non-traditional academic path (such as extended leavers), there are systemic failures in the academic and research system to support the mentor-mentee relationship during extended leave periods. Oftentimes, the level of support during extended leave is based on the mentor-mentee relationship and as such there are various levels of support available that are relationship-dependent.

This needs to change and as such organizational culture needs to change to ensure that the same level of support is offered to **everyone** during the time away. We need formal support frameworks across institutions that are specific to an individual needs, are flexible and innovative. Examples of formal support could include setting up casual contracts or honorary positions for the mentee. It might include a phased return to work or having protected funding pots for carers to fund childcare support at conferences. Examples of informal support could include setting up parent and carer networks and various platforms to share experiences [3]. My plea to HE and research institutions is to realize the benefits of this approach in retaining talent, provide choice and flexibility in your support and facilitate an open, ongoing, and constructive dialogue.

3.3.7 What Is Your View or Experience on Reverse Mentoring?

Michael

There was a very nice comment from Sir Godfrey at the *Ask Me Anything* session. He pointed out that it's important for experienced researchers to also be mentored by young junior researchers. And I think that addresses quite well this concept of reverse mentoring. I think there are mutual learning benefits, and especially the pace at which science advances nowadays. I feel it's important and it's something that can provide beneficial to mentoring as well. And, apart from that, I think there is always something to learn from each personal interaction that you have.

Reverse mentoring is something I believe becomes integrated into the relationship, it happens through questions asked by the mentee which can bring new perspectives in the mentor's thinking, also through sharing news related to the mentee's learning experience, participation in events and so forth. I don't think having a formal structure for it helps in any way.

Theresa
You learn something from everyone that you have interactions with. So in my teaching role as a Programme Director and Module Leader, I am involved in teaching management and delivery. Here I am thinking of the interactions I have had with current and previous cohorts of students. We've got these really multicultural cohorts with different experiences, age ranges, and I learn something new from students and early career researchers every day. To give you an example, when I was a postdoc, we used to use paper-based A-Z maps to get to all our field sites where you use the co-ordinates to find your way around. It took an informal chat with a postdoc recently to say, 'well should I send you a google pin to get to a fieldsite?' I responded with 'what is a google pin?' So it turned into a lesson for me, learning about what google pins are, how they work and what they can be used for. This doesn't always have to be formal and that reverse mentoring happens a lot of the time without me even realizing it, from students and early career researchers. It does help to go in with an open mind and to be willing to learn from anyone you interact with.

3.3.8 Switching Careers/Disciplines or Approaching Highly Interdisciplinary Research

Theresa
I ultimately took 5 years out of academia and when I returned I went down a teaching route, so I'm slowly getting back into the research over time. I found it quite difficult at the start to get into teaching. I have always been driven by the science and research but I quickly made mentors around teaching in my previous institution. You tend to get communities of practice for example around teaching, similar to having research networks and I found mentors that supported me along the teaching route in that way. So finding people that work in the same sort of area and are passionate about, for example in my case, the teaching and then developing that community. So I didn't have a formal mentor as such. It was more of an informal mentor but they helped guide me through the process of what I needed to do if I was going to follow the teaching route through academia.

Also, for those that have taken extended leave, or considering it, you also pick up really valuable skills in that time out which helps you to become a better mentor and arguably better researcher and/or lecturer when you return. You pick up skills in academia through research and through teaching but similarly you pick up skills in the time away for example project management, resourcefulness, time management, problem-solving and teamwork that can all add to your skills base.

Michael
It is not only about changing fields but it is also about new fields. We have new interdisciplinary areas such as cancer neuroscience, which is neither typical cancer biology nor typical neuroscience. There is something like environmental neuroscience now being termed and most probably will emerge as a research area

by itself. So what happens when you want to jump into an area like this? It's a novel area. Who would you go to talk to now? I think this is a bit, a bit more challenging because there are few experts in general who will take first steps, pioneering steps in developing these areas. So when I looked into cancer neuroscience, there were maybe 6–7 research groups around the world, not more than that focusing on this, so, there are in general, a few people you can talk to. I think then it's important to expand your outreach and do not look only at academics but look also at entrepreneurs for example. They're usually among the first to also take more risky steps into new directions. Possibly researchers from some industry departments too.

I also wish to relate my experience taking an independent study course during my PhD studies at KAUST. I told my PI one day that I am really interested to learn more about lactate metabolism in gut microbiota (because the focus of my project was on deciphering the role that lactate plays in signaling in cortical neurons) as I was reading lots of new articles focused on lactic acid bacteria in the context of the gut-brain axis. And that I have considered signing up for an independent study course. The curriculum allowed me to find a second supervisor that coordinated the course together with my PI. So under the guidance of an expert in microbiology and my PI, I designed this independent course focused on lactic acid bacteria metabolism in the gut and brain physiology. It was an amazing learning experience and something I encourage all those who have this opportunity offered by their university to go for it.

3.4 Future Outlook of mentoring

Theresa

I would like to stress the importance of mentors and thank all the mentors out there, especially for those supporting mentees that are taking a non-standard route through academia or research. The support that you provide can be completely transformative and life-changing. For mentees out there who are thinking about taking extended leave or are currently taking extended leave, speak to your university or research institute support services about what is available in terms of support. In your time away, continue to nurture your networks and if you are so inclined, keep your hand in with papers and developments in your field as it can really help when you return.

Michael

Try to always dream big, aim high, and, do not limit yourself by, let's say, the capacity around you or the people around you. You can find the path. You can find the way to achieve what you dream of and what you're capable of. So always push for that, ask for help, connect with like-minded people.

References

1. Homer, Wilson ER (2018) The Odyssey, 1st edn. W.W. Norton & Company, New York
2. Mercer T (2023) The support I needed. Science 382(6675):1198
3. Mercer T, Harris J, Corstanje R, Kerai-Jones C (2024) Support for faculty on long-term leave is a career lifeline. Times Higher Education Campus

4 Discussion on Game Changers in Science and Scientific Work Enabled Via an AI-Supported Real-Time Survey

Karin Hrovatin and Ulrich A. K. Betz

Contents

Abstract

The knowledge about important upcoming technologies and scientific challenges is key for developing research management strategies. However, while general game changers and issues are described in different reviews, they often lack individuals' perspectives. A scientific team of high diversity, composed of academic and corporate researchers from all over the world, was surveyed for key emerging game changers in science and scientific work during the alumni meeting of the renowned Merck Innovation Cup. The obtained results were analyzed with the help of ChatGPT, which enabled us to analyze the survey data on the fly and thus further navigate discussion during the meeting. This led to new insights about game changers as perceived by a diverse group of researchers.

Keywords

Innovation · Game changer · Artificial intelligence · Survey · Science · Talent · Foresight

K. Hrovatin · U. A. K. Betz (✉)
Future Insight e.V., Darmstadt, Germany

Merck KGaA, Darmstadt, Germany
e-mail: karin.hrovatin@merckgroup.com; ulrich.betz@merckgroup.com

U. A. K. Betz (ed.), *Science for a Better Tomorrow*,
https://doi.org/10.1007/978-3-031-93623-4_4

The progression of science relies on integrating novel discoveries and technologies into future research. In line with this, previous work has discussed upcoming trends that hold the potential to fundamentally reshape research and its practical applications [1–3]. Concurrently, new advances that will enhance the scientific workflow were likewise described [4]. This research encompasses overarching trends with broad-reaching implications spanning diverse disciplines and work aspects, such as the rise of powerful deep-learning models or new sustainability approaches, as well as field-specific advancements, such as new in vitro model systems in medicine and hard-drive paradigms in computer science [2]. Nonetheless, papers are often outdated by the time of publication [5] and cannot capture all emerging technologies due to the rapid growth in the number of scientific publications [6]. Therefore, we used the opportunity of the Merck Innovation Cup Alumni Symposium in June 2023 to gain first-hand insights into what may be the emerging trends and which problems still need solving.

Merck Innovation Cup Alumni Symposium stands as an ideal platform for the discourse on the future of science and research work. A diverse ensemble of attendees spans generations, ranging from postgraduate students to active researchers with varying levels of experience and to retirees who stayed connected to the scientific community. The participants have backgrounds in different disciplines encompassing STEM fields, such as medicine, life sciences, chemistry, and computer sciences, as well as social sciences, such as economics and law. This leads to the generation of diverse ideas and enables discussion from multiple perspectives. Furthermore, as Merck Innovation Cup focuses on bringing together highly innovative individuals with a collaborative spirit, such brainstorming discussions can catalyze truly ground-breaking ideas [7, 8]. The industrial setting of the cup also facilitates the translation of promising concepts into practical implementations. Here, we discuss how we collected inputs from the participants and summarize the obtained results.

4.1 Conducting an Open-Ended Survey On-the-Fly

The design of our session included two phases: an open-ended survey followed by a subsequent discussion guided by the survey outcomes. Employing an open-ended survey as the basis for our discussion provided multiple advantages. First, we wanted to give voice to everyone, which otherwise would not have been possible during the discussion due to time constraints as well as could be inhibited by the different cultural and personality characteristics of participants affecting willingness to speak before an unfamiliar audience with hierarchical disparities [9]. Second, to obtain new ideas it was necessary to give the participants the opportunity to express themselves freely. Third, responding to open-ended questions can promote introspection, which in our case presented a beneficial preparation for the follow-up discussion.

We supported survey preparation and analysis with ChatGPT, which can be used in multiple ways for survey design. It can enhance initial brainstorming about questions that cover the desired topics. Furthermore, it can aid in refining these queries to increase comprehensibility across a diverse audience, for example, by considering different user personas, while also detecting framing biases [4, 10–12]. Lastly, ChatGPT can serve as a proofreading tool for the questions [13].

Particularly for survey analysis, the use of ChatGPT was necessary as responses needed to be prepared for the group discussion within a limited time frame of a few hours. Consequently, we employed ChatGPT to propose answer categories and classify answers accordingly. ChatGPT's capabilities encompass text summarization, sentiment analysis, and topic annotation [13, 14], which makes it promising for survey analysis, as was also proposed for other NLP models [15–17]. Nevertheless, our experience revealed that complete automation of response analysis was elusive, with recognized ChatGPT limitations expanded upon below. Instead, we employed ChatGPT to facilitate manual analysis by suggesting response topics that were subsequently refined to better align with the discussion audience for the discussion. Notably, ChatGPT has exhibited susceptibility to forgetting [18], which also presented a challenge in our analyzes as only some, but not all of the answers from a list, were assigned to automatically recommended categories in the final output. However, this issue could potentially be ameliorated by modifying the task, for example, initially employing ChatGPT to define categories and subsequently tasking it with categorizing individual responses into these categories. In our experience, the suggestion of the most relevant categories or interesting answers needs manual refinement. This phenomenon could partially be attributed to recognized biases of ChatGPT toward progressive views [19], which could have led to the prioritization of progressive responses over the identification of the most relevant categories or most interesting answers. These issues may potentially be mitigated by more precise prompts, which are pivotal in steering ChatGPT toward desired outputs [20].

Hence, while ChatGPT shows great promise for automated real-time open-ended survey analysis, inherent limitations persist. Consequently, forthcoming efforts should concentrate on delineating optimal workflows for ChatGPT utilization in this context. This would be of great use, given that conventional audience questionnaires either compromise expressiveness by relying on predefined answers or hamper interpretability through the utilization of word clouds. More sophisticated audience questionnaire tools are thus needed to improve outcome representativeness.

Lastly, a significant drawback of ChatGPT lies in its lack of privacy, rendering its application in sensitive settings untenable. However, in the future large language models may be migrated on personal devices [21] and more organizations may establish local instances for their members. Notably, Merck already provides a private version of ChatGPT to its employees.

4.2 Digitalization and Teamwork as Key Targets for Research Optimization

To identify participants' viewpoints on the future of science and its potential enhancement, we posed four questions: (1) Emerging trends: "Which emerging trends or technologies (or old ones that have been overlooked) are you most excited about? How would you apply them in your work/research area? (2) Overarching challenges: "What is the most important problem we should join forces on in the future?" (3) Pain-points: "What are the key challenges or pain-points in your daily work/research that need to be addressed?" (4) Future skills: "What skills or competencies do you think will be crucial for professionals in science to thrive in the future?" We below provide a brief overview of the most interesting survey results. This summary was generated specifically for this publication and does not directly match the more exhaustive synopsis used during the brainstorming session. In the latter, we summarized answers to each question into 3–5 themes alongside several thought-provoking examples.

Numerous responses reflected broader trends in society or specific research fields, including medicine, computer science, and sustainability. Among emerging trends were: personalized medicine, large language models, and different AI applications ranging from task automation to data analysis; and among overarching challenges were: climate change and inequality. Nevertheless, we also identified answers that are rarely broadly discussed outside of specialized domains but are nevertheless of great value. For example, in the emerging trends question, one of the participants suggested "3D printing for tailored lab equipment" and another participant reflected on the importance of metabolomics, which is not among the most widely profiled omic layers due to technological challenges [22]: "Metabolites are the immediate signature of the metabolic status of the organism, something proteomics or genomics cannot match. Metabolites may therefore prove novel biomarkers or therapeutic targets. Furthermore, being small molecules, they could be delivered more easily than e.g. proteins." In the overarching challenges question, the following responses provided a new perspective: "disconnect between the development of technologies and the ways in which they can/will be used" and "better interdisciplinary communication as we all become more specialized so that the big picture is not lost." Seemingly non-standard answers surfaced especially within pain-points and skills categories, which prompted the participants to reflect more deeply about themselves rather than their work. For example, participants pinpointed diverse time-consuming and repetitive tasks that impede their daily work, such as manual annotation of scientific protocols and laboratory results, email correspondence, slide preparation, grant applications, and last-minute changes of weekly plans due to experimental results or tool availability. Similarly, many concerns regarding knowledge and data accessibility emerged, spanning the literature search, data extraction from papers, and the digitization of historical data and existing supply chains. Many of these challenges could be resolved through advancements in computational tools and digitalization. Notably, this was also suggested among emerging trends by some respondents, such as "automated graphic generation to make presentations" and "democratization of access to knowledge and technology through large language models." Such dynamic exchange between problem owners and technology experts harbors potential for innovative solutions that are unlikely to be proposed by any of the participants individually. This aligns with the spirit of the Merck Innovation Cup where participants with different expertise work together to create an innovative business proposal.

In general, there was an optimistic sentiment toward emerging deep-learning tools as enhancers of productivity. This was especially clearly evident from the future skills topic, where many answers nominated the use of the so-called AI and computational approaches as critical skills: "to adjust to technological change" and "big data and AI training for everyone." Additionally, the importance of the awareness about the limitations of emerging deep-learning tools was also exposed.

Despite the overarching theme of digitalization, an emphasis on social skills and team dynamics emerged in the pain points and future skills topics. Some answers underscored the significance of cross-disciplinary communication, including with funding bodies and patients, pooling of knowledge within teams, and other collaborative competencies.

We also observed ideas that likely originated from different participant groups, such as researchers in academia and industry, but were applicable to both sectors. This emphasizes the need for inter-sector discussion for collective solutions. Examples of such answers included concerns over "loss of knowledge when PhD students leave the laboratory; knowledge transfer/documentation" and "digitization of supply chains, even digital savviness of employees; the capacity of industry to move at the same speed as research."

4.3 Advancement of Science by Identifying Common Challenges

The execution of the open-ended survey presented a unique opportunity for trend identification, illuminating pivotal challenges and potential game changers as perceived by individuals. This unveiled two key themes—digitalization and teamwork—which resonated strongly across participant groups. The universality of these challenges suggests an urgent need for innovative solutions, which can further evolve into new business applications. Moreover, the survey revealed sought-after skills in both industry and academia. This can help junior scientists cultivate expertise that align with the evolving job market, paving the way for informed career decisions. Overall, the survey and the subsequent brainstorming session provided new insights across participant groups, fostering the exchange of ideas and the potential for novel cross-disciplinary collaborations.

Acknowledgments K.H. acknowledges past financial support from Joachim Herz Stiftung via Add-on Fellowships for Interdisciplinary Life Science.

Conflicts of Interest U.B. and K.H. are members of Future Insight e.V.; U.B. is Senior Vice President of Innovation at Merck KGaA and Chairman of the Board of Future Insight e.V.; K.H. consulted for Merck KGaA in the scope of the Merck Innovation Cup and is currently employed by Merck KGaA.

Declaration of Generative AI in Scientific Writing During the preparation of this work the authors used ChatGPT in order to spellcheck the text. After using this tool, the authors reviewed and edited the content as needed and thus take full responsibility for the content of the publication.

References

1. Betz UAK (2018) Is the force awakening? Technol Forecast Soc Change 128:296–303
2. Betz UAK, Arora L, Assal RA, Azevedo H, Baldwin J, Becker MS, Bostock S et al (2023) Game changers in science and technology—now and beyond. Technol Forecast Soc Change 193:122588
3. Wang H, Fu T, Du Y, Gao W, Huang K, Liu Z, Chandak P et al (2023) Scientific discovery in the age of artificial intelligence. Nature 620(7972):47–60
4. Noy S, Zhang W (2023) Experimental evidence on the productivity effects of generative artificial intelligence. Science 381(6654):187–192
5. Nguyen VM, Haddaway NR, Gutowsky LFG, Wilson ADM, Gallagher AJ, Donaldson MR, Hammerschlag N, Cooke SJ (2015) How long is too long in contemporary peer review? Perspectives from authors publishing in conservation biology journals. PLoS One 10(8):e0132557
6. Park M, Leahey E, Funk RJ (2023) Papers and patents are becoming less disruptive over time. Nature 613(7942):138–144
7. EBR (2018) An innovator's midsummer dream come true. Eur Bus Rev. https://www.europeanbusinessreview.com/an-innovators-midsummer-dream-come-true/
8. Matos ÂP, Kauffman A, Vickery C, Naegeli K, Strittmatter L, Sybirna A, van Blijswijk J et al (2020) Synthetic biology category wins the 350th anniversary merck innovation cup. Trends Biotechnol 38(1):1–4
9. Tiferes J, Bisantz AM (2018) The impact of team characteristics and context on team communication: an integrative literature review. Appl Ergon 68:146–159
10. Baki Kocaballi A (2023) Conversational AI-powered design: ChatGPT as designer, user, and Product. arXiv [cs.HC]. arXiv. http://arxiv.org/abs/2302.07406
11. Chen Y, Wang R, Jiang H, Shi S, Xu R (2023) Exploring the use of large language models for reference-free text quality evaluation: a preliminary empirical study. arXiv [cs.CL]. arXiv. http://arxiv.org/abs/2304.00723
12. Junprung E (2023) Exploring the intersection of large language models and agent-based modeling via prompt engineering. arXiv [cs.AI]. arXiv. http://arxiv.org/abs/2308.07411
13. Kocoń J, Cichecki I, Kaszyca O, Kochanek M, Szydło D, Baran J, Bielaniewicz J, et al (2023) ChatGPT: Jack of all trades, master of none. arXiv [cs.CL]. arXiv. http://arxiv.org/abs/2302.10724
14. Kuzman T, Mozetič I, Ljubešić N (2023) ChatGPT: beginning of an end of manual linguistic data annotation? Use case of automatic genre identification. arXiv [cs.CL]. arXiv. http://arxiv.org/abs/2303.03953
15. Beeferman D, Gillani N (2023) FeedbackMap: a tool for making sense of open-ended survey responses. arXiv [cs.CL]. arXiv. http://arxiv.org/abs/2306.15112
16. McGillivray B, Jenset G, Heil D (2018) Extracting keywords from open-ended business survey questions. arXiv [cs.CL]. arXiv. http://arxiv.org/abs/1808.10685
17. Schonlau M, Weiß J, Marquardt J (2023) Multi-label classification of open-ended questions with BERT. arXiv [stat.AP]. arXiv. http://arxiv.org/abs/2304.02945
18. Laskar MTR, Bari MS, Rahman M, Bhuiyan MAH, Joty S, Huang JX (2023) A systematic study and comprehensive evaluation of ChatGPT on benchmark datasets. arXiv [cs.CL]. arXiv. http://arxiv.org/abs/2305.18486

19. Rutinowski J, Franke S, Endendyk J, Dormuth I, Pauly M (2023) The self-perception and political biases of ChatGPT. arXiv [cs.CY]. arXiv. http://arxiv.org/abs/2304.07333
20. White J, Fu Q, Hays S, Sandborn M, Olea C, Gilbert H, Elnashar A, Spencer-Smith J, Schmidt DC (2023) A prompt pattern catalog to enhance prompt engineering with ChatGPT. arXiv [cs.SE]. arXiv. http://arxiv.org/abs/2302.11382
21. Zhu X, Li J, Liu Y, Ma C, Wang W (2023) A survey on model compression for large language models. arXiv [cs.CL]. arXiv. http://arxiv.org/abs/2308.07633
22. Alseekh S, Aharoni A, Brotman Y, Contrepois K, D'Auria J, Ewald J, Ewald JC, Fraser PD, Giavalisco P, Hall RD, Heinemann M, Link H, Luo J, Neumann S, Nielsen J, de Souza LP, Saito K, Sauer U, Schroeder FC, Schuster S, Siuzdak G, Skirycz A, Sumner LW, Snyder MP, Tang H, Tohge T, Wang Y, Wen W, Wu S, Xu G, Zamboni N, Fernie AR (2021) Mass spectrometry-based metabolomics: a guide for annotation, quantification and best reporting practices. Nat Methods 18(7):747–756. https://www.nature.com/articles/s41592-021-01197-1

Part II

Healthy Lives

The U.S. National Institutes of Health: Inspiring Innovation in Biomedicine and Health

5

Steven M. Ferguson and Michael L. Salgaller

Contents

Abstract

With its unique system of intramural and extramural research programs, funding for academic and corporate product development, the U.S. National Institutes of

S. M. Ferguson (✉)
Department of Health and Human Services, Office of Technology Transfer, National Institutes of Health, Bethesda, MD, USA
e-mail: fergusos@od6100m1.od.nih.gov

M. L. Salgaller
Department of Health and Human Services, Technology Analysis and Marketing Unit, Technology Transfer Center, National Cancer Institute, National Institutes of Health, Bethesda, MD, USA
e-mail: michael.salgaller@nih.gov

U. A. K. Betz (ed.), *Science for a Better Tomorrow*,
https://doi.org/10.1007/978-3-031-93623-4_5

Health (NIH) has grown from very simple roots changing not only the face of healthcare, but also leading to the creation of the biotech industry in the USA. Whether your interest in biomedicine and health innovation from a scientific, medical, educational, or commercial perspective, the NIH should be a part of your future.

Keywords

Innovation · Biomedicine · NIH

5.1 Introduction

The start of National Institutes of Health (NIH), and thus the origins of its "innovation ecosystem" begins in 1887, when a one-room laboratory was created within the Marine Hospital Service (MHS), predecessor agency to the U.S. Public Health Service (PHS). The MHS itself had been charged by Congress in the 1880s to examine passengers on arriving ships for clinical signs of infectious diseases—especially for dreaded diseases of cholera and yellow fever—and prevent epidemics. Joseph J. Kinyoun, a young MHS physician trained in the new bacteriological methods being reported in Europe, was chosen to set up a one-room laboratory in the Marine Hospital at Stapleton, Staten Island, New York (Photo 5.1). Dr. Kinyoun (essentially the first NIH Director), called this facility a "laboratory of hygiene" to indicate the laboratory's purpose was serving the public's health. Within only a few months, Kinyoun identified the cholera bacillus in suspicious medical cases and used his Zeiss microscope to confirm his colleagues' clinical diagnoses. In stimulating and assisting other parties to improve healthcare, we see the very beginnings of this unique innovation ecosystem built around NIH.

Besides being the founding NIH Director, Dr. Kinyoun also focused on what we could call today bioentrepreneurship and technology transfer. Working first as a federal employee and later in the private sector, Kinyoun invented and patented multiple industrial disinfecting machines used in quarantine operations—such as the "Kinyoun Portable Bed Disinfectors." He developed the first smallpox immune serum; his "Kinyoun Method" of smallpox vaccination used until the 1960s. His "Kinyoun Stain," discovered for tuberculosis, is still in use today. Late in his career, he even worked in pharma for a firm that became a predecessor to Merck [1]. Clearly, Kinyoun led by example in founding NIH both an institution and innovation ecosystem.

Photo 5.1 Dr. Joseph J. Kinyoun, NIH Founder

5.2 NIH Today

Despite his own remarkable vision and activities, Dr. Kinyoun could hardly have imagined the size and scope of the NIH's present programs and the supportive environment for biomedical research and product development fostered today. From its humble beginnings as a single laboratory, the NIH has evolved into a comprehensive program of 27 institutes and centers (ICs) both national and international in scope.

As a result of the numerous scientific opportunities and funding programs that make up today's NIH, the environment it that NIH nurtures continues to accelerate even more significant contributions to human health, innovative medical products and economic development. Using innovations from federal labs to spur both technological and economic development arose from the 1986 Federal Technology Transfer Act. This act codified and promoted partnerships between NIH intramural research and the private sector to develop new medical products.

Around 90% of NIH's $48 billion FY 2024 budget went to more than 300,000 researchers at over 2500 universities, medical schools, companies and other research institutions in every state in the USA and throughout the world. The 1980 Bayh-Dole Technology Act codified and fostered partnerships between NIH-funded extramural research and private-sector development of new medical products [2]. The remaining funding (~10%) was spent on internal NIH R&D projects (Intramural Research Program; IRP) carried out by the approximately 6000 scientists employed by the NIH. Dozens of NIH-supported scientists from its own faculty and around the world have received Nobel Prizes for their groundbreaking achievements in Physiology or Medicine, Chemistry, Physics, and Economic Sciences. To date, 174 NIH supported researchers have been sole or shared recipients of 104 Nobel Prizes. This number includes Noble Prize winners serving as NIH scientists in the IRP.

The continuous process of biomedical research and product development requires a supportive environment and an innovative ecosystem. For new research to truly yield new drugs, devices and reagents, both public and private sector institutions need to use innovation to refine and build upon basic knowledge to enable the development of even better products. Uniquely for NIH, it does not matter whether an idea originates in a supported university laboratory, its own intramural research program or even the private sector. Nor does it matter the geographical origin of the innovation. Each new medical idea can be evaluated and supported based upon its own scientific and product merits—regardless of its origin. Collaborations, publications, and research tool sharing help ensure that important findings percolate through and invigorate the entire scientific community. For NIH's innovation ecosystem, new findings serve as a building block for establishing a deeper understanding of human health and disease and are supported through a wide variety funding, educational, training and developmental programs.

5.3 Structure of the NIH Innovation Ecosystem

To truly function as the foundation of an ecosystem, an institution or organization must realistically be able to help stimulate and sustain two primary functions—for biomedicine these two would be both new research as well as product development. Most biomedical products have of their research and development history that can be traced back to basic research institutions with the original research often funded by NIH or other governmental programs. Licensing and technology transfer programs at these federal labs or other non-profit research organizations then provide a means for getting new inventions to the market for public use and benefit. From a research institution's perspective, this portion of the innovation ecosystem is quite desirable. Public and commercial use of inventions typically comes with new recognition of the value of basic research programs at the university or organization in which it originated. These inventions also serve as helpful means to attract new R&D resources and partnerships within the ecosystem to these laboratories. Through licensing or other technology-transfer mechanisms, these institutions also receive a "return on investment"—whether that is measured in terms of financial, educational or societal parameters—or some combination thereof. An example of this would be the recently concluded *Public Health and Economic Impact Study of NIH Intramural Technology Transfer Licensing,* which measured and assessed the significant impact the technology licensing program of the NIH IRP had in innovation, healthcare, and economic development [3].

5.4 NIH Innovation Keystone: Bayh-Dole and the Birth of Technology Transfer

In 1980, continuing the momentum of the policies of Presidents John F. Kennedy and Richard Nixon, Senators Birch Bayh and Robert Dole enacted legislation that gave universities, nonprofits, and small-businesses the right to own inventions made by their employees for federal government-funded research. The Bayh-Dole Act of 1980 (P.L. 96-517) reversed the presumption of title ownership by NIH in NIH grants and permitted a university, small business or nonprofit institution to elect and pursue ownership of an invention in preference to the government. The underlying spirit of this important piece of legislation was to maximally utilize the outstanding research at these universities and other recipients for the good of the public who funded the research through their tax dollars. This set the stage for explosive growth of a new system of innovation built around government biomedical funding agencies, such as NIH.

Though the ownership right that universities and other funding recipients have to these inventions comes with obligations, these also have stimulated activity in the ecosystem. The primary obligation for these institutions is to actively market and attempt to commercialize the invention, preferably through U.S.-based business enterprises (including start-ups) to benefit the public. Thus, was born the field of "technology transfer" and the establishment and growth of technology-transfer offices (TTOs) now part of every research campus. Prior to Bayh-Dole, 28,000 patents were owned by the U.S. government—less than 5% of which were commercialized. Since the enactment of Bayh-Dole, more than 7000 newly created companies are still operational. The combination of innovation and entrepreneurship resulted in billions of dollars of direct economic impact within the USA and more than 700 new products put in the market during those years—all largely based upon NIH or other agency funded research [4].

Similarly in the 1980s, federal intramural laboratories—including NIH—were also given a statutory mandate under the Stevenson-Wydler Technology Innovation Act (P.L. 96-480), the Federal Technology Transfer Act (P.L. 99-502), and Executive Order 12591. These actions ensured that new technologies developed in federal laboratories were similarly transferred to the private sector and commercialized. One metric of the success of NIH IRP innovations licensed by the private sector is the resulting 47 new FDA-approved drugs and vaccines [5].

Within the system of innovation, NIH and NIH-funded universities developed a more strategic focus for their technology-transfer activities that more fully supports working with entrepreneurs. Although licensing revenue is collected, maximization of such funds is not the goal. Instead, research organizations find themselves primarily looking for increased product launches, company formation and new job creation based upon NIH-funded inventiveness. This process also supports faculty recruitment and retention, enhanced access to follow-on research funding and, in general, creating an entrepreneurial culture to attract venture investment. The economic development aspects of research are now recognized as a "fourth mission" for such institutions—along with education, research, and public service. Entrepreneurs play a key role in this fourth mission by establishing companies driven by new research discoveries—thus building out the innovation ecosystem.

5.5 Accessing Technologies and Collaborations in the NIH Innovative Ecosystem

Generally, bioentrepreneurs can directly access NIH-supported research and inventions for product development from three main sources, as given in Table 5.1. For research funded by grants and contracts from NIH (extramural research), the individual university or small business controls commercial rights. Biomedical research conducted by NIH itself (intramural research program) is licensed directly through the individual IC technology transfer offices or their service centers at NIH [6]. The full spectrum of NIH intramural technology transfer activities is given in Table 5.2.

Both NIH and NIH-supported research institutions have a robust research program "pipeline" providing novel, fundamental research discoveries available for commercial applications. NIH, for instance, as both a large-scale provider and consumer, represents a sort of "supermarket" of research products or tools for its commercial partners and suppliers. Additionally, overall product sales of all types by NIH licensees generally are around $7 billion annually. As previously mentioned, most NIH intramural technology transfer activities date from the Federal Technology Transfer Act of 1986 which first allowed federal laboratories to keep their license royalties and share them between the individual inventors and further, internal reinvestment.

Research collaborations and assistance from NIH or NIH-funded institutions can take several forms as these researchers and clinicians can work with industry under different collaborative agreements. For example, research institutions may seek to access technologies developed by industry for research studies—an imaging tool, a sequencing platform, or a drug discovered and in development by a company. The technology transfer office then works with companies and clinical partners to memorialize the understanding between the scientists and/or clinicians to allow the collaborations to happen. The key components of these collaboration agreement are terms related to inventions, rights to inventions, confidentiality versus publication,

Table 5.1 Sources for accessing NIH-funded research in the innovation ecosystem

• Intramural Research (from institute technology transfer offices)
• University Grantee Research (from individual university technology transfer offices)
• SBIR and STTR Programs (from individual small business awardees)

Table 5.2 Intramural NIH technology transfer innovation ecosystem activities

managing conflicts of interest and indemnification—especially for work involving patient care.

5.6 Industry Collaborations in the NIH Innovation Ecosystem

There are several types of research or collaboration-related agreements that companies commonly encounter in working with NIH and NIH-funded institutions:

Confidential Disclosure/Nondisclosure Agreements (CDA/NDA) Prior to engaging in any collaboration, each party may need to disclose to the other party some proprietary information that—if passed on to third parties—might be detrimental to the interest of the disclosing party. Such a discussion is a necessary first step to determine the interest in, and the breadth and scope of any potential collaboration. The parties negotiate a CDA/NDA ensuring the information disclosed is held confidential, only used for establishing the collaboration, stipulates a term of how long the information needs to be held confidential and describes the consequences of nonadherence to agreement terms.

Material Transfer Agreement (MTA), Sponsored Research Agreement (SRA), Research Collaboration Agreement (RCA), Clinical Trial Agreement (CTA), and Cooperative Research and Development Agreement (CRADA) Companies, both small and large, typically need to invest a significant research and development funds toward developing drugs or other biomedical products. NIH and NIH-funded research institutions have several programs key towards understanding the fundamental biology underlying a wide variety of commercial products. When companies and research institutions seek to collaborate, they often have very different focuses. A company seeks to get through regulatory requirements and onto the market as expeditiously as possible. Their aim is better understanding the mechanism-of-action to penetrate as many verticals as possible and have discoveries arising from collaborations improve the addressable market for their eventual product. In the case of collaborations with NIH-supported clinical programs, access to patient samples may be an option. Samples provide valuable insights the company hopes will guide them through clinical validation of their product—be it a potential drug, medical device, or diagnostic. In contrast, NIH or university investigators are often interested in testing experimental solutions from companies to build a scientific insight or medical knowledge that is publishable. The good news is that agreements are crafted to satisfy the needs of both parties. And it is possible under CRADAs (for NIH) or SRAs (for NIH-funded entities) for the investigator to receive additional funding support from the company for basic or clinical research programs, in turn, enhancing commercial development.

MTAs and SRAs are agreements dictating the terms of the transfer of material and/or money from the company to the academic institution. Similarly, at NIH, joint projects with companies for basic research or clinical studies can be formalized as CRADAs. RCAs are more appropriate when no IP options or provided funding are anticipated. Because of their clinical hospitals and centers as well as other networks and facilities, the NIH and at least some of its supported universities can also take some medical discoveries (or those of their partners) into early clinical trials through CTAs.

5.7 Licensing Technologies from the NIH Innovation Ecosystem

Basic Licensing Principles of University and Federal Laboratories Compared to technology licensing with corporations, NIH and NIH-supported institutions bring a different focus and perspective to the table when negotiating technology transfer agreements. These agreements are used to further overall institutional missions. Therefore, representatives from such nonprofit institutions consider the public consequences of such licenses as their priority—not the financial terms involved. For example, when compared with their peers in industry, NIH-funded institutions have the mandate to make new technology as broadly available as possible. This means that there is a strong preference to limit the scope of a license to only what is needed to develop specific products. Exclusive licenses are quite typical for biomedical products such as vaccines, therapeutics, and others. The underlying technologies require substantial private risk and investment (and a prior public notice and comment period in the Federal Register in the case of NIH). In their agreements, NIH laboratories and universities would also typically expect to retain the right to permit further research use of the technology whether to be conducted either in the NIH intramural program, universities or companies. Because the commercial rights granted represent institutional (and public) assets, these agreements have enforceable performance benchmarks to ensure that the public will eventually receive the benefit (through commercialized products) of the research it funded. Regulations governing the license negotiation of federally owned technologies and their mandated requirements are described in more detail at 37 Code of Federal Regulations (CFR), Part 404, while those for federally funded technologies can be found at 37 CFR Part 401.

In a license agreement, the research entity essentially grants rights to a company to make, use and sell products that were it not for the license, would infringe on the patent rights that the research center owns and/or controls. In some instances, the research center also grants the company rights to use technological information/know-how or materials that goes together with the information in the patent application and that is valuable to the company as it hopes to commercialize the technology into products. Licensing is at the heart of operations of a technology transfer office since NIH or NIH-funded universities function as nonprofits. They do not and cannot have a product commercialization arm. NIH or NIH-funded universities cannot convert inventions into commercial products and processes. They must partner with industry to do that—as is often the case with NIH-funded U.S. small businesses under the Small Business Innovation Research (SBIR) programs. Thus, these out-licensing activities are the key for research programs to fulfil the core of the Bayh-Dole Act and other federal mandates to commercialize arising from NIH funding.

Licensing from NIH and NIH-Funded Laboratories Commercializing technologies (for example, vaccines or drugs) and then entering world-wide markets cannot be the responsibility or mission of research institutions or government agencies. As is the case with its funded universities, the NIH is not able to commercialize its discoveries even with its considerable size and resources. Instead, it relies instead upon industry partners. Companies with access to the needed expertise financial resources are needed to undertake continued development of these inventions from the NIH or other research institutions into final products. Typically, a royalty-bearing license agreement with the right to sublicense is given to a company from NIH (if NIH-owned) or the university (if university-owned) to use patents, materials or other assets to bring a therapeutic, vaccine, or other product concept to market. Exclusivity is almost always the norm for the U.S. Food and Drug Administration (FDA)-regulated products due to the risk involved in time, money and regulatory pathways involved for companies and their investors. Financial terms of the license agreement are negotiable but typically reflect the nascent, high-risk nature of the discovery. Technologies coming from NIH or NIH-funded research are often early-stage; many years prior to generating revenue as a commercialized product. Consequently, most licensees are early-stage companies or start-ups. Larger firms typically want assets further along in the product development continuum. In addition to the license agreement, there will often be research collaborations between the licensee and the NIH or university to assist with additional work needed. When the NIH licensee can sufficiently "de-risk" the technology, these companies then sublicense, partner or get acquired by larger biotech or pharmaceutical firms. Extensive resources are needed for the final, most expensive stages of development with the large company expected to sell the product once it achieves regulatory marketing approval.

Start-Ups as Licensing Vehicles in the NIH Innovation Ecosystem Since the 1980s, federally funded health research institutions developed an active but increasingly strategic focus on improving public health through technology-transfer activities. As such, they are particularly interested in working with start-ups and other early-stage companies looking to develop and deliver innovative products. Rather than just seeking a financial return through revenue generation, these institutions are looking to utilize licensing of nascent inventions to increase new company formation, support faculty recruitment and retention, enhance research funding and create in general a more entrepreneurial culture within the organization. Successful start-ups attract venture investment and develop the product for as many indications as possible.

The licensing practices for most NIH-funded nonprofit research institutions changed significantly over time with respect to biomedical inventions [7]. Until the last 20 or so years ago, most of the important medical products based on licenses from university or federal laboratory research came from direct agreements with large pharmaceutical firms. With its ever-increasing consolidation, large pharmaceutical firms are typically no longer looking to directly license early-stage technologies for commercialization. In contrast, the number of licenses signed with start-ups as well as small- to medium-sized biotechnology companies is rising. Typically, around 70% of the total licenses are executed with start-ups and small biotech firms. Most success stories tend to be from those originally partnered with biotech or other smaller companies at the time of the original license agreement. Some examples from the NIH licensing program are:

- Kepivance® (a human growth factor used to treat oral sores arising from chemotherapy licensed to Amgen).
- Velcade® (a small molecule proteasome inhibitor used to treat multiple myeloma from Millennium).
- Synagis® (a recombinant monoclonal antibody for preventing serious lung disease caused by respiratory syncytial virus in premature infants from Med-Immune).

- Prezista® (an HIV protease inhibitor used to treat drug-resistant AIDS patients from Tibotec).
- Taxus Express® (a paclitaxel drug-eluting coronary stent used to prevent restenosis from Angiotech).

Although these firms or their successors are all now late-stage, well-resourced companies, they were early, small companies when the underlying technology was licensed to them.

5.8 Funding in the NIH Innovation Ecosystem

NIH is well known as the largest public funder of biomedical research in the world and invests roughly $43 billion a year to support outside institutions to enhance life and reduce illness and disability. This level of funding supports a strong research ecosystem that has led to breakthroughs and new treatments, helping people live longer, healthier lives, and building the research foundation that drives discovery. NIH offers funding for many types of grants, contracts, and even programs that help repay loans for researchers.

While perhaps best known for grants to academic scientists throughout the world, NIH also provides private sector U.S. entities with nondilutive funding through the Small Business Innovation Research (SBIR) and Small Business Technology Transfer Research (STTR) programs which are now known collectively as the NIH Seed Fund [8]. The NIH Seed program is perhaps the most valuable and stable funding source for new companies. These grant funds do not need to be repaid—unlike small business loans or convertible notes.

Other noteworthy advantages of the NIH Seed Program for small companies (eligibility as defined by the U.S. Small Business Administration) include:

- retention by the company of any intellectual property rights from the research funding,
- receipt of early-stage funding that doesn't impact stock or shares in any way (e.g., non-dilutive capital),
- national recognition,
- verification and visibility for the underlying technology,
- generation of a leveraging tool that can attract other funding from private venture capital or angel investors.

The SBIR program itself was established in 1982 by the Small Business Innovation Development Act to increase the participation of small, high technology firms in federal R&D activities. Under this program, departments and agencies with R&D budgets of $100 million or more are required to set aside 3.2% of their R&D budgets to sponsor research at small companies. The STTR program was established by the Small Business Technology Transfer Act of 1992 and requires federal agencies with extramural R&D budgets over $1 billion to administer STTR programs using an annual set-aside of 0.45%. In FY 2023, NIH's combined SBIR and STTR grants totaled over $1.3 billion [9].

The STTR and SBIR programs are similar in that both seek to increase small business participation and private-sector commercialization of technology developed through federal support of R&D. The SBIR program itself funds early-stage research and development at small businesses. The unique feature of the STTR program is the requirement for the small business applicant to formally collaborate with a research institution in Phase I and Phase II.

However, the SBIR and STTR programs at NIH differ in two major ways. First, under the SBIR program, the principal investigator must have their primary employment with the small business concern at the time of the award and for the duration of the project period. However, under the STTR program, primary employment is not so stipulated. Second, the STTR program requires research partners at universities and other nonprofit research institutions to have a formal collaborative relationship with the small business concern. At least 40% of the STTR research project is to be conducted by the small business concern and at least 30% of the effort is to be conducted by the single "partnering" research institution.

As a major mechanism at the NIH for achieving the goals of enhancing innovation and public health via commercialization of new technology, the NIH SEED Program presents an excellent funding source for start-up and other small biotechnology companies. They are structured in three primary phases: Phase I (feasibility), Phase II (development), and Phase III (commercialization).

In addition to receiving funding through the NIH SBIR and STTR programs, small companies may also be eligible for technical and management assistance programs designed to increase their chances for successful commercialization of the funded technology. Initially, the SBIR and STTR programs provided little or no assistance beyond the funds itself. More recently, many initiatives have been established to educate, mentor and connect awardees to valuable stakeholders. These are a key part of the NIH innovation ecosystem and include:

NIH Entrepreneurship Bootcamp—The NIH Entrepreneurship Bootcamp is designed to equip life science investigators and nascent companies with specialized innovation and entrepreneurship training. The course requires no prior experience. It uses a life science-focused customer discovery process to assess customer and stakeholder needs, and teaches participants to develop stronger business models, market strategies, and commercialization plans prior to their initial SBIR/STTR application.

Innovation Corps (I-Corps™) at NIH—The I-Corps program provides funding, mentoring and networking opportunities to help SBIR Phase I awardees commercialize promising biomedical technology. During this 8-week, hands-on program, companies learn how to focus their business plans and get the tools to bring their treatment to market. Program benefits include funding up to $55,000 to cover direct program costs; training from biotech sector experts; expanding professional networks; creating a comprehensive business model; and gaining entrepreneurial skills.

Technical and Business Assistance Needs Assessment—The Needs Assessment Report provides a third-party, unbiased assessment of an NIH Phase I project's progress in technical and business areas that are critical to success in the competitive healthcare marketplace. This no-cost report helps companies strategize a project's next steps.

5.9 Using NIH Basic and Clinical Research Assistance to Enhance Innovation

Basic and clinical research assistance from the NIH institutes is also available to companies or other partners through specialized services such as drug candidate compound screening and preclinical and clinical drug development and testing services—offered by several programs. These initiatives are particularly targeted towards developing and enhancing new clinical candidates in the disease or disorders strategically aligned with NIH's mission. The largest and perhaps best-known of these programs are found in the National Cancer Institute (NCI) [10]. The NCI has played an active role in the development of drugs for cancer treatment for over 50 years. This is reflected in the amazing fact that approximately one half of the chemotherapeutic drugs currently used by oncologists for cancer treatments were in some form discovered and/or developed with NCI. Its Developmental Therapeutics Program (DTP) promotes all aspects of drug discovery and development before testing in humans (preclinical development) as part of the Division of Cancer Treatment and Diagnosis (DCTD). NCI also funds an extensive clinical (human) trials network to ensure that promising agents are tested in humans. NCI's Cancer Therapy Evaluation Program (CTEP), also part of the DCTD, administers clinical drug development. Compounds can enter at any stage of the development process—even those needing extensive testing before applying for human trials. Drugs developed through these programs include well-known products such as cisplatin, paclitaxel, and fludarabine. Clinical, translational and pre-clinical research collaborations are also available with intramural investigators in all 27 NIH institutes.

Established in, the National Center for Advancing Translational Sciences (NCATS) is designed to assist companies with the many costly, time-consuming bottlenecks that exist in translational product development [11]. In partnership with both the public and private organizations, NCATS develops innovative ways to reduce, remove or bypass such bottlenecks to speed the delivery of new drugs, diagnostics, and medical devices to patients. NCATS does not conduct its own drug development—as would a company. Instead, it focuses on using science to create powerful new tools and technologies for wide adoption by translational researchers in all sectors. NCATS-supported programs and projects also produced numerous tools to help basic and clinical researchers advance translational science.

Programs of note for the NIH innovation ecosystem from NCATS include:

- *Bridging Interventional Development Gaps (BrIDGs)* which enables research collaborations to advance candidate therapeutics for both common and rare diseases into clinical testing.
- *Clinical and Translational Science Awards (CTSA)* support a national network of medical research institutions that work together to improve the translational research process to get more treatments to more patients more quickly.
- *Therapeutics for Rare and Neglected Diseases (TRND)* offers collaborative opportunities to access rare and neglected disease drug-development capabilities, expertise, and clinical/regulatory resources.

There is additional assistance available from other institutes in a wide range of unmet medical needs—including infectious diseases, drug abuse, and others; too many to be highlighted here. All in all, such efforts can provide a wide variety of technical assistance (typically at modest or no cost) for preclinical and even clinical development of novel therapies or other biomedical products by a variety of partners within the NIH innovation ecosystem.

5.10 Contracting Opportunities with NIH and NIH-Funded Institutions

One of the most overlooked opportunities by biomedical-focused companies is the ability to sell products and services to the NIH and NIH-funded centers. This opportunity is relevant and highly valuable to companies with a business model viewing NIH as a customer rather than development partner. Indeed, the NIH may be the first customer for start-up companies looking to develop new products used in conducting basic, translational or clinical research. With an intramural staff of about 18,000 employees, laboratories in several regions of the country (most at the main campus in Bethesda, Maryland), and an annual intramural budget of over $4 billion, the NIH is perhaps the largest individual institutional consumer of bioscience research reagents and instruments in the world. A variety of mechanisms for selling products and services to the NIH are possible, including stocking in government storerooms and general contracting opportunities. Companies providing products and services to NIH laboratories and programs generate cash flow and revenues to fuel their own R&D. Moreover, they start to demonstrate commercial acumen to buy-side stakeholders. As the world's largest biomedical research organization, the NIH has hundreds of contracting opportunities. Specific information on such opportunities is found on the NIH Office of Acquisition Management and Policy website [12].

A series of annual research festivals are also an excellent starting point for companies hoping to sell products to the NIH [13]. These events are held at the Bethesda and Frederick, Maryland campuses. Part scientific, part social, part informational and part inspirational, such events draw a variety of small- to medium-sized bioscience firms to exhibit their product and services available to NIH. They are cost-effective since NIH does not profit from exhibitors. Lastly, they are quite egalitarian since all allotted exhibition space is identical regardless of a company's size and resources.

5.11 Training and Education in the NIH Innovation Ecosystem

In addition to traditional scientific training supported at all educational levels, NIH, and NIH-funded universities have set up or have access to educational programs that train scientists and engineers to greater appreciate the importance of commercialization. These programs are often funded and supported at NIH institute training offices. Furthermore, the NIH Office of Intramural Training and Education (OITE) provides resources and information to enhance the educational experience of NIH trainees. It assists with finding appropriate workshops, arranging individual career counseling and identifying other NIH resources to meet trainee needs. OITE resources are also available for trainees in the extramural NIH community. Other options for education and training include entrepreneurship centers and small business assistance programs at many NIH-funded universities as well as the Federal Laboratory Consortium for Technology Transfer that is also funded by NIH and other federal labs.

5.12 NIH Innovation Ecosystem Has Spurred Biotechnology Industry Growth

As previously noted, the economic development potential of biomedical research is being recognized as a "fourth mission" for research institutions such as the NIH—along with education, research, and public service. Thus, it is in this fourth mission that bioentrepreneurs and NIH find themselves again sharing the common goal of having new companies established based upon developing innovative research discoveries.

The economic importance of licensing and technology transfer has become better recognized in recent years and some of the figures can be quite striking. For example, the overall product sales of all types by licensees of NIH intramural research reported by the NIH Office of Technology Transfer are nearly $7 billion annually—equivalent to mid-tier Fortune 500 companies. Economic development also was the focus of a U.S. Presidential Memorandum entitled "Accelerating Technology Transfer and Commercialization of Federal Research in Support of High-Growth Businesses" [14]. This White House directive recognizes the economic aspects of innovation and technology transfer for federal research in the way it fuels economic growth as well as creating new industries, companies, jobs, products, and services, and improving the global competitiveness of U.S. industries. The directive requires federal laboratories such as the NIH to support high-growth entrepreneurship by increasing the rate of technology transfer and the economic and societal impact from federal R&D investments.

Looking at the university and academic medical center figures reported by the Association of University Technology Managers (AUTM), there are similar economic indications for the impact of technology transfer and the initial funding of research from NIH and other federal programs [15]. For 2023, AUTM reported 9299 new license agreements and new research expenditures of $104.9 billion by reporting universities. That same year, more than 7214 start-ups remained operational from prior years. By the end of 2023, 714 new products had been introduced into the marketplace.

5.13 NIH Innovation Ecosystem: Results to Date

With leading-edge research programs and focus on the healthcare market, NIH and NIH-funded research programs have an exemplary record in providing opportunities for bioentrepreneurs to develop both high-growth companies and high-impact medical products. Indeed, an early study from 2007 described over 100 drug and vaccine products approved by the U.S. FDA based at least in part on technologies directly licensed from university and federal laboratories. Federal labs (e.g., NIH) provided nearly 20% of the total [16]. A later study from 2009 showed that university-licensed products commercialized by industry created at least 279,000 jobs across the United States during a 12-year period. Further, an increasing share of the United States GDP each year was attributable to university-licensed products [17]. A 2011 study published in the *New England Journal of Medicine* [18], based upon their 2007 preliminary study, showed the intramural research laboratories at the NIH as by far the largest single nonprofit source of new drugs and vaccines approved by the FDA. Finally, a 2017 study from the NCI SBIR Development Center showed that out of 690 SBIR grant awards, 368 (53%) had already resulted in sales. Total cumulative sales were $9.1 billion, which equates to average sales of approximately $24.8 million for each of the 368 awards [19].

These strong sales underscore the significant impact of the NIH innovation ecosystem. It is strong and will be increasingly effective and important going into the future. Although new knowledge and product development has been a model in showing the value of the NIH innovation ecosystem from NIH and NIH-funded institutions, it is not the entire story. The final tally must include not only the full societal value and economic impact both of new companies, but more importantly the life-saving or impactful therapeutics, vaccines, diagnostics and other biomedical products on the market having origins in federally funded research. This is believed to be the truest measure of a system of innovation as well demonstrating the value and importance of having the growth of the intramural and extramural research programs of the NIH since its humble origins in 1887 [20].

References

1. Morens DM, Fauci AS (2012) The forgotten forefather: Joseph James Kinyoun and the founding of the National Institutes of Health. mBio 3(4):e00139–12
2. https://www.nih.gov/about-nih/what-we-do. Accessed 26 Dec 2024
3. https://www.techtransfer.nih.gov/reports/public-health-and-economic-impact-study. Accessed 26 Dec 2024
4. https://autm.net/AUTM/media/SurveyReportsPDF/AUTM-Infographic-23-DIGITAL.pdf. Accessed 26 Dec 2024
5. https://www.techtransfer.nih.gov/reports/hhs-license-based-vaccines-therapeutics. Accessed 26 Dec 2024
6. https://www.ott.nih.gov/tdcs. Accessed 26 Dec 2024
7. Ben-Menachem G, Ferguson S, Balakrishnan K (2006) Doing business with NIH. Nat Biotechnol 24(1):17–20
8. https://seed.nih.gov/. Accessed 26 Dec 2024
9. https://report.nih.gov/nihdatabook/category/8. Accessed 26 Dec 2024
10. https://dtp.cancer.gov/ and https://ctep.cancer.gov/. Accessed 26 Dec 2024
11. https://ncats.nih.gov/. Accessed 26 Dec 2024
12. https://oamp.od.nih.gov/. Accessed 26 Dec 2024
13. https://researchfestival.nih.gov/ and https://technicalvendor.com/. Accessed 26 Dec 2024
14. https://federallabs.org/about/who-we-are/history. Accessed 26 Dec 2024
15. https://autm.net/surveys-and-tools/surveys/licensing-survey/2023-licensing-survey. Accessed 26 Dec 2024
16. Jensen J, Wyler K, London E, Chatterjee S, Murray F, Rohrbaugh M (n.d.) The contribution of public sector research to the discovery of new drugs. In: Personal communication of poster at 2007 AUTM annual meeting
17. https://ipwatchdog.com/2009/10/28/bio-study-us-economy/id=6943/. Accessed 26 Dec 2024
18. Stevens A, Jensen JJ, Wyller K, Kilgore P, Chatterjee S, Rohrbaugh M (2011) The role of public-sector research in the discovery of drugs and vaccines. N Engl J Med 364:535–541
19. https://sbir.cancer.gov/impact. Accessed 26 Dec 2024
20. https://pmc.ncbi.nlm.nih.gov/articles/PMC8045787/. Accessed 26 Dec 2024 has been revised and updated for this chapter

Building Pandemic Resiliency in the Wake of COVID-19

6

Maximilian Marshall and Lauren Gardner

Contents

M. Marshall (✉)
Systems Engineering, Johns Hopkins University, Baltimore, MD, USA
e-mail: mmarsh29@jhu.edu

L. Gardner
Department of Civil and Systems Engineering, Johns Hopkins Whiting School of Engineering, Baltimore, MD, USA
e-mail: l.gardner@jhu.edu

U. A. K. Betz (ed.), *Science for a Better Tomorrow*,
https://doi.org/10.1007/978-3-031-93623-4_6

6.1 Introduction

COVID-19 affected the lives of almost everyone on the planet. It killed millions, crashed economies, and upset political power balances. Now imagine something similar happens again, today. A novel, highly transmissible virus emerges and starts to spread. We have no vaccine, no effective drugs, and don't fully understand how the virus affects people and is transmitted. How do you think the world would react? What would governments do? What would businesses do? What would you do? We ask these questions to contextualize the systemic danger that is still presented by a pandemic and to motivate an honest reflection about how to prepare for future pandemic threats in a world still grappling with the last one.

COVID-19 stressed almost every aspect of society, prompting a wide-ranging and varied response. Elements of this response, such as the historically rapid development of vaccines, were important successes and are vital for future resiliency. Additionally, COVID-19 was the first global pandemic to emerge in the age of the internet, leading to unprecedented and widespread gathering and dissemination of epidemiological data. However, the pandemic also brought us to the point of fracture and failure. This had profound social effects, from deaths in overcrowded hospitals to unexpected school closures, and rampant vaccine misinformation to inflation and increased the cost of living. These systemic failures and disruptions have created and exacerbated disconnections between the institutions responsible for public health and the general populace. Governmental and medical institutions are distrusted, their policies and recommendations limited by a lack of compliance. Emerging scientific advancements face a lack of scientific literacy, limiting the implementation of new technologies. Vaccines save lives and reduce harm, but their population-level efficacy is constrained by individual-level uptake. Looking ahead, we confront an uncertain combination of two paradigms inherited from COVID-19. On the one hand, we have significant technological and medical capabilities, along with a massive amount of relevant data and accumulated experience with pandemic response. On the other hand, many of our systems did not hold up to the pressure of the pandemic. We need to address both the structural reasons for this lack of resiliency and reckon with the ramifications of these failures on the public psyche, because SARS-COV-2 will not be the last dangerous pathogen to emerge in the twenty-first century. Therefore, we need to work toward a pandemic response system that fosters individual-level participation in higher-level response efforts, as conceptually illustrated in Fig. 6.1.

In this chapter, we consider the path forward for applied epidemiology in the twenty-first century. To do so, we draw on our experience creating and managing the JHU COVID-19 Dashboard, which was the preeminent worldwide source for COVID-19 data during the peak of the pandemic. The breadth of the Dashboard's use was well summarized by its description in Time Magazine's best inventions of 2020: the "data has been downloaded billions of times, helping governments to decide where to dispatch resources and when to reopen—and individuals to suss out the safety of hosting a socially distant backyard barbecue."[1] This description

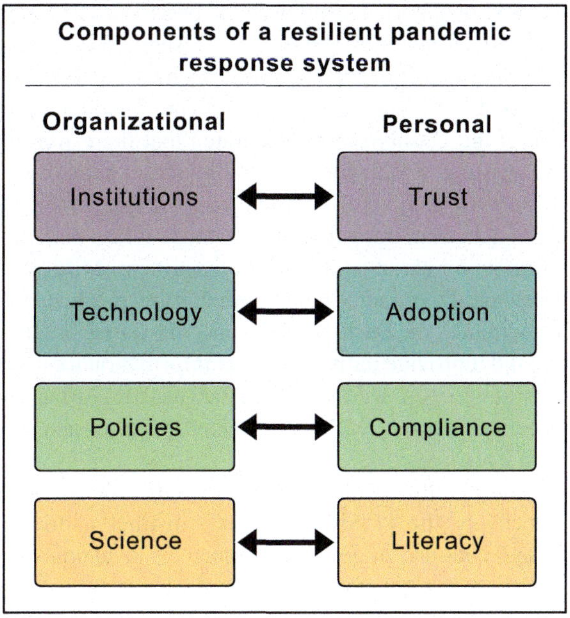

Fig. 6.1 The co-dependency between organization and personal components of a resilient pandemic response system

is particularly appropriate for our discussion in this chapter. It includes both the actual service that the Dashboard provided (making information available) and the uses to which it was put (governmental- and individual-level choices). The need for innovation in this space, the epidemiology landscape that begins with data and ends in decision-making, is the subject of our discussion in this chapter. First, it's worth a slightly more detailed discussion of the Dashboard, showing why it's a relevant basis for a forward-looking discussion about improving the information space around epidemiology.

6.2 What We Learned from Building the COVID-19 Dashboard at Johns Hopkins University

The Johns Hopkins COVID-19 Dashboard was a dynamic, interactive map and fully open platform for accessing all publicly reported global cases, deaths, and vaccination data for COVID-19. Launched in January of 2020, the Dashboard quickly became the most authoritative, centralized, real-time reporting, and open data sharing system for tracking the pandemic around the world. Within a week of its launch, the Dashboard and its data were receiving over one million hits per day, and by early March 2020, it was receiving over a billion hits per day. Cumulatively there have been over 220 billion queries (interactions and data requests) on the Dashboard and over 3.5 billion page views. Heads of state relied on the Dashboard to initiate stay-at-home orders, close schools and businesses, and impose travel restrictions. Dashboard data was used to decide where to hold COVID-19 vaccine trials. Several international corporations used the Dashboard to determine when and where to close and reopen factories and offices. Dashboard data also fed pandemic visualizations for almost every major media organization worldwide (including the New York Times, CNN, and The Guardian). The immediate global adoption and widespread use of the Dashboard demonstrated that there were (and still are) serious holes in the existing worldwide epidemiological data ecosystem. However, the Dashboard is not just indicative of what was missing from institutional and governmental data-gathering infrastructure. It tapped into the modern expectation for instant information about something so important. Millions of individual people used the Dashboard to monitor their communities' risk levels and inform their own behavior. Importantly, the Dashboard did not editorialize or make recommendations to people, not did it provide individual-level information like contact tracing. Nevertheless, by providing open access to granular, near real-time data on infections, deaths, and vaccinations, the Dashboard helped people make informed decisions about their own lives.

In sum, the Dashboard is the foundation of our discussion because it was effectively the prototype for two distinct pathways for the future of science-based, data-driven decision-making for pandemic preparedness and response. The first path (or the top-down approach) is the modernization and harmonization of existing epidemiological data and modeling infrastructure. There are reasons why a third-party data aggregator like Johns Hopkins University became the trusted source for pandemic data. What can we do to promote efficiency, gain trust, and build resiliency in such a fragmented system? The second path (or the bottom-up approach) is directly connecting with the public, recognizing that individual engagement with public health practices is a necessary subcomponent of aggregate, population-level outcomes. Therefore, it is critical to strategically implement public health communication that is resilient to a media environment rife with misinformation, especially when individuals are their own policymakers.

For brevity and clarity, we will generally refer to these two pathways as the "top-down" and "bottom-up" approaches during this chapter. Before diving into their specifics below, we identify here some aspects of the paths that we encourage the reader to bear in mind. First, both paths have the same intended starting and finishing points. Their shared start is epidemiological data, and their shared finish is, ideally, mitigating harms related to pandemic threats. However, the paths differ in the routes they take from beginning to end. The top-down path seeks to improve extant public health practice, and thus is grounded in more concrete recommendations based on our experience with public health institutions and systems throughout the pandemic. The bottom-up path aims to improve public health engagement and enable more scientific decision-making practices amongst the general public. Its tenets are slightly more speculative than the top-down path, as its goal is a sustainable inculcation of community values and trust in science. In fact, we believe that moving forward, both approaches will be mutually requisite. We need to invest in and strengthen both our epidemiological infrastructure and the public's receptivity toward public health. Finally, we note that in exploring these two paths, we are not identifying an *exhaustive* set of necessary innovations

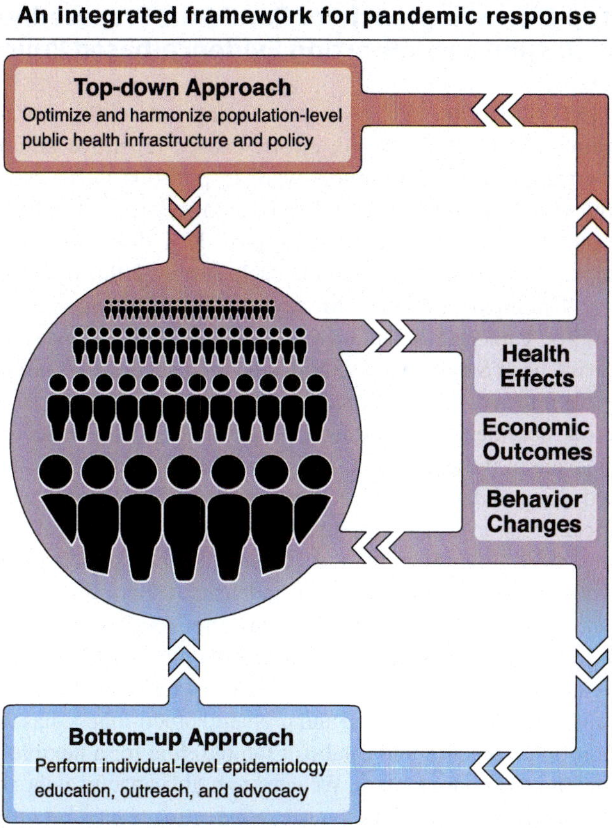

Fig. 6.2 Conceptual framework for an integrated pandemic response system, including both top-down and bottom-up pathways. The cyclic structure captures the feedback loops between policies and interventions, individuals' behavior and population-level outcomes

for epidemiology in the twenty-first century. We draw from our experience with the Dashboard (and other pandemic response efforts) to identify key improvements within the landscape of gathering, communicating, and taking effective action based on epidemiological data. In essence, our recommendations for the two pathways can be summarized as follows. The top-down approach is framed around optimizing the larger public health environment, whereas the bottom-up approach is intended to connect with people at the individual level and enable thoughtful and informed decision-making. As shown in Fig. 6.2, these approaches build toward a more harmonious, and therefore efficacious, combination of public health imperatives and personal decision-making.

6.3 A Top-Down Approach for Pandemic Preparedness and Response: Supporting Evidence-based Policy

Throughout COVID-19, the lack of a pre-existing data collection and modeling infrastructure to support public health efforts caused repetitive failures in policy-making and governance. These failures were in part due to the inconsistency of the messaging resulting from generation and publicization of extremely variable predictions and poor quality, biased data sets. This disconnection is a critical shortcoming, because in such a system, each stage of the process is not only dependent on, but highly sensitive to steps taken upstream. For example, bias in procedures for gathering ground-level data can lead to misleading estimates of disease incidence. Such uncertainty leads to inaccurate or highly uncertain epidemiological forecasts. Poorly communicated modeling results, in turn, have the capacity to erode confidence in or even the reliability of public health responses. The more separate each of these different stages are, the less likely relevant information is to be passed along as data is translated into decisions. In essence, a well-informed epidemiological decision-making process, especially in the context of an emerging pathogen, must tread a fine line. On the one side, oversimplification, assumptions, and anecdotes have power to misinform and lead to harmful outcomes. On the other side, an overly complex, detailed, or fastidious system may be too slow or inefficient to implement necessary changes before outbreaks grow out of control or cause significant damage. We believe a unified set of tools and best practices, along with a dedicated group of experienced and prepared practitioners, has the capacity to strike the appropriate balance and establish the baseline for a flexible and effective top-down pandemic response system. We need to build capacity at the intersection of public health, engineering, and data science, connecting researchers to practitioners to boost their collective potential for effective evidence-based decision-making. We organize and summarize the necessary elements of this top-down approach by separating it into the following three categories:

1. ***Data***—Gather and disseminate more of it at higher quality. Invest in expanding disease surveillance efforts, standardizing methodologies, and ensuring accessibility.
2. ***Modeling***—Develop new and improve existing modeling tools for early outbreak detection, short-term forecasting, and long-term scenario analysis.
3. ***Translation***—Connect researchers and practitioners to enable the translation of key data and modeling insights into decision support tools and timely evidence-based policy.

The inherent uncertainty of pandemic threats means that modeling is the core technical component of this pathway. Generally, the purpose of modeling is to fill gaps in existing knowledge and make informed predictions about unknown information. When responding to an emerging pathogen, models help fill holes in patchy real-time information. Models generate forecasts of the future. They are also used to test and refine policies by simulating their effects before implementation. However, in order for modeling tools to meaningfully contribute to prospective and real-time decision-making, they must be wrapped in a system that ensures their optimal functionality. This means that this system is capable of dealing with the fundamentally heterogeneous and uncertain data streams that define real-time observation of the pandemic landscape, and that the system must ensure that key decisions are able to be made with knowledge of this uncertainty. No matter the objective, a robust decision in the pandemic response space exists in this paradigm, requiring effective documentation, confrontation, and management of uncertainty.

Within the translation category, we encourage both testing and implementation of decision support tools, emphasizing the need for feedback between researchers and practitioners. This prospective process of cyclically working to improve outbreak preparedness is a key feature of our top-down approach. During "peace times" (i.e., no known imminent epidemic threats), prospective work is traditionally and shortsightedly devalued. This belies the fact that the epidemic risk landscape is in perpetual motion. Viruses constantly evolve in their host populations and human and animal migration is in perpetual flux. We should constantly be investigating, modeling, and evaluating different hypothetical outbreak scenarios. This is particularly relevant for governmental or healthcare entities with significant power to set policy or allocate resources, because the other side of the preparedness coin is the real-world reaction to an emerging threat. As the widespread failure to contain COVID-19 demonstrated, a pre-existing framework with a well-defined system for incorporating real-time and uncertain data is an essential tool for rapid response to a novel pathogen. Only a comprehensive system that incorporates both short-term awareness and response alongside longer term preparedness and optimal policy is suitable for robust pandemic prevention. The following sections detail each of the three aims above, including technical and practical methodologies needed for each step.

6.3.1 Aim 1: Expand Disease Surveillance and Implement Data Standardization

A critical first step in the development of modeling tools for pandemic preparedness and response is the collection and aggregation of data. Disease surveillance sets the tone for the rest of the process because data inputs have a profound effect on model utility and validity. In short, this means that increasing the quality and quantity of epidemiological data collection will directly improve decision-making capabilities. This is no small process and deserves significant investment and planning, as effective surveillance requires substantial resources and coordination. Before further discussion of this topic, we provide a general introduction to the current landscape of epidemiological data:

1. *Traditional epidemiological data* directly relate to disease status or outcomes, of which cases, deaths, vaccinations, and hospitalizations are common examples.

These variables have underpinned the majority of data-driven pandemic response efforts for COVID-19 and other outbreaks, and are used to generate the parameters that guide disease dynamics models. Despite the fact that these ostensibly measure health outcomes directly, it's important to take the sampling and sourcing of these data into careful consideration. Data reported from hospitals, for example, reflects the subset of the population who choose or need hospital admission. Trends in this subset might not reflect the larger population.

2. *Emerging epidemiological data* include new ways of measuring disease prevalence and characteristics. One example that is coming into extensive use is wastewater-based epidemiology (WWBE). Humans shed virus into the sewer via bodily waste, bathing/handwashing, and even laundry. WWBE assesses the disease prevalence at the community level by tracking the amount of pathogens in sewer systems. This type of active surveillance can be used to determine prevalence of a disease within a community, and has been implemented as an early warning system for a multitude of diseases including SARS-CoV-2, poliovirus, Hepatitis A, Norovirus, non-polio enterovirus, and more [2]. Another important emerging data source is genomic surveillance. Both COVID-19 and flu variants clearly demonstrate the necessity for accounting for disease variants in response planning, with vaccination as a key example application. However, we include genomic surveillance in the "emerging" category because of the increasing dissemination and access to this data online [3]. These systems are still confronted by inconsistency, lag times, and other barriers, but have the possibility to be effective early warning systems as well as sources of information about the pathogen itself [4].
3. *Non-epidemiological data* are sources of information that are not explicitly related to disease dynamics but are still of (potential) use in characterizing risk and outcomes. An important example of this is human mobility data from cell phones, which has a significant and well-established correlation with infectious disease transmission [5–9]. Another example is web search datasets, which have shown that increased COVID-19 related search activity can be a warning sign of early spread [10]. Other information sources worth considering include climate/hydrometeorological data and demographic data, as climate change will play a significant role in both human migration and vector transmission of diseases for the foreseeable future. Demographic information is also vital because diseases like COVID have been demonstrated to disproportionally affect minorities and historically underserved populations [11, 12]. Whenever possible, these disparities should always be accounted for.

Fundamentally, the purpose of gathering data for decision-making is to provide a meaningful and useful representation of reality. This does not always look the same, depending on where, when, and for what the data are meant to be used. For example, surveillance efforts in under-resourced locations might simply not have the capacity to capture data from their entire population. Instead, collectors might focus on surveilling wastewater locations that are most representative of the demographics of their populations. A different option (again depending on the use case for the data) might be to focus on direct outreach in areas with more vulnerable people, where earlier warnings have the capacity to save more lives. This sort of heterogeneity in need and flexibility in application of data collection means that it is necessary to support local, regional, and national surveillance efforts. Different groups with different priorities need different data. However, for the purposes of larger-scale and generalizable outbreak detection and modeling efforts, maintaining consistent data standards and availability across space, time, and hierarchies is vital. Thus, support for surveillance itself should also come with institutional support to manage, curate, and disseminate the resulting data.

Throughout COVID-19 a number of independently managed data catalogs were created, but they remained disparate, and none were fully optimized for data science applications. These datasets often suffered from inconsistent naming conventions, uneven quality control, and lack of alignment between disease information and potential predictors. Faulty data, often the result of sampling or aggregation problems, makes a poor source for robust modeling and analysis [13]. If data are incomplete, inconsistent, or do not reflect reality, models have limited reliable ground truth from which to learn or be evaluated. "Garbage in, garbage out" may be a cliché, but it is overused for a reason. In order to make effective decisions for the reality around you, you need a good understanding of the reality around you. In this aim, we are not naïve enough to suggest a single database formulation or data standard that will solve all of these problems in any context. Planning for a single hospital is entirely different than planning for a nation. Even so, there are essentials for constructing a useful and robust epidemiology dataset, no matter the context.

A quality epidemiology dataset provides a flexible and accessible framework that integrates different information streams for the purpose of analysis and disease modeling. Repositories should include data quality checks, address biases and uncertainty, and include new and updated data as they become available. Cataloging and evaluating metadata is a necessity, particularly documentation of limitations and biases. Spatial data should be harmonized using a consistent hierarchy and clearly documented formatting. Any spatial aggregation or disaggregation performed for harmonization needs to be explicitly documented in the metadata, along with any assumptions baked into the spatial statistics. Putting in this effort will yield a rich dataset of covariates that can be revised iteratively and deployed quickly, if necessary.

6.3.2 Aim 2: Develop New Modeling Tools

An important part of why we refer to the necessary innovations in epidemiology as "pathways" from data to decision-making is to remind people that this work has an end goal, and that the goal is to be useful. An efficacious mapping of epidemiological data to decision-making is, by definition, worthwhile. It seems ludicrous to say, but research in public health should be intended to help the health of the public. We bring this point up here in the modeling and data analysis section because of a review we conducted that quantitatively assessed prospective, data-driven modeling studies of COVID-19 [14]. In our review, we found significant gaps in model performance and transparency, especially on reporting aspects of models that are crucial for translating the work to real-world applications. Published modeling studies did not consistently state their precise objectives, detail their methodology, express uncertainty, evaluate performance across a long and varied timespan, or clearly list their limitations. Without this information, studies are more vulnerable to misinterpretation, which can have serious consequences during a global health crisis in which the public is paying attention to scientific literature [15]. In addition, poor reporting limits the ability to synthesize insights from research and determine best practices.

In essence, an effective institutional (top-down) pandemic modeling system is accomplished via successfully performing the following three objectives: early outbreak detection, short-term forecasting, and long-term scenario analysis. We argue that while early outbreak detection might instinctively seem the most important piece, the other two are equally valuable. Why? It comes down to answering a simple question. What is the point of a pandemic early warning system? Our answer is to try as best as we can to stop a pandemic in its tracks, to prevent it from ever happening. Doing this does not just mean detecting a threat, it also means responding to it. Therefore, a meaningful pandemic early warning system must not just warn that there is a risk, it must also inform (or ideally, suggest) policies to mitigate the risk. Short-term forecasting helps with critical decision-making over smaller timescales, and long-term scenario analysis allows for policy planning to prevent, for example, a local epidemic from becoming a worldwide pandemic. Each of these objectives requires unique versions of and interconnections between data, modeling, and communication. Additionally, it is not useful to assume that epidemic modelers will be able to predict reality perfectly, because such an assumption will lead to flawed expectations by policy makers and incorrectly applied decisions. Instead, the inherent uncertainty in underlying data and prediction performances must be carefully incorporated through all steps of the modeling process.

6.3.2.1 Early Outbreak Detection (Modeling Objective 1): "What Is Happening Right Now?"

High quality surveillance data are needed to identify outbreaks prior to significant spread. This requires not only real-time data reporting, but tools that are capable of rapidly processing this generally messy and incomplete information and identifying the true, underlying risks. Unfortunately, perfect data are hard or impossible to come by, especially in the context of an emerging pathogen. Addressing this need requires understanding the current situation given limited or poor-quality data. For example, biased testing or underreporting requires inferring the true number of cases. This is known as nowcasting. These transformed data must then be processed for signals that indicate specific growth patterns to be communicated to key decision-makers, while explicitly accounting for the possible error present due to potentially flawed data. These efforts require various forms of statistical models and anomaly detection methods.

Historically, passive surveillance has been the most widely available data source for alerting outbreak detection systems, assessing epidemiological risk and establishing situational awareness; however, it is unlikely to be sufficient at the spatial and temporal scale required for timely response. It is therefore promising that new active surveillance systems are becoming more readily available. As indicated in the above section, wastewater-based epidemiology (WBE) data, genomic surveillance and random population sampling are more likely to enable early detection on the actionable timescale required. Further investment in these strategies is worthwhile, along with development of standardized deployment methods for them.

Successful early detection of a novel pathogen enables (but does not guarantee) containment before significant infection of the adjacent population. It is therefore critical for reducing the impact of epidemics and preventing pandemics. The development of real-time outbreak detection tools, building on novel, timely data sources is an essential component of the top-down pathway to effective institutional decision-making. Decision efficacy could improve tremendously given distributed testing and processing routines that accurately detect the true risks of outbreaks in their earliest states.

6.3.2.2 Short-Term Forecasting (Modeling Objective 2): "What Is About to Happen?"

Even in a scenario with successful early warnings, short-term forecasting is still necessary. Forecasting aims to predict near term (days to weeks) epidemiological dynamics and health outcomes. On this timescale, it is generally assumed that epidemic outcomes cannot be significantly altered by human intervention. Thus, the utility proposition of forecasts is generally one of harm mitigation, not (for example) longer term policy development. For example, a short-term prediction of incident cases might be used for determining quarantine protocols in the context of an emerging pathogen, and a hospitalization forecast could help with optimal allocation of limited medical resources. Forecasting models, therefore, should focus on predicting near term disease outcomes that burden the healthcare system, such as hospitalizations and ICU occupancies, which can thus be applied to support real-time pandemic response such as optimal resource allocation.

Different combinations of forecast methodologies and research objectives (e.g., continuous versus discrete categorical risk targets) should be considered for optimal utility. Two critical components of this work are the incorporation of uncertainty into the predictions, and the development of novel evaluation tools to assess the performance of the models. Methods should be refined using simulated and historical pandemic scenarios to both increase the accuracy of predictions and infer the most important short-term factors driving outbreaks. We note that current forecasting and evaluation methods have generally proven insufficient for guiding evidence-based policy because of significant decreases in model performance under highly varying temporal and spatial disease dynamics. Putting this issue bluntly, it's really hard to accurately forecast surges. However, as more and more genomics data becomes available closer to real time, and as wastewater surveillance is scaled up, and given the ongoing advances in artificial intelligence, we believe that short-term forecasting is likely to become much more useful and is thus worth including in our top-down decision-making pathway [4, 16, 17].

Future forecasting work should focus on the development of accurate, high-resolution, short-term prediction methods. They should be designed to support governmental and healthcare decision-making in times of crisis and be transferrable across different types of diseases. Significant improvement in short-term forecasting utility would be a huge advancement toward consistently being able to prevent localized epidemics from growing into worldwide pandemics.

6.3.2.3 Long-Term Scenario Analysis (Modeling Objective 3): "What Will Happen If We Do X vs. Y?"

Scenario analysis refers to outbreak projections over longer temporal horizons (months to years) to evaluate the impact of different possible future contexts, i.e., intervention strategies, availability and uptake of vaccines, or new variants. Meaningful longer-term projections are an important part of pandemic mitigation because they allow for preemptive and proactive development of policy and deployment of resources. However, as the varied public response to COVID-19 demonstrated, changes in behavior occur due to new policies, changes in risk perception, political manipulation, or economic concerns. These reactive behaviors are traditionally excluded from modeling, severely limiting the utility of the best available tools [18]. Effective scenario analysis should integrate models of human behavior and infectious disease dynamics. By developing joint models for the spatiotemporal course of an infectious disease and the individual-level health and economic decisions that are made in response to the epidemic, scenario modelers could simultaneously predict the time-evolution of economic/behavioral and health/epidemic outputs under a given set of assumptions or scenarios. If appropriately calibrated and representative of real-world behavioral and epidemiological mechanics, both model accuracy and utility for evaluating the effectiveness of public health policies should improve. This allows for more comprehensive comparisons of the potential impact of different interventions under consideration.

6.3.3 Aim 3: Communication and Translation of Data and Models into Decision Support Tools

Enabling institutions to make assured and scientifically based decisions is the final, vital element of our first "top-down" pathway. This needs to be accomplished through capacity-building with key stakeholders in public health and policymaking as well as iterative development of communication strategies. The generally inadequate public health response to the emergence of COVID-19 demonstrates the necessity of this element. The scattershot, heterogeneous response generated in real time failed to contain, characterize, or effectively communicate the risk of COVID-19. However, the pandemic eventually galvanized broad governmental and public health efforts, but not in time to prevent a global pandemic and a stop a new virus becoming endemic in humans.

In general, modelers have a responsibility to facilitate appropriate stakeholder interpretation of their work. They should be explicit in transparently reporting how assumptions and limitations shape the interpretation of their results [19]. However, optimal support for decision-makers is often the result of a mutually beneficial, direct collaboration with modelers [15, 20]. On the one hand, these relationships allow modelers to better understand the needs of on-the-ground users of their products and tailor their work for maximum efficacy. On the other hand, stakeholder involvement in the modeling process facilitates better understanding of the utility and limitations of infectious disease modeling. This paradigm is also true for people and organizations involved with on-the-ground gathering of data. In other words, deepening the connection between those that generate the input for models is also important, not just those who use the output. This is an important aspect of our "top-down" approach that is worth emphasizing. By "top-down" we mean institutions, policymakers, governments, etc. acting in their aggregate capacities to try to improve public health outcomes. We do not mean that there is a linear flow from data, to modeling, to communication and translation of results. Iterative, prospective improvement of data/surveillance, modeling, and policymaking is a necessity. In turn, this will allow institutions to optimally react to and minimize the destructive impact of emerging epidemic threats.

6.4 The Bottom-Up Approach for Pandemic Preparedness and Response: Engaging the Individual

Now, we come to the transition point between our description of the "top-down" path and the "bottom-up" path. Before switching completely to our description of the latter, it is essential to consider why we are discussing both pathways, and not merely emphasizing one or the other. Until now, we've discussed how incorporating modern and emerging technology can be used to improve the efficacy of existing approaches in epidemiology. We call this a "top-down" approach because by and large, these efforts manifest as assertive, imposed, aggregate policies. For example, consider governmental action during COVID-19, such as imposing limitations on restaurants and closing schools. From a slightly different perspective, consider the massive investment in developing and distributing vaccines. Even though these are quite different examples, one being fundamentally restrictive while the other makes beneficial resources available, both attempt to alter the landscape of transmission by making large scale interventions that then "trickle down" to affect individual people. ***It's essential to consider this endpoint of individual people interacting with these systems, because these interactions are really where the rubber meets the road for public health outcomes***. Whether or not people choose to get vaccinated, whether or not they wear masks, if they choose to get tested when they feel a little sick, these individual choices are the components of larger, aggregate outcomes. Top-down interventions are essential, but they only help up to a certain point. Past that point, individual people must opt-in to risk reduction behavior or to engage with interventions. This is why the bottom-up path is necessary. Personal behavior will always have an effect on the efficacy of public health measures designed to limit disease spread through populations. Therefore, it is necessary to not only design appropriate (top-down) interventions, but to work additionally to promote engagement with these interventions and behavior that mitigates the spread of infection (bottom-up).

Before discussing how we would think this bottom-up approach would work, it is worth it to consider why individual-level engagement about epidemiology was and still is difficult. First of all, epidemiological interventions rarely have outright, easily discernable, inherently positive short-term benefits that can be observed at the individual level. Success in applied epidemiology is primarily an absence of negative outcomes. For illustrative purposes, consider the ideal COVID-19 risk reduction scenario for a specific individual, in terms of actions taken and outcomes incurred. They take all the appropriate measures they can (vaccination, masking, testing, etc.), and they do not get COVID-19. This is problematic in terms of building and maintaining sustainable, population-level impetus to behave in epidemiologically desirable ways because even in this ideal scenario, the most tangible benefit for the participant is the *absence* of illness. Humans are simply not built to find this consistently rewarding. Additionally, because we tend to learn heuristically, any deviation in behavior and/or change in outcome either reinforces irresponsibility, yields a negative outcome, or both. The options are as follows:

1. Behavior remains the same (caution), but the result changes (they get sick). Bad for short-term public health and also bad for long-term public health because it indicates heuristically that precautions don't work.
2. Behavior changes (riskier), but the result stays the same (they don't get sick). Bad for long-term public health because it indicates heuristically that precautions are not necessary.
3. Behavior changes (riskier) and the outcome change (they get sick). Bad for short-term public health.

As you can see, this is a tough situation from which to operate. It is hard, outside the context of an extremely deadly disease, to imagine these short-term heuristic incentives motivating emergent, population-level reductions in risky behavior. Therefore, we need to work to break people out of this paradigm of short-term optimization. Pandemics are significant threats to people's lives and livelihoods, and thus will always have the potential to cause significant social grievance. Under intense pressure, presented with the frightening and chaotic reality of an emerging pandemic, people crave clear explanations that reflect their frustration and affirm their existential fear. Thus, aggregate social grievance is a fertile ground for the spread of generalized suspicion. Individual people can then absorb this suspicion and apply it to aspects of their own lives, generating heuristic justifications for disengaging with best public health practices. This leads to apathy, skepticism, and engagement with predatory misinformation. However, this skepticism should not necessarily be thought of as an intrinsic barrier to efficacious public health practice. Instead, it can represent an opportunity to directly engage with the personal stakes people have in these matters.

Skepticism is, at least partially, a function of being historically underserved and exploited. It is incumbent upon current and future public health practitioners to work to build this trust back. This means educating people, demonstrating that their individual contributions to public health are worth it to the communities that they care about, even if it's hard to perceive the benefits for themselves. It means being proactive in convincing people that public health systems are operating in good faith. History is filled with predatory actors using people's legitimate grievances to advance their own agendas. As public health practitioners, what can we do to erode suspicion and push back against misinformation? What should we be doing to help people make the decisions that will help themselves? This is the purpose of our bottom-up pathway, which has three major tenets:

1. **Education**: Involve people in and educate people about epidemiology outside the chaotic context of an emerging pandemic.
2. **Personalization:** Customize the way we connect to and communicate with individual people, honoring their specific needs, backgrounds, and opinions.
3. **Transparency**: Demonstrate that public health agendas are intended to benefit the public and not to empower or enrich other entities.

6.4.1 Tenet 1: Education

As mentioned earlier in this chapter, we directly experienced the public demand for information in this space with the widespread use of the Dashboard during peak COVID-19 times. This was true despite the fact that the Dashboard was limited to providing real-time pandemic data. With this bottom-up path, we advocate for an extension and a deepening of what the Dashboard offered, with the goal of purposeful education to increase societal preparedness for pandemic response efforts. This preparedness has three different purposes. First and simplest, increasing knowledge about epidemiology increases the corresponding likelihood of informed decision-making. Second and by extension, a secure relationship with scientific fact is the best way to inure people against misinformation. As expressed above, misinformation during a chaotic time has immediate power and appeal that is unmatched by sober, scientific reality in the short-term. Thus, a longer-term effort ahead of time is required, leading to third purpose of education. Besides providing important information, we conceive of this effort as a form of inoculation: mental toughening and preparedness gained by exposing people to the realities of epidemiology outside the context of a pandemic. This way, in an urgent scenario, communications about public health priorities and policies will not seem as new, strange, or scary as they might otherwise. Broadly speaking, education gets people "in shape" for both engaging in effective pandemic response and resisting misinformation at an individual level.

6.4.2 Tenet 2: Personalization

While education for the sake of maximizing preparedness is the goal of the bottom-up pathway, personalized outreach is how this objective is implemented. Before delving into this application as it relates to our bottom-up pathway, it's worth briefly recognizing the larger context of the information space online. Hyper-specific personalized algorithms deliver people much of their information. Targeted advertising efforts cull huge amounts of personal data, predicting users' individual preferences in order to maximize the likelihood of successfully making a sale. In general, we are moving into an information ecosystem that is defined by the following relationship. For individual users, the media environments encountered are conditioned for them, are recognizably *theirs*. In exchange, actors (such as advertisers or political campaigns) wishing to route specific information to users can do so with extremely high precision. Public health outreach, therefore, must contend with and confront this personalized and algorithmic consumption of information. By "contend with" we mean matching the level of personalization that people have come to expect, and by "confront" we mean taking on the exploitative and divisive misinformation in the public health space. Misinformation efforts play on the divisions between people, using personal data to target messaging that emphasizes social discord. Effective public health outreach must therefore personalize messaging and recommendations without otherizing or blaming. At its root, this means meeting people where they are, recognizing the validity of each individual's needs, opinions, and lived experiences. This refers to both how messaging is delivered, as well as the actual content of the communication. Vaccine misinformation, for example, has propagated through wellness-focused social media [21]. By providing a target for people's anger and suspicion, anti-vax sentiment validates social frustration. Such scapegoating is a destructive shortcut to actionizing and taking advantage of legitimate social grievance during pandemics (and other times of unrest). Bad actors in this space stoke a general sense of frustration and distrust, which individual people apply to their own specific circumstances. Building capacity for a positive public health response (mental inoculation) must work the opposite way, by encouraging the individual choices that aggregate up to making a meaningful difference.

6.4.3 Tenet 3: Transparency

Mental inoculation and personal outreach are the first-order components of the bottom-up pathway. As the goal of the process and the methodology required to attain it, they are the direct action and outcomes that practitioners are directly engaging with. Transparency, however, refers to an emergent property of the bottom-up pathway that is necessary for its outcome to sustainably take root. In the age of algorithmic advertising, constant connectivity, and data mining, we are relentlessly bombarded with communication intended to get something from us.

Media networks have political biases. Social media influencers constantly promote their sponsors. Even the algorithms that present us information are designed to play with our emotions. We have become conditioned to expect an underlying extractive motivation behind so much of the content that we absorb. Therefore, an absolutely essential part of working toward achieving significant societal pandemic resiliency is to show people that public health policies are *not* doing this same thing. This is tremendously difficult, not least because a significant amount of current and recent public health messaging fails this test. For example, during the initial stages of the COVID-19 pandemic, public health modeling results were sensationalized by the media and bent to serve predetermined political purposes. From a different perspective, much of the "outreach" to promote COVID-19 vaccination and antiviral drugs is just advertising by their manufacturers, the pharmaceutical companies. It is tough to wholeheartedly believe public health messaging that so clearly has a profit motive baked in. Engaging in the individual-level communication space with the *clearly demonstratable, transparent goal of helping people,* not profiting off of them or benefiting from their support, is a prerequisite for credibility in this sort of effort. Transparency must emerge from a good faith demonstration of public health principles by public health practitioners.

6.4.3.1 Lessons from an Existing Application: What Public Health and Weather Have in Common

Now that we have broken down and explored the core components of the bottom-up pathway, let's consider it as a whole. What might real-world applications that follow this framework actually look like, and are there any comparable efforts from which we can draw inspiration? Weather forecasting is an illustrative example that synthesizes the three tenets of our bottom-up path. Here, we work our way through several aspects of weather prediction and forecasting that can be mapped onto our proposal.

6.4.3.2 Effectively Conveying Uncertainty

A foundational issue of operationalizing models and data-driven decision-making is the translation/explanation of uncertainty. This is a constant frustration for modelers. When presenting our work, people normally want a definite and specific prediction. For complex problems, it's impossible to make correct predictions all the time. Instead, models are useful because they present a range of possible outcomes or a set of different scenarios that could potentially happen. Weather is complex, it's a chaotic system that can't be perfectly predicted. Nevertheless, weather forecasts are able to effectively capture and explain some of this uncertainty to the general public. Take an example weather forecast: "there is a 50% chance of rain today." The uncertainty is expressed in a way that conveys the possible outcomes to the user, allowing them to plan for different potential scenarios. This might seem trivial, but it is actually an extremely effective translation of a highly uncertain forecast. A different, equally technically valid, but significantly worse way to communicate this would be "we have no idea if it's going to rain or not today." Fundamentally, these two statements convey the same information. However, the former expression clearly grounds the concept of uncertainty in the user's possible future experience. It expresses a clear characterization of two possible scenarios to plan for, instead of general confusion about what might happen. Bottom-up epidemiology communication needs to work in the same way. Instead of expressing individual-level risk with statements like "COVID cases have doubled in your city this week" or "you are twice as likely to get COVID this week as last week", it should be done with something like "2 out of every 10 people you see in public today are likely to have COVID, up from 1 out of 10 last week." Uncertainty is conveyed, grounded in the experience and potential risk to the individual, without making or implying an outcome ("you are going to get sick") that could be invalidated and reduce the authority of the message. In general, a successful adoption of this aspect for epidemiology messaging would help people understand that even though the future is not certain, it's still possible to take appropriate action oneself to minimize the risk from infectious disease.

6.4.3.3 Normalization of Use

Weather forecasts are a commonly used everyday resource. This leads to their second aspect that can be ported to our bottom-up pathway: normalization and trust-building from consistent lower-stakes communication. Most of the time, weather predictions help people make relatively minor decisions, like whether or not to bring an umbrella to work, or how warmly to dress. These are expected, run-of-the-mill daily choices. However, paired with the effective use of uncertainty referenced above, this means that strategic planning based on variable weather forecasting scenarios is incorporated into peoples' everyday routines. This consistent use of forecasts, along with their general utility, gives them a foundational level of trust and reliability. Importantly, weather forecasts are not always so low-stakes. During extreme events like hurricanes and tornadoes, weather services provide urgent and life-saving information. They make use of the fact that they already have a pre-existing way to route information to users in order to deliver critical messages like warnings and evacuation notices. Additionally, because users have a history with these weather service providers and have broad associations with them providing useful information, extreme warnings are less jarring and inherently rejectable. It is easier to internalize urgent and disruptive warnings when they come from a familiar and trusted source. This benefit is easy to see in an epidemic context, as it would encourage people to believe public health warnings and take recommendations seriously. However, this requires an equivalent normalization of day-to-day engagement with epidemiology. This, as of now, is not common practice. For most people, daily disease tracking is an unpleasant memory associated with peak COVID. Altering this paradigm means bringing epidemiology into people's lives outside the hysterical context of an emerging pandemic. People could be kept informed about the ratios of prevalent illnesses based on their location, and what they or their loved ones are most likely to have if they develop symptoms. This could also mean keeping people abreast of foodborne illnesses, updating them when new vaccines are released, or modeling disease prevalence in their communities based on wastewater data. Risk profiles relating to endemic diseases like influenza and COVID-19 could be customized for individual people, based on their own comorbidities or those of vulnerable people that they have regular contact with. Bringing epidemiology into more people's lives this way could also have the additional benefit of increasing awareness of positive community level outcomes, which as we described above is harder to feel at the individual level. This everyday experience and discourse will acclimatize people to the realities of epidemiology. In turn, this will help with the uncertainty and turbulence of emerging pandemic threats.

In sum, we see all three of our bottom-up pathway tenets present in weather forecasting. Mental inoculation takes the form of the choices people make every day based on predicted conditions. This consistent, low-level scenario analysis helps develop their ability to respond to uncertain (but much higher stress) situations. Personalized outreach is, of course, present in the specific location-based weather forecasting that people are given. However, it also manifests in the way weather forecasts are presented, ensuring relevance and utility for their users. Public advocacy is demonstrated by the alignment of the goals of the providers and users of weather forecasting. While undoubtedly monetized in ways like advertising or data collection, weather prediction services are incentivized to provide information that is as accurate and useful as possible. This builds a sense of trust in their user base. In general, this demonstrates the necessity and potential of our bottom-up approach for epidemiology communication. No matter what the top-down policy is, broad and aligned individual-level decision-making will make it easier to accomplish.

6.4.3.4 A Proposed Paradigm Shift for Public Health Communication

What, then, would it actually look like to implement this bottom-up approach for epidemiology? There is not a single answer to that question. In effect, we view it as a proposed paradigm shift for public health communication at large. It is an information ecosystem that is still emerging and mutable, with room to be filled in many different ways. New epidemiology dashboards are an option. Mobile phone applications that capture and reflect infection possibilities based on personal risk tolerance and location-based disease prevalence could also help. Personal biosecurity devices and IOT-based contact tracing could help engender more effective, community-based data collection. Social media is a massive and fertile ground for this kind of outreach and should be used to show people that epidemiology is a necessary part of our health and security infrastructure that must be maintained and invested in, even when pandemics are not in the news. Individual-level pandemic preparedness should be invested in at the local, provincial, national, and international levels. As we mentioned in the introduction of this chapter, the top-down pathway exists, but the bottom-up effort is more speculative. It's an information ecosystem that does not yet exist, at least not as a cogent effort with direct and widespread support from public health institutions and practitioners. It is vital for this investment to happen, because if the public health community does not positively engage in the individual-level communication/outreach space, we are effectively ceding that ground to actors who have other (generally more misinformative, divisive, and self-serving) interests. If we want our societies to be better prepared for the strain and challenge of the next emerging pandemic threat, this must change.

6.5 Conclusion

The genesis of our conception of both the top-down and bottom-up pathways was the role that the Johns Hopkins COVID-19 Dashboard played during the emergence and peak disruption of COVID-19. It represents both the promise and peril of future pandemic response efforts. On the positive side, epidemiology is a fundamentally data-driven discipline and we are moving into a world of increasing data availability and granularity. The Dashboard is powerfully indicative of this as it was the standard, ground truth source for pandemic information used worldwide to track, understand, and predict COVID-19. However, data alone are not a magic bullet, especially not in epidemiology. Sourcing, sampling, bias, and (mis)representation must always be considered. The Dashboard was not immune to these problems. Since it only gathered and reported data from local governments, Dashboard data was subject to wildly different baseline conditions. For example, countries with less testing infrastructure (often under-resourced countries) tended to report less COVID-19 activity than countries with more testing capacity. Without understanding this context, this gave the impression that, at times, the global south was relatively less affected, which was not the case. The potential upside of the data-driven world we're moving toward is significant, but we must remember that the picture painted by data is not always real. Despite the fragmented nature of our public health systems, the knowledge and expertise within them remains vital to interpret and operationalize new data resources. This leads to the improvements suggested in our top-down pathway, which are necessary in order to make optimal use of new data sources, emerging technology, and institutional experience. The Dashboard also showed us a different public health mechanism. Again, it only reported government data with no additional modeling or analysis for contextualization and interpretation. Nevertheless, it was still used by individual people worldwide because it was filling an important demand for real-time, location-specific pandemic information. Moving forward, the way individual people access epidemiological information must be contextualized and interpreted in a responsible and scientifically valid way, and this is where our bottom-up approach is applied. Sadly, many political and cultural figures are not friendly toward or supportive of public health practice. This is unfortunate, as the public should not have an adversarial relationship with public health. Nevertheless, public health practitioners need to try to engage with people to overcome hesitancy and suspicion, and to encourage positive engagement with best practices, recommendations, and healthier choices.

Finally, we reflect on our proposal with a common theme that emerges from both pathways. For provinces, countries, companies, and public health institutions, collaboration can be difficult or even dangerous. Entities may look to limit culpability, save money, or even save face by not being truthful about epidemic outcomes or removing resources from public health efforts. This instinct is mirrored for individual people. The discomfort of letting one's coworkers or friends know about a positive COVID test after spending time together reflects this same tendency. Working toward the collective good is difficult from both perspectives. This is why, at their core, both of our approaches promote collaboration as a pathway to epidemic resiliency. From the top-down perspective, we advocate for a harmonization and incorporation of data-driven decision-making that we think will improve the response capabilities of public health institutions en masse. From the bottom-up, we encourage the inculcation of individual responsibility, building the understanding that (sometimes difficult) personal choices aggregate up to positive community level outcomes. Collaboration also defines the relationship between the two pathways, which are not fundamentally separable. Public health institutions require a receptive population in order to implement efficacious policy. In turn, populations require effective public health practice for their own well-being. In the modern world, sustaining this relationship requires a useful exchange of data and ideas between individuals and institutions. This is a significant challenge, but we believe our approach provides a piece of the framework to get there.

References

1. Johns Hopkins Coronavirus Resource Center. The 100 best inventions of 2020 | TIME. https://web.archive.org/web/20241224172852/https://time.com/collection/best-inventions-2020/5911434/johns-hopkins-coronavirus-resource-center/
2. Shah S, Gwee SXW, Ng JQX, Lau N, Koh J, Pang J (2022) Wastewater surveillance to infer COVID-19 transmission: a systematic review. Sci Total Environ 804:150060
3. Elbe S, Buckland-Merrett G (2017) Data, disease and diplomacy: GISAID's innovative contribution to global health. Glob Chall 1(1):33–46
4. Du H, Dong E, Badr HS, Petrone ME, Grubaugh ND, Gardner LM (2023) Incorporating variant frequencies data into short-term forecasting for COVID-19 cases and deaths in the USA: a deep learning approach. EBioMedicine 89:104482
5. Badr HS, Du H, Marshall M, Dong E, Squire MM, Gardner LM (2020) Association between mobility patterns and COVID-19 transmission in the USA: a mathematical modelling study. Lancet Infect Dis 20(11):1247–1254
6. Ilin C, Annan-Phan S, Tai XH, Mehra S, Hsiang S, Blumenstock JE (2021) Public mobility data enables COVID-19 forecasting and management at local and global scales. Sci Rep 11(1):13531
7. Chang S, Pierson E, Koh PW, Gerardin J, Redbird B, Grusky D et al (2021) Mobility network models of COVID-19 explain inequities and inform reopening. Nature 589(7840):82–87
8. Guan G, Dery Y, Yechezkel M, Ben-Gal I, Yamin D, Brandeau ML (2021) Early detection of COVID-19 outbreaks using human mobility data. PLoS One 16(7):e0253865
9. Badr HS, Gardner LM (2021) Limitations of using mobile phone data to model COVID-19 transmission in the USA. Lancet Infect Dis 21(5):e113
10. Mavragani A, Gkillas K (2020) COVID-19 predictability in the United States using Google Trends time series. Sci Rep 10(1):20693
11. Tai DBG, Shah A, Doubeni CA, Sia IG, Wieland ML (2021) The disproportionate impact of COVID-19 on racial and ethnic minorities in the United States. Clin Infect Dis 72(4):703–706
12. Yaya S, Yeboah H, Charles CH, Otu A, Labonte R (2020) Ethnic and racial disparities in COVID-19-related deaths: counting the trees, hiding the forest. BMJ Glob Health 5(6):e002913. https://gh.bmj.com/content/5/6/e002913

13. Dong E, Ratcliff J, Goyea TD, Katz A, Lau R, Ng TK et al (2022) The Johns Hopkins University Center for Systems Science and Engineering COVID-19 Dashboard: data collection process, challenges faced, and lessons learned. Lancet Infect Dis 22(12):e370–e376
14. Nixon K, Jindal S, Parker F, Reich NG, Ghobadi K, Lee EC et al (2022) An evaluation of prospective COVID-19 modelling studies in the USA: from data to science translation. Lancet Digit Health 4(10):e738–e747
15. Nixon K, Jindal S, Parker F, Marshall M, Reich NG, Ghobadi K et al (2022) Real-time COVID-19 forecasting: challenges and opportunities of model performance and translation. Lancet Digit Health 4(10):e699–e701
16. Rankin N, Saiyed S, Du H, Gardner L (2024) A multi-city COVID-19 categorical forecasting model utilizing wastewater-based epidemiology. medRxiv, p 2024.09.16.24313752. https://www.medrxiv.org/content/10.1101/2024.09.16.24313752v1
17. Du H, Zhao J, Zhao Y, Xu S, Lin X, Chen Y et al (2024) Advancing real-time pandemic forecasting using large language models: a COVID-19 case study . arXiv. http://arxiv.org/abs/2404.06962
18. Darden ME, Dowdy D, Gardner L, Hamilton BH, Kopecky K, Marx M et al (2022) Modeling to inform economy-wide pandemic policy: bringing epidemiologists and economists together. Health Econ 31(7):1291–1295
19. Pollett S, Johansson MA, Reich NG, Brett-Major D, Valle SYD, Venkatramanan S et al (2021) Recommended reporting items for epidemic forecasting and prediction research: the EPIFORGE 2020 guidelines. PLoS Med 18(10):e1003793
20. Reich NG, Lessler J, Funk S, Viboud C, Vespignani A, Tibshirani RJ et al (2022) Collaborative hubs: making the most of predictive epidemic modeling. Am J Public Health 112(6):839–842
21. Wellness communities and vaccine hesitancy. https://journals.sagepub.com/doi/epub/10.1177/1329878X241270526

Human Longevity: Bringing Radical Disruptive Ideas into Light 7

Vaibhav Tiwary, Baishnabi Kovvali, Musab Bin Umair, and Shahaf Peleg

Contents

Abstract

Human aging is a puzzling and convoluted phenomenon. It is a gradual and lengthy process, which we all eventually experience. From a molecular standpoint, it is complex and multifactorial. Scientists are divided as to the causes of aging and how to define aging. While many accept aging as a natural process, we currently see a surge in longevity research in an effort to substantially increase

Baishnabi Kovvali and Musab Bin Umair contributed equally with all other contributors.

V. Tiwary · B. Kovvali · M. B. Umair · S. Peleg (✉)
Research Group Energy Metabolism and Epigenetics, Research Institute for Farm Animal Biology (FBN), Dummerstorf, Germany
e-mail: peleg@fbn-dummerstorf.de

U. A. K. Betz (ed.), *Science for a Better Tomorrow*,
https://doi.org/10.1007/978-3-031-93623-4_7

life span. Indeed, the quest to increase our human life span has been embedded in our history and literature—and it is the most formidable and potentially impossible human quest, which is surpassed by difficulty only to human travel in outer space.

Although our basic understanding of aging is rather fragmented at best, scientists have nonetheless discovered, mostly in the recent three decades, various interventions to extend the life span in animal lab models. These interventions are mainly clustered into three approaches—life style, drugs, and genetic interventions. However, while they work on certain animals, the translation of these approaches to human is limited at best. Perhaps partially due to our lack of understating of the core roots of aging, we likely face a defiant barrier to meaningfully extend human health span and an even more firm barrier to extend it beyond the current mark of 120 years. As such, it becomes increasingly evident that we should not only expend our research on current human interventions on aging, but also on interventions that are more radical.

In recent years, we observed an improvement of biological technologies and in particular in synthetic biology. This enables scientists to generate novel proteins and essentially upgrade the biological capacities of an organism to combat aging. One recent fascinating line of such synthetic upgrade is the introduction of light harvesting proteins from distant organisms, such as bacteria and fungi, into animals. This ultimately enables respective animals to utilize the energy of light, much like the transduction of energy from the food we are eating, directly into electrochemical energy. Indeed, several lines of evidence support the notion that introducing light-activated pumps inside the mitochondria of yeast and worms mimics the activity of the electron transport chain, thus promoting the synthesis of ATP by light and independently from oxygen. Preliminary work shows promising utilization of such approach to extend life span, although the road into translating such radical approach into human health and aging is still ahead of us.

Keywords

Aging · Lifespan · Interventions · Opto-energetics · Reprogramming · Caloric restriction

7.1 The Search for the Elusive Youth Fountain

Qin Shi Huang (259–210 BCE), the first Emperor of China, is a notable historical figure in the quest for immortality. Driven by a desire to extend life span, he dispatched expeditions in search of a mythical elixir, believed to grant eternal youth. According to some sources, he may have eventually died after drinking a so-called elixir of life containing mercury. Another famous historical tale involves the Spanish explorer Juan Ponce de León. According to legend, he set out in 1513 to find the Fountain of Youth, hoping to rejuvenate himself and gain eternal life, though he never found the mythical fountain [1, 2]. Their efforts and others, undertaken over a span of 2000 years, highlight the long-standing human fascination with overcoming the limitations of aging and the pursuit of longevity [3]. The search for the elixir or Fountain of Youth is often seen as a metaphor for humanity's eternal quest for immortality and the desire to reverse the aging process. Such stories underscore the timeless and universal aspiration to explore ways of enhancing and prolonging life. Even today, there are individuals that self-experiment their bodies in hope to increase their life span.

In ancient civilizations like Greece and Rome, aging was often viewed as a natural biological process of decline. Aristotle, for example, saw aging as the result of a gradual loss of innate heat and moisture, leading to the weakening of the body's vital forces [4]. Later during the Renaissance period, influenced by the revival of classical knowledge, humans continued to view aging through the lens of humoral theory, where aging was seen as an imbalance of the body's four humors (blood, phlegm, yellow bile, and black bile), leading to a natural decline in health [5, 6]. In eighteenth and nineteenth Centuries, the enlightenment and the advent of modern science began to shift perspectives on aging. Aging began to be viewed more as a pathological process that could potentially be studied, treated, or even delayed. This period saw the emergence of geriatrics as a field of study [7].

With advances in biology and genetics, aging is now often defined in terms of cellular and molecular changes. The accumulation of damage to DNA, proteins, and cells over time is considered central to the aging process, leading to a decline in the function of tissues and organs [8]. In recent years, there has been growing interest in longevity research and anti-aging interventions, with a broad effort in exploring novel means to slow down, reverse, or even "cure" aging. This has led to debates about the ethics and feasibility of such pursuits [9].

The concept of aging is ever-evolving from early philosophical musings to a nuanced scientific inquiry in modern times to determine the underlying mechanism of aging and influence these mechanisms in favor of longevity. Today, aging is understood not only as a process of physical decline but also as a complex phenomenon shaped by evolutionary pressures, cellular processes, and social factors. Modern research continues to refine our understanding of aging, offering insights into how we might delay its onset or mitigate its impacts on health and wellbeing.

7.2 How Do We Define Aging?

The modern definition of aging can be broadly simplified to "the time-dependent functional decline of biological function that affects living organisms". Another popular view of aging is "accumulation of a wide variety of molecular and cellular damage, leading to a gradual decrease in physical and mental capacity, increased risk of disease, and ultimately, death". This perspective emphasizes the role of damage at the cellular level, such as DNA mutations, oxidative stress, and telomere shortening, all of which contribute to the aging process [8].

From an evolutionary perspective, aging is a conserved biological phenomenon, which can be defined as "the progressive accumulation of deleterious mutations and trade-offs that favor early life reproductive success at the cost of late-life survival and function". The evolutionary perspective focuses on why aging occurs, suggesting that aging is a byproduct of natural selection, which tends to favor traits that enhance reproductive success early in life rather than longevity. The question raised here is that *"Is Aging programmed?"*—In spite of the disadvantages that aging poses for an individual to survive, it might favor the survival of the species [10] reasoned, "Worn out individuals are not only valueless to the species, but they are even harmful, for they take the place of those which are sound" [11].

A consensus has yet to be achieved by scientists regarding how aging should be explained. There are many theories, describing the probable causes that may drive aging associated deterioration of the body. The majority of the theories put forth to explain aging are somewhat intertwined indicating the intricate pathophysiology of the aging process [12].

In an attempt to understand aging, some modern theories are mentioned as follows:

7.2.1 Entropy and Aging

The premise of this theory revolves around the fact that there are parallels in the thermodynamics concept "entropy" and biological aging. As entropy is the measure of the disorder in a system, aging has been referred to as the process characterized by entropy accumulation and irreversibility [13]. According to Hayflick, there is significant evidence that changes associated with age are defined by increase in entropy and loss of molecular fidelity, which consequently leads to the deterioration of the biological systems [14]. Mesenchymal stem cell (MSC) aging is a multifaceted phenomenon as it leads to the dysfunction of a plethora of essential biological processes. In his insightful review, Tatullo uses the example of MSCs to explain the concept of disorder and entropy in biological systems and mentions that combined deterioration of the efficiency of biological systems because of inflammaging along with other factors leads to an irreversible state of biological entropy [15].

7.2.2 Ground Zero of Organismal Aging and Its Reprogramming

A new perspective to look at aging was put forth by Gladyshev in 2021, where he proposed that there is a "ground zero" of organismal life where an organism comes to life and at the same time the aging starts as well. He posited that this ground zero stage is mid-embryonic stage and this model provides clarity regarding the relationship between aging, de-differentiating, rejuvenation and development. He suggests there is a preparatory phase during and after conception for some time during the early embryogenesis, which is marked with a rejuvenated state in which the telomeres are extended, epigenetic marks are removed and right after this stage is a ground zero, where the organismal life and aging begins. This is the stage where telomeres start to shorten, body plan formation is established, self-recognition is established, and neural activity starts along with the development of body systems like the immune system. According to the author, the cellular life begins with conception and ground zero is the origin of organismal life [16].

7.2.3 Damage Accumulation and Disposable Soma Theory

Disposable soma theory incorporates the concept of damage accumulation as a reason for aging within it, because it is based on two postulates, one being the concept of damage accumulation and the second pertaining to repair mechanisms. The concept of damage accumulation comes from "error catastrophe" theory, which states that, with age, randomly occurring molecular errors, resulting during biosynthetic processes, keep increasing at an exponential rate. The damage keeps on accumulating with subsequent new damage. The second postulate is pertaining to the fact that repair mechanisms have high energy demands and there is a trade-off between the resources of somatic maintenance and reproduction which leads to aging [17].

7.2.4 Antagonistic Pleiotropy

Certain genes that are beneficial and enhance fitness in early life become the cause of deterioration later in life leading to aging. It is hypothesized that these genes are favored by natural selection because of their crucial nature in early life characterized by development and reproduction [18, 19]. It shows a trade-off between the utilization of genes essential for the reproductive age and become detrimental later on in adult life when the chances of reproduction are less. The lessened chances of reproduction later in life protects these genes from being eliminated [20]. Although there is currently less evidence for this, there are still a few examples observed in favor of the theory. The *Drosophila* males that have mated with more females reduce their ability of clearing out infections caused by bacteria. Similarly, the red squirrels that breed early in life have shorter life spans [19].

7.2.5 ROS and PMF

A significant source of damage to cells and cellular components is reactive oxygen species (ROS). The major source of ROS are mitochondria and when the delicate balance between ROS production and the antioxidant system of the body is disturbed due to several reasons, ROS wreak havoc on the macromolecules and damage the mitochondria itself, the organelle that generates critical metabolites for the function of cells [21]. Mitochondrial dysfunction is one of the hallmarks of aging and there is agreement on the fact that ROS dependent mitochondrial dysfunction drives aging. Protonmotive force (PMF) is generated in the mitochondria Electron Transport Chain (ETC) and this is used by the ATP synthase to generate ATP. Although mitochondrial dysfunction is a hallmark of aging, the specific role of PMF decline observed in the mitochondria of cells reaching replicative senescence is not yet resolved. Studies in old mice and cells used to model aging disorders have suggested a link between PMF and aging [22]. Similarly, it was observed that treating old stem cells with MitoQ, an antioxidant that targets the mitochondria, caused those stem cells to return to a young phenotypic state [22, 23].

7.3 Biomarkers of Aging

As mentioned above, aging can be viewed as the result of "the accumulation of damage to biomolecules, organelles, and cells, leading to loss of homeostasis and functional decline in tissues and organs" [24]. The updated mechanistic view of aging includes the hallmarks of aging: genomic instability, telomere attrition, epigenetic alterations, loss of proteostasis, disabled macroautophagy, deregulated nutrient-sensing, mitochondrial dysfunction, cellular senescence, stem cell exhaustion, altered intercellular communication, chronic inflammation, and dysbiosis [8]. To accurately identify and define aging, we rely on specific biomarkers, which are measurable indicators that reflect the biological processes underlying aging. These biomarkers help us discern what constitutes inside the process of "aging".

There are three criteria for each hallmark that are identified, one of which is **Time-Dependent Manifestation**: a hallmark must exhibit changes that naturally occur as the organism ages. These changes are not random but are systematically associated with the aging process, corresponding with the biological deterioration that occurs over time. For instance, the shortening of telomeres (protective caps at the ends of chromosomes) is a change that is associated to aging and cellular replication limits, making it a recognizable hallmark. Another criterion is **Acceleration of Aging When Hallmarks Enhanced**—if we are experimentally enhancing a hallmark it should lead to a noticeable acceleration in the aging process. For example, if the function of mitochondria (the powerhouse of the cell) is further damaged in an experimental context, it often results in rapid aging and more substantial age-related diseases, showing the importance of mitochondrial function to overall longevity. Lastly, **Therapeutic Reversibility of Hallmark**, or in other words modulating or correcting a hallmark should have the potential to slow down, halt, or even reverse aspects of the aging process. It is the most defining characteristic of a hallmark of aging because it offers a target for therapeutic intervention. This criterion underlines that these hallmarks are more than just aging markers; they are dynamic factors which can be altered to influence the wide aspects of aging [8].

Genomic Instability—It denotes disruptions in the normal pathways that maintain genomic integrity, leading to errors in the production of biological molecules and overall cellular function. Genomic instability arises from a variety of factors that can broadly be categorized into exogenous (external) and endogenous (internal) agents. Both types damage the integrity of nuclear and mitochondrial DNA, leading to damage that manifests as mutations, chromosomal rearrangements, and structural alterations. These genetic disruptions contribute significantly to aging and the development of age-related diseases. The burden of mutations increases with age, particularly in stem cells, causing their exhaustion and a decline in tissue renewal capacity, which accelerates aging and vulnerability to diseases. Since fully correcting all mutation is an energy intensive process, survival takes precedence over complete DNA integrity [25]. Research indicates that shorted lifespans are associated with higher mutation rates, while longer lifespans are also associated with improved DNA repair systems [26].

Telomere attrition—Telomeres are the protective caps on the ends of chromosomes, which prevent DNA damage during cell division. However, because the DNA replication machinery is unable to fully replicate the ends, telomeres shorten with each cell division, resulting in genomic instability, cellular senescence, or apoptosis. Telomerase, a reverse-transcriptase enzyme, can extend telomeres to maintain their length but is rarely expressed in most somatic cells, resulting in telomere erosion over time. While telomere shortening can reduce the lifespan of malignant cells, slowing cancer formation, it also leads to premature aging diseases such as pulmonary fibrosis, aplastic anemia, and dyskeratosis congenita, which hinder tissue regeneration [27].

Epigenetics Alteration—The term Epigenetics refers to the study of heritable changes in gene expression that occur without alterations to the underlying DNA sequence. Aging is influenced by a broad spectrum of epigenetic changes, including alterations in DNA methylation, histone modifications, chromatin remodeling, and non-coding RNA regulation. These modifications impact gene expression and contribute to various age-related diseases, such as cancer, neurodegeneration, and metabolic disorders.

DNA Methylation is the addition of methyl group (CH_3) to cytosine bases in cytosine-phosphate-guanine (CpG) sites. This chemical alteration does not alter the DNA sequence itself, but it impacts gene expression by influencing how genes are switched on or off. Studies show that as human age, the DNA methylation patterns undergo numerous alterations over time. In recent times, scientists are developing DNA methylation-based epigenetic clocks, which focus on specific sites in the genome to estimate chronological age and assess mortality risk. These clocks are also used to measure the effectiveness of interventions aimed at prolonging human lifespan [28].

Histone modifications are the addition or the removal of specific chemical groups at histones. Histones are the proteins that wrap around DNA in the cell nucleus while these modifications influence how tightly or loosely the DNA is bundled thereby governing gene expression. There are various types of post-translational modifications like—acetylation, methylation, phosphorylation, ubiquitination and sumoylation. Indeed, many studies from different animal models have demonstrated that aging is accompanied by epigenetic alterations and that targeting these alterations result in impact of lifespan.

Degradation of proteostasis—Proteins are the foundation of all biological systems, and therefore proteostasis, or protein homeostasis, is essential for cellular health. Disruptions in proteostasis lead to the accumulation of misfolded or damaged proteins that form aggregates, contributing to cellular dysfunction. Research has shown that enhancing the accuracy of protein translation extends lifespan in multiple organisms, while error-prone translation accelerates aging [29]. Impaired proteostasis can result from slowed translation, oxidative damage, or the failure of quality control systems like the unfolded protein response (UPR) and degradation pathways, including the proteasome and lysosome [30].

Disturbed Macroautophagy—Macroautophagy is a catabolic process preserved throughout evolution and is primarily responsible for breaking down intracellular components to provide cells with metabolites and energy. It involves multiple vesicle transport activities, typically initiated under stressful conditions like nutrient deprivation [31]. As organisms age, there is a significant decline in autophagic activity, leading to impaired organelle turnover, which is a key factor in aging. Studies show that the genetic inhibition of autophagy accelerates aging. For example, mice with doxycycline-inducible shRNA-mediated knockdown of *Atg5* exhibit premature aging, degeneration, and senescence across multiple organ systems, which can be reversed by restoring autophagy [32].

Deregulated Nutrient-Sensing and Aging—The nutrient-sensing network plays an important role in regulating cellular activity in response to nutrient availability and stress. When nutrients are abundant, the nutrient-sensing network promotes anabolic processes, including ribosome biogenesis and mRNA translation. In contrast, during nutrient scarcity or stress, it triggers cellular defense mechanisms such as autophagy to maintain cellular homeostasis. The decreased activity of the nutrient-sensing network affects various processes during aging, enhancing stress resistance, triggering repair mechanisms, promoting autophagy, and managing inflammation [8].

Cellular Senescence—Senescence is a cellular condition triggered by various stimuli, marked by permanent cell cycle arrest, a secretory phenotype, damage to macromolecules, disrupted metabolism, and distinct changes in cellular structure [33]. Cellular senescence is a biological response to acute or chronic cellular damage and stress. In humans, senescent cells progressively accumulate in various tissues with age, increasing by 2- to 20-fold in older individuals compared to younger ones [34]. This phenomenon primarily affects fibroblasts, endothelial cells, and immune cells, although all cell types can become senescent. A key factor in this process is telomere shortening, which triggers senescence. Even tissues with slow or non-dividing cells, such as the brain and heart, can harbor senescent cells, while localized senescence is also common in many age-related diseases [8]. The most compelling evidence linking cellular senescence to aging is derived from studies in mice, where continuous genetic or pharmacological removal of senescent cells has been shown to extend both health span and lifespan [35]. Furthermore, eliminating senescent cells has proven to be therapeutic in a wide array of mouse models of disease [36].

Stem Cell Exhaustion—Stem cells can be viewed as the body's raw materials, capable of generating specialized cells such as muscle, nerve, or blood cells through differentiation. For example, skeletal muscle relies on satellite cells, while the skin epidermis uses multiple stem cell niches that can adapt during injury for repair. With age, there is a gradual decline in their regeneration capacity. Organs like the liver, lung, and pancreas typically exhibit low renewal rates but demonstrate injury-induced stem-like plasticity through de-differentiation and reactivation of embryonic programs, a capacity that declines with age [8].

Chronic Inflammation—Chronic inflammation hastens the aging process of immune cells, impairing their functionality and preventing the clearance of senescent cells and inflammatory molecules. Certain factors, such as senescence-associated secretory phenotypes (SASPs), contribute to chronic inflammation, which, in turn, induces senescence in normal cells. If constant high levels of inflammation in organs like the bone marrow, liver, and lungs remain unresolved, this can cause tissue damage and contributes to age-related diseases. As a result, inflammation is identified as an internal driver of aging [37].

Dysbiosis and Aging—The gut microbiome plays a vital role in aging, influencing nutrient metabolism, pathogen defense and the production of essential metabolites. Disruption of the microbiome–host communication, known as dysbiosis, has been linked to obesity, diabetes, neurodegenerative diseases, cardiovascular diseases, and cancer. Age-related changes in the microbiome include a decline in microbial diversity and an increase in harmful taxa [38]. Studies of centenarian populations reveal a decrease in core bacterial species like *Bacteroides* and *Roseburia* but an increase in longevity-associated bacteria like *Akkermansia* and *Bifidobacterium* [8].

Altered intercellular communication—Intracellular communication is important to maintain the harmony across the body. Aging involves disruption in intercellular communication. The fundamental cause of such disruption is cell-intrinsic factors such as SASP, which connect cellular and systemic aging hallmarks, resulting in chronic inflammation, impaired immune surveillance, and altered genome–microbiome interactions, resulting in dysbiosis. Several studies have investigated blood-borne factors which alter the process of aging or can help in exploring longevity, examined various cell communication systems, and assessed the impact of extracellular matrix dynamics damage on aging [8].

Mitochondrial Dysfunction—Mitochondria are essential for cellular energy production, which also play a critical role in aging as they can trigger inflammation and cell death when ROS or mitochondrial DNA (mtDNA) leak from the organelle. Although the specific molecular mechanisms can vary depending on the disease context, mitochondrial age-related dysfunctions often involve impaired energy production, increased oxidative stress, disrupted calcium balance and alterations in mitochondrial dynamics, as well as quality control processes [39]. With age, mitochondrial function declines due to a combination of factors, including mtDNA mutations, impaired proteostasis, destabilization of respiratory chain complexes and reduced mitochondrial turnover. These changes impair cellular energy production, increase ROS levels and promote mitochondrial membrane permeabilization, contributing to inflammation and cellular death. The progressive deterioration of mitochondrial function is considered as a hallmark of aging [40].

7.4 Current Lifespan Interventions

There are already interventions that at least in model organisms, increase life and heath span. They mainly cluster under three categories: **life style, drug therapy,** and **gene therapy**.

The impact of lifestyle such as diet has been long explored. Caloric restriction (CR) is known to effectively delay the aging process. The term "Caloric Restriction" refers to the deliberate reduction in the intake of high caloric diet in the range of 20–60%, without causing any malnourishment. CR has been found to exhibit reduction in occurrence of diverse age-related diseases like diabetes, cardiovascular diseases, neurological disorders, cancer and obesity. CR can influence nutrient-sensing mechanism, which leads to an improved metabolic profile, increased ability to handle stress, reduced oxidative stress and a more robust inflammatory response. These improvements in health eventually lead to a higher life expectancy in a wide range of organisms [41].

CR mimetic can be defined as the class of molecules that induce physiological effects comparable to the Caloric restriction state. These effects include decrease in blood glucose levels, insulin secretion and triglyceride concentrations, alongside an increase in blood ketone bodies. Many examples of CR mimetics include Nicotinamide mononucleotide, Glucosamine, Dimethoxychalcone, Spermidine, Rapamycin, Metformin and Aspirin [42]. Research has been focused on exploring new compounds that can be used as CR mimetics due to the potential impact of caloric restriction on aging disorders and the aging process [42].

Here we review some of examples of repurposing disease specific **FDA approved drugs** for that have been in-use at our bedsides for different specific diseases and repurposing them for aging would be a faster way to make it accessible to larger populace as a longevity interventions:

Metformin is primarily used as an anti-diabetic drug, metformin activates AMPK, which helps maintain cellular energy balance and mimics caloric restriction effects. It also reduces hepatic glucose production and may influence gut microbiota, contributing to systemic anti-inflammatory effects and improved insulin sensitivity [43–45]. In studies on various organisms, metformin has demonstrated lifespan-extending effects, and human observational studies suggest it may reduce risks of cancer, cardiovascular disease, and neurodegenerative conditions. The Targeting Aging with Metformin (TAME) trial aims to assess its effects on delaying age-related diseases in humans [46]. Potential side effects include vitamin B12 deficiency, gastrointestinal issues, and lactic acidosis. Long-term use may require monitoring of vitamin B12 levels due to risks of cognitive dysfunction from deficiency [47].

Rapamycin and Rapalogs (e.g., Everolimus, Temsirolimus) inhibits mammalian Target of Rapamycin (mTOR), a central pathway regulating cellular metabolism, catabolism, immune responses, autophagy, survival, proliferation, and migration, to maintain cellular homeostasis [48]. mTOR inhibition reduces cell senescence and inflammation, delays cellular aging and mimics effects of caloric restriction. Rapamycin has shown positive results in *C. elegans and Drosophila melanogaster*, delaying onset of age-related diseases like cardiovascular conditions, cognitive decline, and cancer [49]. It may also mitigate tendon and joint degeneration [50], liver damage [51] and muscle weakening with age [52]. Contradictory studies indicate long-term rapamycin use can induce insulin resistance and Type 2 diabetes-like symptoms. Combining rapamycin with metformin might help mitigate these effects [53]. Nonetheless, its immunosuppressive properties may complicate its use as a widespread anti-aging drug [47].

Non-Steroidal Anti-Inflammatory Drugs (NSAIDs)—Aspirin or its derivatives is capable of prolonging longevity *C. elegans*, Drosophila and mice [54–56]. Aspirin inhibits cyclooxygenase (COX) enzymes reducing the production of pro-inflammatory mediators like prostaglandins. Aspirin may reduce risks of cancer, cardiovascular disease, and may even Alzheimer's and Parkinson's due to its anti-inflammatory and anti-oxidative effects [57, 58]. A downside of Aspirin is increased risk of gastrointestinal bleeding and hemorrhage. Aspirin's benefit/risk ratio suggests cautious, individualized use, especially in elderly patients [58]. Similar to aspirin, ibuprofen is a COX inhibitor but tends to have fewer gastrointestinal side effects. In model organisms, ibuprofen extends lifespan by up to 15% [59]. Observational studies suggested potential benefits in reducing risks of Alzheimer's and Parkinson's [60, 61]. Clinical trials in humans have been less conclusive, with ibuprofen showing limited cognitive benefits and an association with certain risks, such as renal impairment and cardiovascular issues [47, 60].

Statins inhibit 3-hydroxy-3-methylglutaryl-CoA reductase, the key enzyme of the cholesterol synthesis pathway, reducing cholesterol levels and impacting the mevalonate pathway, which plays roles in inflammation and cellular health. Statins are associated with lower risks of cardiovascular diseases and may even improve survival in elderly populations, particularly those with diabetes [62]. Drosophila and progeroid mice models show lifespan extension, with human studies reporting increased survival among older adults, including those in nursing homes [63–66]. They are also associated with new-onset diabetes, cognitive impairment, and muscle damage in certain cases. Their use in the elderly requires careful balancing of benefits and side effects [47].

Senolytics (e.g., Dasatinib and Quercetin)—The strong association between cellular senescence and age-related diseases has spurred the development of senolytic therapies—compounds that selectively kill senescent cells. Senolytics selectively clear senescent cells, which accumulate with age and contribute to inflammation and tissue dysfunction. Senescent cells are a key source of the inflammatory cytokines, chemokines, and matrix-degrading enzymes that characterize the senescence-associated secretory phenotype. Several senolytic agents have been identified, including navitoclax, dasatinib, quercetin, fisetin, and cardiac glycosides [67]. In mice, senolytics improve physical function, reduce frailty, and delay age-related diseases [35, 68]. Dasatinib and quercetin combination (DQ) therapy has shown benefits in patients with idiopathic pulmonary fibrosis and diabetic kidney disease [69]. Recently, cyclic treatment of ABT263, a senolytic drug, was shown to increase lifespan of giant mice [35].

Navitoclax, for example, targets proteins from the BCL2 family, which are crucial for the survival of senescent cells, making these cells particularly vulnerable to the drug [70]. Clinical trials are currently evaluating navitoclax for its potential to treat senescence-associated diseases [71].

Safety concerns exist, as senolytics might impair tumor suppression by removing senescent cells. Other potential side effects include impaired wound healing and pro-inflammatory effects due to inefficient clearance of cell debris from senescent cell death [47].

Antioxidant Vitamins (e.g., Vitamins A, C, and E) neutralize reactive oxygen species (ROS), reducing cellular damage from oxidative stress, and a key factor in aging. While antioxidants show health and longevity benefits in animal models [72], health benefits in human studies are less conclusive [73]. In certain cases, high doses of antioxidants (like vitamin E and beta-carotene) were linked to increased cancer and mortality risks [73, 74]. Vitamins are treated as dietary supplements and not as drugs by FDA and other authorities so the access to them is easier. The "antioxidant paradox" suggests that while oxidative stress is implicated in aging, antioxidant supplementation has not consistently prevented or treated age-related diseases in humans, possibly due to issues like the pro-oxidant effect of excessive doses and interference with beneficial ROS signaling [47, 75].

Melatonin regulates sleep and circadian rhythms, with additional roles as an antioxidant and anti-inflammatory agent. It can cross the blood–brain barrier and protects cells from oxidative damage. Studies support melatonin's potential lifespan-extending effects in Drosophila melanogaster [76, 77], and mice [78, 79] as well as potential neuroprotective effects against Alzheimer's and Parkinson's disease [80]. It may reduce inflammation, modulate immune responses, and promote autophagy. Possible side effects, especially in elderly users, include headaches, nausea and prolonged drowsiness [47, 81].

Edible medicinal plants contain a variety of bioactive compounds that could have the potential to reduce the risk of age-related chronic diseases, delay the aging process, and prolong the health span. These compounds include *chlorogenic acid, caffeic acid, quercetin, rutin, anthocyanins, fisetin, kaempferol, ellagic acid,* and others, which have been recognized for their potential to counteract the aging process and extend lifespans in various model organisms such as fruit flies and worms [82, 83].

Anti-aging drug therapy is a promising strategy to combat aging. However, every drug has only a limited set of targets, which limits it effects. Therefore, to increase the effectiveness of anti-aging drug therapy and significantly extend lifespan, it might be preferable to use drug combinations targeting several anti-aging pathways simultaneously. In a right composition of multiple drugs, the result might not just be additive along different pathways but synergistically multiplicative. Multiple principles need to be considered when selecting such compounds for combination to guarantee an advantage of the cocktail over therapy with individual compounds, for example, minimal off-target effects, negligible additive toxicity, and little difference in absorption of the combination compared to individual compounds. Various diseases including age-related disorders like type-2 diabetes, cancer, obesity, and hypertensions already use combination drug therapy for treatment [84].

7.5 Gene Therapy for Aging

The tweaking of various genes have been shown to increase lifespan in model animals. For example, the inhibition of the nutrient-sensitive and growth-promoting insulin/IGF-1 and mTOR signaling pathways results in the largest and most reproducible increases in health span and lifespan [85]. Classic studies by Kenyon et al. in 1993 involved a **Daf-2** mutation of the hermaphrodite worms *C. elegans* that lived more than twice as long than wild type while being fertile and active [86]. An earlier study by Friedman et al. demonstrated that the mutation of **age-1** gene in *C. elegans* lengthens lifespan while reducing the fertility in the hermaphrodite worm [87]. The daf-2 gene and age-1 were later found to encode parts of the worm's homolog of the insulin/insulin-like growth factor-1 (IGF-1) receptor. Reduction of insulin/IGF-1 signaling throughout life was the cause of the slowdown in the progression of aging [88].

Telomerase gene therapy, i.e., expression of telomerase reverse transcriptase (TERT) is another promising intervention. Studies show that increased TERT expression by germ line modifications found that this favors the expansion of cancerous cells and the development of cancer later in life [89]. Recently, modifications to this approach involve an adeno-associated virus (AAV) based delivery in later life. In mice, TERT delivered via AAV leads to telomere lengthening in late life and improves several molecular markers of aging resulting in increased healthspan. The treatment also extended the median lifespan by 24% in 1-year-old mice and 13% in 2-year-old subjects [89]. In another study, hyper-long telomeres in mice resulted in reduced DNA damage, better glucose and insulin tolerance, lower the low-density lipoprotein (LDL) levels, decreased cancer incidence, and increased longevity [90]. Due to long-term safety and ethical concerns, human studies have not been conducted yet, even though gene therapy could be a very promising approach to target aging [89, 91].

In recent years, using OSKM factors (Oct4, Sox2, Klf4, and c-Myc) or chemical cocktails for partial reprogramming became an exciting opportunity to revert aged cells to a younger state without completing the reprogramming cycle to pluripotency, thus retaining their cellular identity in a younger epigenetic state [92, 93]. Partial Reprogramming has significant potential applications for cellular rejuvenation and finally organism level rejuvenation. Organismal level rejuvenation can be achieved theoretically in two ways. First, by direct editing of germline to arm each cell in the adult body with the reprogramming factors OKSM, which is currently prohibited due to safety and ethical regulations. Secondly, by supplying the factors directly to an adult organism in the form of DNA or mRNA with gene therapy techniques. However, due limitations of efficacy and specificity of this technique, its uses remain limited [92].

Full organismal level reprogramming experiments have been conducted in chimeric OKSM-inducible mice. In this study by Ocampo et al. [94], doxycycline-inducible expression of polycistronic OSKM cassette in progeria LAKI mice led to a median lifespan increase of 33% compared to control mice that received no treatment. Other effects of this treatment included rejuvenation of certain cellular phenotypes, including the reduction of mitochondrial ROS and restoration of H3K9me levels [94]. Notably, later study by the same group used in wild type mice induced teratoma formation and there was no lifespan extension. Interestingly, the transcriptome, lipidome, and metabolome of multiple tissues were reverted to a younger state. Moreover, this treatment increased skin regeneration capacity in mice [95].

Another recently emerging approach is to use chemical cocktails for reprogramming. It is an attractive approach as it is non-genetic and supports easy delivery of molecules to the whole body. In *C. elegans,* a lifespan increase by 42.1% accompanied by a reduction of DNA damage and an amelioration of epigenetic age-related marks such as H3K9me3 and H3K27me3 was observed in a study of two-chemical reprogramming procedures [96]. In another study, human somatic cells were transformed to pluripotent stem cells using chemical factors instead of biological transcription factors [97].

A synergistic treatment approach is the combination of cellular reprogramming and removal of senescence cells in the same individual. Kaur et al. [98] tested the interaction between cellular reprogramming and senolytic removal of senescent cells in an adult fruit fly model with inducible overexpression of the mammalian OKSM factors and Senolytic factor (Sen), which is a peptide based *Drosophila foxo (forkhead box, sub-group O)* gene, by either inducing OKSM or Sen or both. They observed increased number of intestinal stem cell, which usually declines with age leading to poor barrier function in flies and also in mammals. All three treatments led to a decrease in senescent cells. Constant and ubiquitous expression of both OKSM and Sen led to significant detriment in mean lifespan. Interestingly, upon a change in approach for the cyclical expression of 24 or 12 h per week, Sen treated flies produced a substantial increase in mean but not in maximum lifespan. This showed the compression of mortality with increase in later deaths as a compensatory mechanism for early life protective effects. Contrarily, OKSM flies showed a statistically significant maximum lifespan extension with both 24 and 12 h expression induction [98]. Strikingly, expression of the OKSM-Sen combination, especially for 12 h expression induction, resulted in a survival curve that combines beneficial effects of both the interventions resulting in increased mean and maximum lifespan [98].

Nonetheless, it is unlikely that these interventions alone can ever enable a human to live meaningfully past 120 years as a hard limit for such extension has been suggested [99]. Instead, more radical approaches which alter the very basic biological pathway of an organism are needed.

7.6 Examples of Long Lived Animals and Toward Radical Lifespan Intervention

There are animals that exhibit unusual long lifespans. Some, such as the *Turritopsis dohrnii* (Immortal Jellyfish), may even be able to turn back the clock from mature medusa (adult form) to its polyp form (juvenile stage) in response to adverse environments like physical damage or aging. This process called as transdifferentiation allows the jellyfish to escape biological aging. In this process senescent, weakened, or damaged medusa revert back to a cluster of poorly differentiated cells (the cyst stage) which metamorphose back into a preceding life cycle stage (the polyp) [100].

The Greenland shark (*Somniosus microcephalus*) is the longest-lived vertebrate known, with an estimated lifespan of ~400 years. On chromosome level it has a genome assembly of 6.45 Gb; one of the largest non-tetrapod genome sequenced till date. Based on radiocarbon dating, females show an extraordinary lifespan estimated at 392 ± 120 years [101]. Longevity in this species is coupled to a very slow growth rate of <1 cm/year. Most of genome expansion is largely due to the expansion of transposable elements. Intriguingly, when compared to the public shark genomes; genes specifically duplicated in the Greenland shark form a functionally connected network enriched for DNA repair function [102].

Naked mole rat is one of the most interesting and relatively newer models of aging research. It is an extremophilic rodent species found in the arid and semi-arid regions of equatorial Africa. It leads a eusocial, subterranean lifestyle living in an extensive maze of poorly ventilated, hot, and humid sealed burrows where the conditions could get extreme including hypoxia, low food availability and no access to free water. Naked mole rats feature numerous adaptations to survive in these extreme conditions [103].

Interestingly, it is extremely long-lived based on its body-size, living approximately 30 years [104] and also have been seen to live till 40 in lab environments [103]. At the cellular and molecular level, several factors contribute to their exceptional longevity. Naked mole rats possess enhanced genomic stability, characterized by lower mutation rates and efficient DNA repair mechanisms, likely due to increased expression of DNA repair genes and higher levels of protective proteins like HSP70, HSPA12A/B, and HSP27. They also display unique proteomic adaptations, such as highly stable proteins resistant to oxidation, misfolding, and degradation. Their cells exhibit heightened resistance to oxidative stress and maintain efficient protein quality control mechanisms. Furthermore, naked mole rats possess unique metabolic adaptations, including a lower metabolic rate and efficient energy utilization, potentially contributing to their longevity. Further investigation of the naked mole rat's unique biology promises to provide critical insights into the mechanisms of law defying longevity and may ultimately inform the development of interventions aimed at promoting human health span and life span [103].

The bowhead whale (*Balaena mysticetus*) is a warm-blooded species has not only been estimated to live over 200 years (estimated age of one specimen 211 SE ±35 years), suggesting it is the longest-lived mammal, but also exhibits very low disease incidence until an advanced age compared to humans. Remarkably, large whales with over 1000 times more cells than humans do not exhibit an increased cancer risk suggesting the existence of natural mechanisms that can suppress cancer more effectively in these animals [105].

Perhaps collectively, one key aspect toward meaningfully increasing lifespan and break the glass ceiling of 120 years for human lifespan is to be able, just like these unique animals, to alter and upgrade key and essential biological process. In other words, we need to seek radical intervention to alter our canonical biology in a more modern pursuit of human life extension. One potential avenue of radical intervention is 'upgrading' an animal and enable it new to harness a completely different biological idea. In other words, incorporating, via synthetic biology, proteins from entirely different organism into a target animal model.

One promising and recent idea is enabling an animal to harness the energy of light and transform it directly into chemical energy inside their mitochondria. While plants, bacteria and fungi are able to harness the energy of light and generate chemical gradient in their cells, animals do not have such ability. However, with modern synthetic tools, it is possible to re-design a protein from plants/bacteria/fungi and express it in animals. One example is light-activated proton pumps, converting light to electrochemical potentials. It was previously shown by Bernhanu et al. that artificial organelle containing light-activated proton pumps can make ATP using the energy from the light. The organelle consisting of bacteriorhodopsin and ATP synthase generated PMF and produced ATP. After combination with PURE (protein synthesis using recombinant elements) system, this ATP was then utilized as a substrate for transcription of mRNA and to drive tRNA aminoacylation and GDP phosphorylation [106].

Similarly, our mitochondria break down macromolecules that are generated from the food we eat, to eventually support the PMF across the mitochondrial membrane. One attractive idea was to introduce a light-activated proton pump inside the mitochondria to support the PMF. The potential introduction of such a synthetic tool could have several advantages. It could for example, increase the levels of ATP in an old organism that harbors dysfunctional mitochondria. Furthermore, it would generate so-called clean ATP, which would not involve complex biochemical reactions that may produce toxic by-products such as advanced glycation end products AGE, lipfuscin or ROS. Another potential advantage is to supply cells sufficient ATP in instances where oxygen levels are low (hypoxic state) that can be observed during aging and vascular maladies. Thus, using light for 'partial energy replacement' and maintenance of potentiated PMF inside mitochondria could be a radical way to slowdown aging.

One major obstacle is to guide the light-activated pump to the mitochondria. A major breakthrough has been shown in 1994 were scientists designed such light-activated proton pump to translocate to the mitochondrial membrane of yeast [107]. This upgraded yeast needed less glucose to survive and proliferated as they could utilize also the energy of light. Years later, Berry et al., showed a similar proof of concept in worms [108]. Demonstrating that optogenetically increasing PMF during adulthood using a light-activated proton pump targeted in the inner membrane of mitochondria improves age-associated phenotypes and extends lifespan in *C. elegans* [109].

Light-dependent partial energy replacement is still in early stages of development and its use in humans needs many more questions answered. One of the biggest challenges to applications of energy replacement using light would be the delivery of light to the inner organs which are not exposed to light [110].

Another approach that involves using the energy of light was shown in 2022, when Chen et al. coated plant photosynthetic system, nanothalakoid units (NTU), with chondrocyte membrane (CM) and showed that CM-NTUs entered chondrocytes and upon light activation enhanced ATP and NADPH levels in the cell [111]. It was also observed that after the penetration of CM-NTU into the cartilage explant, NTU's were secreted as extracellular vesicles to other chondrocytes for even distribution throughout the cartilage. The question remained if light could penetrate the joints of mice. For this purpose, red light penetration was tested and it was demonstrated that it effectively attenuated cartilage degradation, and hindered the progression of osteoarthritis by activating the CM-NTU construct [111].

Further potential approach is Photosynthetic Artificial Organelles with Dynamic ATP Control. Lee et al. in 2019 developed a dynamic light-dependent ATP producing organelle utilizing plant derived photosystem II, responsible for light-induced water oxidation and electron donation to the electron transport chain, and bacterial proteorhodopsin along with an ATP synthase. The nature of this organelle is switchable, as synthesis of ATP is impeded by green light and promoted by red light. In vitro, they demonstrated that light-dependent ATP synthesis induced actin polymerization in the giant lipid vesicle where they introduced the artificial organelle reactants (ADP and G-actin) and magnesium ionophores which led to the growth of the three dimensional actin filaments [112] (Fig. 7.1).

7.7 Conclusions and Perspectives

How far are we from witnessing the first human to reach 100 years old with adequate physic and mental health, leading them to live pass 150 years? This question is still difficult to answer.

Large amount of funding effort is still channeled toward basic research focused on the canonical rather than radical lifespan interventions. This allocation of effort may well hinder meaningful progress toward witnessing such radical lifespan extension. Nonetheless, we also see more and more effort toward supporting out-of-the-box ideas. Such ideas are, for example, synthetic organ replacement, organelle transplantation (such as engineered mitochondria), the introduction of genes from long-lived animals, stem cell therapy, cryopreservation, synthetic gene engineering and others. While these ideas have yet to have sufficient scientific basis to be considered for trials in human, they show the promise of potential alterations of the basic biological mechanisms that may delay aging. Therefore, more resources to such moonshot ideas may facilitate seeing human lifespan increase.

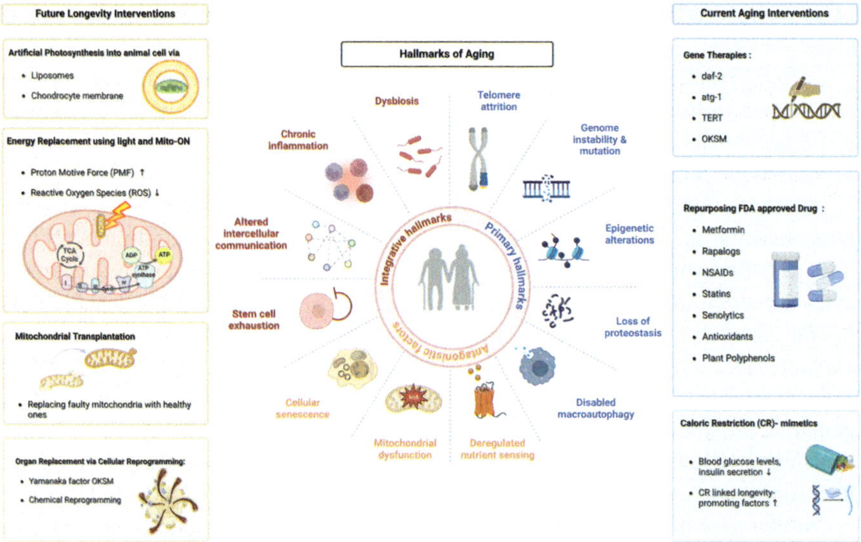

Fig. 7.1 Hallmarks of aging, emerging, longevity, and interventions. A consolidated view of aging hallmarks—classified into primary, antagonistic, and integrative categories—alongside current interventions (e.g., gene therapy, senolytics, CR mimetics) and future strategies such as mitochondrial transplantation, light-driven energy restoration, and cellular reprogramming. (Adapted from López-Otín et al., Cell (2023); created with BioRender (Tiwary, 2024). https://BioRender.com/q76w504)

We discussed the usage of opto-energetics as means to transduce the so-called cleaner energy inside animal mitochondria by using synthetic biology. We envision that such technology could be used in the near future to treat age-associated maladies in the eye, particularly in cases where the ability to synthesize ATP in the age/diseased eye is declining. Similarly, such technology may be applied for the skin. In the next years, we expect successful testing in mice prior to translating this technology into human health. Nonetheless, one eventual obstacle for using opto-energetics for inner organs is that light cannot readily reach them and therefore require further out-of-the-box approaches.

Interestingly, there is a substantial number of people who regularly take drugs to extend their lifespan. Few individuals go even further and use more drastic measures such as injecting genes in their bodies in designated locations that are less subjected to the regulation of health care. It is apparent that this may have opposite outcomes, as without sufficient human longitudinal studies, some if not most untested treatments would be harmful and cause a decrease in health and lifespan.

Lastly, we predict an integration of novel approaches, techniques, and ideas streaming into the longevity field in the years to come. Such synergy and cooperation is critical for progress in a field that had limited scope in the last few decades. This has to go hand in hand with proper funding which support taking risky ideas that may fail. Thus, we soon hope to see an unusual spike in disruptive approaches to break the human lifespan barrier that has never been considered before.

Acknowledgements We would like to thank Anne-Marie Galow for her critical input.

References

1. Weber DJ (2009) The Spanish frontier in North America: the brief edition. Yale University Press, New Haven
2. Milanich JT (2018) Florida Indians and the invasion from Europe. University Press of Florida, Gainesville
3. Lewis ME (2009) The early Chinese empires, vol 1. Harvard University Press, Cambridge
4. Parkin TG (2003) Old age in the Roman world: a cultural and social history. JHU Press, Baltimore
5. Siraisi NG (2009) Medieval and early Renaissance medicine: an introduction to knowledge and practice. University of Chicago Press, Chicago
6. Ramsey M (2013) The popularization of medicine. Routledge, New York, pp 97–133
7. Bynum WF (2006) The western medical tradition: 1800–2000. Cambridge University Press, Cambridge
8. López-Otín C, Blasco MA, Partridge L, Serrano M, Kroemer G (2023) Hallmarks of aging: an expanding universe. Cell 186:243–278. https://doi.org/10.1016/j.cell.2022.11.001
9. Fontana L, Partridge L, Longo VD (2010) Extending healthy life span—from yeast to humans. Science 328:321–326
10. Weismann A (1891) Essays upon heredity and kindred biological problems, vol 1. Clarendon Press, Oxford
11. Kirkwood TBL (2005) Understanding the odd science of aging. Cell 120:437–447. https://doi.org/10.1016/j.cell.2005.01.027
12. Jin K (2010) Modern biological theories of aging. Aging Dis 1:72–74
13. Hershey D (1989) Another way of looking at entropy: entropy and aging, evolving systems. Chem Eng Educ 23:154–160
14. Hayflick L (2007) Entropy explains aging, genetic determinism explains longevity, and undefined terminology explains misunderstanding both. PLoS Genet 3:e220. https://doi.org/10.1371/journal.pgen.0030220
15. Tatullo M (2023) Entropy meets physiology: should we translate aging as disorder? Stem Cells 42:91–97. https://doi.org/10.1093/stmcls/sxad084
16. Gladyshev VN (2021) The ground zero of organismal life and aging. Trends Mol Med 27:11–19. https://doi.org/10.1016/j.molmed.2020.08.012
17. Lemaître J-F, Moorad J, Gaillard J-M, Maklakov AA, Nussey DH (2024) A unified framework for evolutionary genetic and physiological theories of aging. PLoS Biol 22:e3002513. https://doi.org/10.1371/journal.pbio.3002513
18. Wu D et al (2022) An antagonistic pleiotropic gene regulates the reproduction and longevity tradeoff. Proc Natl Acad Sci 119:e2120311119. https://doi.org/10.1073/pnas.2120311119
19. Austad SN, Hoffman JM (2018) Is antagonistic pleiotropy ubiquitous in aging biology? Evol Med Public Health 2018:287–294. https://doi.org/10.1093/emph/eoy033
20. Chistyakov VA, Denisenko YV (2019) In: Danan G, Dupre ME (eds) Encyclopedia of gerontology and population aging. Springer International Publishing, Berlin, pp 1–7

21. Auten RL, Davis JM (2009) Oxygen toxicity and reactive oxygen species: the devil is in the details. Pediatr Res 66:121–127. https://doi.org/10.1203/PDR.0b013e3181a9eafb
22. Berry BJ, Kaeberlein M (2021) An energetics perspective on geroscience: mitochondrial protonmotive force and aging. GeroScience 43:1591–1604. https://doi.org/10.1007/s11357-021-00365-7
23. Mansell E et al (2021) Mitochondrial potentiation ameliorates age-related heterogeneity in hematopoietic stem cell function. Cell Stem Cell 28:241–256.e246. https://doi.org/10.1016/j.stem.2020.09.018
24. Campisi J (2013) Aging, cellular senescence, and cancer. Annu Rev Physiol 75:685–705. https://doi.org/10.1146/annurev-physiol-030212-183653
25. Nik-Zainal S, Hall BA (2019) Cellular survival over genomic perfection. Science 366:802–803
26. Quesada V et al (2019) Giant tortoise genomes provide insights into longevity and age-related disease. Nat Ecol Evol 3:87–95
27. Alder JK, Armanios M (2022) Telomere-mediated lung disease. Physiol Rev 102:1703
28. Horvath S, Raj K (2018) DNA methylation-based biomarkers and the epigenetic clock theory of ageing. Nat Rev Genet 19:371–384. https://doi.org/10.1038/s41576-018-0004-3
29. Shen Z, Hinson A, Miller RA, Garcia GG (2021) Cap-independent translation: a shared mechanism for lifespan extension by rapamycin, acarbose, and 17α-estradiol. Aging Cell 20:e13345. https://doi.org/10.1111/acel.13345
30. Hetz C, Zhang K, Kaufman RJ (2020) Mechanisms, regulation and functions of the unfolded protein response. Nat Rev Mol Cell Biol 21:421–438
31. Hayat MA (2015) Autophagy: cancer, other pathologies, inflammation, immunity, infection, and aging. Academic Press, New York, pp 1–51
32. Cassidy LD et al (2020) Temporal inhibition of autophagy reveals segmental reversal of ageing with increased cancer risk. Nat Commun 11:307
33. Mylonas A, O'Loghlen A (2022) Cellular senescence and ageing: mechanisms and interventions. Front Aging 3:866718. https://doi.org/10.3389/fragi.2022.866718
34. Tuttle CSL et al (2020) Cellular senescence and chronological age in various human tissues: a systematic review and meta-analysis. Aging Cell 19:e13083. https://doi.org/10.1111/acel.13083
35. Gille B et al (2023) Longevity interventions in Titan mice attenuate frailty and senescence accumulation. bioRxiv. 2023.2009.2008.556694. https://doi.org/10.1101/2023.09.08.556694
36. Robbins PD et al (2021) Senolytic drugs: reducing senescent cell viability to extend health span. Annu Rev Pharmacol Toxicol 61:779–803
37. Li X et al (2023) Inflammation and aging: signaling pathways and intervention therapies. Signal Transduct Target Ther 8:239. https://doi.org/10.1038/s41392-023-01502-8
38. Zmora N, Soffer E, Elinav E (2019) Transforming medicine with the microbiome. Sci Transl Med 11:eaaw1815
39. Zong Y et al (2024) Mitochondrial dysfunction: mechanisms and advances in therapy. Signal Transduct Target Ther 9:124. https://doi.org/10.1038/s41392-024-01839-8
40. Amorim JA et al (2022) Mitochondrial and metabolic dysfunction in ageing and age-related diseases. Nat Rev Endocrinol 18:243–258
41. Madeo F, Carmona-Gutierrez D, Hofer SJ, Kroemer G (2019) Caloric restriction mimetics against age-associated disease: targets, mechanisms, and therapeutic potential. Cell Metab 29:592–610. https://doi.org/10.1016/j.cmet.2019.01.018
42. Nassar K et al (2024) The significance of caloric restriction mimetics as anti-aging drugs. Biochem Biophys Res Commun 692:149354. https://doi.org/10.1016/j.bbrc.2023.149354
43. Macneil LT, Schertzer JD, Steinberg GR (2020) Bacteria transmit metformin-associated lifespan extension. Nat Rev Endocrinol 16:9–10. https://doi.org/10.1038/s41574-019-0278-3
44. Markowicz-Piasecka M et al (2020) An investigation into the pleiotropic activity of metformin. A glimpse of haemostasis. Eur J Pharmacol 872:172984. https://doi.org/10.1016/j.ejphar.2020.172984

45. Shintani H, Shintani T, Ashida H, Sato M (2018) Calorie restriction mimetics: upstream-type compounds for modulating glucose metabolism. Nutrients 10:1821. https://doi.org/10.3390/nu10121821
46. Barzilai N, Crandall JP, Kritchevsky SB, Espeland MA (2016) Metformin as a tool to target aging. Cell Metab 23:1060–1065. https://doi.org/10.1016/j.cmet.2016.05.011
47. Vaiserman A, Koliada A, Lushchak O, Castillo MJ (2021) Repurposing drugs to fight aging: the difficult path from bench to bedside. Med Res Rev 41:1676–1700. https://doi.org/10.1002/med.21773
48. Panwar V et al (2023) Multifaceted role of mTOR (mammalian target of rapamycin) signaling pathway in human health and disease. Signal Transduct Target Ther 8:375. https://doi.org/10.1038/s41392-023-01608-z
49. Ehninger D, Neff F, Xie K (2014) Longevity, aging and rapamycin. Cell Mol Life Sci 71:4325–4346. https://doi.org/10.1007/s00018-014-1677-1
50. Zaseck LW, Miller RA, Brooks SV (2016) Rapamycin attenuates age-associated changes in tibialis anterior tendon viscoelastic properties. J Gerontol Ser A Biol Med Sci 71:858–865. https://doi.org/10.1093/gerona/glv307
51. Ge C et al (2023) Rapamycin suppresses inflammation and increases the interaction between p65 and IκBα in rapamycin-induced fatty livers. PLoS One 18:e0281888. https://doi.org/10.1371/journal.pone.0281888
52. Tang H, Shrager JB, Goldman D (2019) Rapamycin protects aging muscle. Aging 11:5868–5870. https://doi.org/10.18632/aging.102176
53. Blagosklonny MV (2019) Fasting and rapamycin: diabetes versus benevolent glucose intolerance. Cell Death Dis 10:607. https://doi.org/10.1038/s41419-019-1822-8
54. Huang X-B et al (2018) Aspirin derivative 5-(bis(3-methylbut-2-enyl)amino)-2-hydroxybenzoic acid improves thermotolerance via stress response proteins in Caenorhabditis elegans. Molecules 23:1359. https://doi.org/10.3390/molecules23061359
55. Song C et al (2017) Metabolome analysis of effect of aspirin on Drosophila lifespan extension. Exp Gerontol 95:54–62. https://doi.org/10.1016/j.exger.2017.04.010
56. Randy et al (2008) Nordihydroguaiaretic acid and aspirin increase lifespan of genetically heterogeneous male mice. Aging Cell 7:641–650. https://doi.org/10.1111/j.1474-9726.2008.00414.x
57. Hua H, Zhang H, Kong Q, Wang J, Jiang Y (2019) Complex roles of the old drug aspirin in cancer chemoprevention and therapy. Med Res Rev 39:114–145. https://doi.org/10.1002/med.21514
58. McNeil JJ et al (2018) Effect of aspirin on cardiovascular events and bleeding in the healthy elderly. N Engl J Med 379:1509–1518. https://doi.org/10.1056/nejmoa1805819
59. He C et al (2014) Enhanced longevity by ibuprofen, conserved in multiple species, occurs in yeast through inhibition of tryptophan import. PLoS Genet 10:e1004860. https://doi.org/10.1371/journal.pgen.1004860
60. Vlad SC, Miller DR, Kowall NW, Felson DT (2008) Protective effects of NSAIDs on the development of Alzheimer disease. Neurology 70:1672–1677. https://doi.org/10.1212/01.wnl.0000311269.57716.63
61. McSharry C (2011) Could over-the-counter treatment protect against Parkinson disease? Nat Rev Neurol 7:244. https://doi.org/10.1038/nrneurol.2011.49
62. Toth PP, Banach M (2019) Statins: then and now. Methodist DeBakey Cardiovasc J 15:23. https://doi.org/10.14797/mdcj-15-1-23
63. Spindler SR et al (2012) Statin treatment increases lifespan and improves cardiac health in drosophila by decreasing specific protein prenylation. PLoS One 7:e39581. https://doi.org/10.1371/journal.pone.0039581
64. Varela I et al (2008) Combined treatment with statins and aminobisphosphonates extends longevity in a mouse model of human premature aging. Nat Med 14:767–772. https://doi.org/10.1038/nm1786

65. Luotola K et al (2020) Statin treatment, phenotypic frailty and mortality among community-dwelling octogenarian men: the HBS cohort. Age Ageing 49:258–263. https://doi.org/10.1093/ageing/afz138
66. Schlesinger A et al (2019) Statins and selective serotonin reuptake inhibitors are associated with longer survival in nursing home residents. Isr Med Assoc J 21:719–723
67. Calabrò A et al (2024) Senotherapeutics to counteract senescent cells are prominent topics in the context of anti-ageing strategies. Int J Mol Sci 25:1792. https://doi.org/10.3390/ijms25031792
68. Xu M et al (2018) Senolytics improve physical function and increase lifespan in old age. Nat Med 24:1246–1256. https://doi.org/10.1038/s41591-018-0092-9
69. Novais EJ et al (2021) Long-term treatment with senolytic drugs Dasatinib and Quercetin ameliorates age-dependent intervertebral disc degeneration in mice. Nat Commun 12:5213. https://doi.org/10.1038/s41467-021-25453-2
70. Ashkenazi A, Fairbrother WJ, Leverson JD, Souers AJ (2017) From basic apoptosis discoveries to advanced selective BCL-2 family inhibitors. Nat Rev Drug Discov 16:273–284. https://doi.org/10.1038/nrd.2016.253
71. He Y et al (2020) Using proteolysis-targeting chimera technology to reduce navitoclax platelet toxicity and improve its senolytic activity. Nat Commun 11:1996
72. Vaiserman AM, Lushchak OV, Koliada AK (2016) Anti-aging pharmacology: promises and pitfalls. Ageing Res Rev 31:9–35. https://doi.org/10.1016/j.arr.2016.08.004
73. Bjelakovic G, Nikolova D, Gluud LL, Simonetti RG, Gluud C (2012) Antioxidant supplements for prevention of mortality in healthy participants and patients with various diseases. Cochrane Database Syst Rev 2012:CD007176. https://doi.org/10.1002/14651858.cd007176.pub2
74. Goszcz K et al (2015) Antioxidants in cardiovascular therapy: panacea or false hope? Front Cardiovasc Med 2:29. https://doi.org/10.3389/fcvm.2015.00029
75. Halliwell B (2013) The antioxidant paradox: less paradoxical now? Br J Clin Pharmacol 75:637–644. https://doi.org/10.1111/j.1365-2125.2012.04272.x
76. Bonilla E, Medina-Leendertz S, DíAz, S. (2002) Extension of life span and stress resistance of Drosophila melanogaster by long-term supplementation with melatonin. Exp Gerontol 37:629–638. https://doi.org/10.1016/s0531-5565(01)00229-7
77. Izmaylov DM, Obukhova LK (1999) Geroprotector effectiveness of melatonin: investigation of lifespan of Drosophila melanogaster. Mech Ageing Dev 106:233–240. https://doi.org/10.1016/s0047-6374(98)00105-5
78. Anisimov V (2003) Dose-dependent effect of melatonin on life span and spontaneous tumor incidence in female SHR mice. Exp Gerontol 38:449–461. https://doi.org/10.1016/s0531-5565(02)00240-1
79. Anisimov VN et al (2001) Melatonin increases both life span and tumor incidence in female CBA mice. J Gerontol Ser A Biol Med Sci 56:B311–B323. https://doi.org/10.1093/gerona/56.7.b311
80. Cardinali DP (2019) Melatonin: clinical perspectives in neurodegeneration. Front Endocrinol 10:480. https://doi.org/10.3389/fendo.2019.00480
81. Posadzki PP et al (2018) Melatonin and health: an umbrella review of health outcomes and biological mechanisms of action. BMC Med 16:18. https://doi.org/10.1186/s12916-017-1000-8
82. Iweala EJ et al (2024) Anti-aging potential of bioactive phytoconstituents found in edible medicinal plants: a review. Sci 6:36. https://doi.org/10.3390/sci6020036
83. Kharat P et al (2020) Ellagic acid prolongs the lifespan of Drosophila melanogaster. GeroScience 42:271–285. https://doi.org/10.1007/s11357-019-00135-6
84. Rybina OY, Symonenko AV, Pasyukova EG (2023) Compound combinations targeting longevity: challenges and perspectives. Ageing Res Rev 85:101851. https://doi.org/10.1016/j.arr.2023.101851
85. Longo VD et al (2015) Interventions to slow aging in humans: are we ready? Aging Cell 14:497–510. https://doi.org/10.1111/acel.12338

86. Kenyon C, Chang J, Gensch E, Rudner A, Tabtiang R (1993) A C. elegans mutant that lives twice as long as wild type. Nature 366:461–464. https://doi.org/10.1038/366461a0
87. Friedman DB, Johnson TE (1988) A mutation in the age-1 gene in Caenorhabditis elegans lengthens life and reduces hermaphrodite fertility. Genetics 118:75–86. https://doi.org/10.1093/genetics/118.1.75
88. Molière A et al (2024) Improved resilience and proteostasis mediate longevity upon DAF-2 degradation in old age. GeroScience 46:5015–5036. https://doi.org/10.1007/s11357-024-01232-x
89. de Jesus BB et al (2012) Telomerase gene therapy in adult and old mice delays aging and increases longevity without increasing cancer. EMBO Mol Med 4:691. https://doi.org/10.1002/emmm.201200245
90. Muñoz-Lorente MA, Cano-Martin AC, Blasco MA (2019) Mice with hyper-long telomeres show less metabolic aging and longer lifespans. Nat Commun 10:4723. https://doi.org/10.1038/s41467-019-12664-x
91. Schellnegger M, Hofmann E, Carnieletto M, Kamolz L-P (2024) Unlocking longevity: the role of telomeres and its targeting interventions. Front Aging 5:1339317. https://doi.org/10.3389/fragi.2024.1339317
92. Yücel AD, Gladyshev VN (2024) The long and winding road of reprogramming-induced rejuvenation. Nat Commun 15:1941. https://doi.org/10.1038/s41467-024-46020-5
93. Simpson DJ, Olova NN, Chandra T (2021) Cellular reprogramming and epigenetic rejuvenation. Clin Epigenetics 13:170. https://doi.org/10.1186/s13148-021-01158-7
94. Ocampo A et al (2016) In vivo amelioration of age-associated hallmarks by partial reprogramming. Cell 167:1719–1733.e1712. https://doi.org/10.1016/j.cell.2016.11.052
95. Browder KC et al (2022) In vivo partial reprogramming alters age-associated molecular changes during physiological aging in mice. Nat Aging 2:243–253. https://doi.org/10.1038/s43587-022-00183-2
96. Schoenfeldt L et al (2022) Chemical reprogramming ameliorates cellular hallmarks of aging and extends lifespan. Cold Spring Harbor Laboratory
97. Guan J et al (2022) Chemical reprogramming of human somatic cells to pluripotent stem cells. Nature 605:325–331. https://doi.org/10.1038/s41586-022-04593-5
98. Kaur P et al (2022) Combining stem cell rejuvenation and senescence targeting to synergistically extend lifespan. Aging 14(20):8270–8291. https://doi.org/10.18632/aging.204347
99. Milholland B, Vijg J (2022) Why Gilgamesh failed: the mechanistic basis of the limits to human lifespan. Nat Aging 2:878–884. https://doi.org/10.1038/s43587-022-00291-z
100. Matsumoto Y, Piraino S, Miglietta MP (2019) Transcriptome characterization of reverse development in Turritopsis dohrnii (Hydrozoa, Cnidaria). G3 (Bethesda) 9:4127–4138. https://doi.org/10.1534/g3.119.400487
101. Nielsen J et al (2016) Eye lens radiocarbon reveals centuries of longevity in the Greenland shark (*Somniosus microcephalus*). Science 353:702–704. https://doi.org/10.1126/science.aaf1703
102. Sahm A et al (2024) The Greenland shark (*Somniosus microcephalus*) genome provides insights into extreme longevity. bioRxiv. 2024.2009.2009.611499. https://doi.org/10.1101/2024.09.09.611499
103. Buffenstein R, Amoroso VG (2024) The untapped potential of comparative biology in aging research: insights from the extraordinary-long-lived naked mole-rat. J Gerontol A Biol Sci Med Sci 79:glae110. https://doi.org/10.1093/gerona/glae110
104. Ruby JG, Smith M, Buffenstein R (2018) Naked mole-rat mortality rates defy Gompertzian laws by not increasing with age. elife 7:e31157. https://doi.org/10.7554/elife.31157
105. Keane M et al (2015) Insights into the evolution of longevity from the bowhead whale genome. Cell Rep 10:112–122. https://doi.org/10.1016/j.celrep.2014.12.008

106. Berhanu S, Ueda T, Kuruma Y (2019) Artificial photosynthetic cell producing energy for protein synthesis. Nat Commun 10:1325. https://doi.org/10.1038/s41467-019-09147-4
107. Hoffmann A, Hildebrandt V, Heberle J, Büldt G (1994) Photoactive mitochondria: in vivo transfer of a light-driven proton pump into the inner mitochondrial membrane of Schizosaccharomyces pombe. Proc Natl Acad Sci 91:9367–9371. https://doi.org/10.1073/pnas.91.20.9367
108. Berry BJ et al (2020) Optogenetic control of mitochondrial protonmotive force to impact cellular stress resistance. EMBO Rep 21:e49113. https://doi.org/10.15252/embr.201949113
109. Berry BJ et al (2022) Optogenetic rejuvenation of mitochondrial membrane potential extends C. elegans lifespan. Nat Aging 3:157–161. https://doi.org/10.1038/s43587-022-00340-7
110. Tiwary V, Galow A-M, Wojtovich AP, Peleg S (2023) Using light to drive energy transduction in metazoan aging. Trends Biochem Sci 48:920–922. https://doi.org/10.1016/j.tibs.2023.08.010
111. Chen P et al (2022) A plant-derived natural photosynthetic system for improving cell anabolism. Nature 612:546–554. https://doi.org/10.1038/s41586-022-05499-y
112. Lee KY et al (2018) Photosynthetic artificial organelles sustain and control ATP-dependent reactions in a protocellular system. Nat Biotechnol 36:530–535. https://doi.org/10.1038/nbt.4140

Individualized Medicine for Cardiovascular Disease: The *curATime* Consortium

8

V. ten Cate and Philipp S. Wild

Contents

Abstract

The *curATime* cluster (Consortium, Tc, CurATime website, 2023) is a regional network that aims to advance research in cardiovascular medicine and achieve new progress and innovation in cardiovascular medicine by combining medical expertise (coordinated by the University Medical Center of the Johannes Gutenberg University Mainz), artificial intelligence (coordinated by the German Research Center for Artificial Intelligence, DFKI), and biotechnological entrepreneurship and drug development (coordinated by TRON gGmbH and BioNTech SE). The network, which is funded by the German Federal Ministry of Education and Research (BMBF) as part of the clusters4future initiative, includes research institutes, small- and medium-sized enterprises (SMEs) and larger corporations in the Rhine-Main area. The development ranges from new diagnostic tools, preventive wearables and apps to new personalized therapeutics

V. ten Cate · P. S. Wild
Preventive Cardiology and Preventive Medicine, Center for Cardiology, University Medical Center of the Johannes Gutenberg University Mainz, Mainz, Germany

Center for Thrombosis and Hemostasis (CTH), University Medical Center of the Johannes Gutenberg University Mainz, Mainz, Germany

German Center for Cardiovascular Research (DZHK), Partner Site Rhine Main, University Medical Center of the Johannes Gutenberg University Mainz, Mainz, Germany

P. S. Wild (✉)
Institute of Molecular Biology (IMB), Mainz, Germany
e-mail: philipp.wild@unimedizin-mainz.de

U. A. K. Betz (ed.), *Science for a Better Tomorrow*,
https://doi.org/10.1007/978-3-031-93623-4_8

that leverage recent biotechnological advances to provide safer and more effective treatment for people with cardiovascular disease or at risk of developing it.

Keywords

Preventive medicine · Drug · Discovery · Development · Artificial intelligence

Cardiovascular diseases, in particular ischemic heart disease and stroke, are the leading causes of morbidity and mortality, accounting for approximately 28% of all deaths worldwide [1]. In recent decades, the numbers of disability-adjusted life years (DALYs) attributable to ischemic heart disease and stroke—which represent the loss of the equivalent of 1 year of healthy life—have risen to 182 and 143 million, respectively [2]. This considerable burden on society has only grown and continues to do so, in part because of the increasing prevalence of common risk factors such as obesity, diabetes mellitus, and arterial hypertension, which have become more commonplace due to sedentary lifestyles and poor diets. A growing body of research indicates that other environmental exposures also contribute to cardiovascular disease risk, including ambient air pollution, ambient temperature, and water, noise and light pollution [3, 4]. The term "exposome" was coined to describe the totality of exposures to which a person is exposed over the course of their life. It is known that cardiovascular disease is the disease category most influenced by the exposome [5]. For the further development of cardiovascular medicine, there is a need to address all causes of cardiovascular disease therapeutically and to tailor treatments to the specific etiology of the individual in order to make treatment more effective and efficient. These more personalized therapies should ideally also imply a reduction in polypharmacy, which is particularly relevant for older patients, who are quite often taking more than ten medications per day [6]. This carries the risk of drug interactions and side effects, which are evident in polypharmacy in cardiovascular disease, for example, through increased susceptibility to bleeding, hospitalization and death [7].

Cardiovascular research has made great therapeutic strides in past decades, with the invention of lipid and blood pressure lowering drugs, but also surgical and increasingly more minimally invasive or interventional treatments like heart valve replacements, defibrillators, and coronary revascularization procedures such as percutaneous coronary intervention (PCI) or coronary artery bypass grafting (CABG), which have greatly improved both life expectancy and health span of those affected. Among more recent innovations are drugs for diabetes and obesity, which are taking the world by storm [8–10]. This has once again highlighted the huge impact of obesity on health. Another recent trend in cardiovascular medicine is the introduction of remote cardiac rehabilitation programs and the incorporation of telemedicine in general, which enables patients to recover increasingly more and quicker in outpatient settings [11, 12]. This has lowered barriers to treatment and made the care itself more comfortable. In the realm of prevention, there are new innovations such as digital wearables coupled with artificial intelligence (AI) that help track heart rhythm and bodily functions over time [13, 14]. These developments can improve detection of cardiac arrhythmias like atrial fibrillation, for example, but also "digital mobility outcomes" (DMOs), which encompass every aspect of mobility, from gait to temporal patterns of starting and stopping during the day [15]. However, most of these innovations are still focused on treating traditional risk factors, particularly obesity, diabetes mellitus, and physical inactivity. This ignores the fact that approximately 12% of patients with acute coronary syndrome have no traditional risk factors [16], indicating the relevance of disease mechanisms that are not addressed by the current generation of therapeutics.

The *curATime* cluster [17] is a regional network that aims to advance research in cardiovascular medicine and achieve new progress and innovation in cardiovascular medicine by combining medical expertise (coordinated by the University Medical Center of the Johannes Gutenberg University Mainz), artificial intelligence (coordinated by the German Research Center for Artificial Intelligence, DFKI), and biotechnological entrepreneurship and drug development (coordinated by TRON gGmbH and BioNTech SE). The network, which is funded by the German Federal Ministry of Education and Research (BMBF) as part of the clusters4future initiative, includes research institutes, small- and medium-sized enterprises (SMEs) and larger corporations in the Rhine-Main area. The development ranges from new diagnostic tools, preventive wearables and apps to new personalized therapeutics that leverage recent biotechnological advances to provide safer and more effective treatment for people with cardiovascular disease or at risk of developing it. Building on the success of the mRNA-based SARS-CoV-2 vaccines developed by the Mainz-based company BioNTech SE, this modular system will be used to develop similar therapeutics for cardiovascular diseases. By optimizing the technology for organ- and cell-specific delivery and by identifying new molecular targets in pathways not yet targeted by current therapeutics, the curATime consortium holds the promise to develop a prototype and—eventually—bring to market a new class of drugs that may render personalized cardiovascular medicine a reality.

Drug development by pharmaceutical companies is a time-consuming, complicated and expensive process, characterized by low success rates in translating preclinical developments into application [18]. New pharmaceuticals have to pass a series of phases before regulatory approval to release a drug onto the market is granted. These phases include target identification and lead compound design, toxicity testing in preclinical models, phase 0 preliminary trials to select lead compounds, initial phase 1 trials for basic safety testing, phase 2 dose-finding trials, and finally large-scale phase 3 trials, in which efficacy and safety are tested among a large number of individuals. The entire process takes an average of 13 years, and the average total costs for development of a new drug amount to $2.6 billion, a figure that doubles every 9 years due to increasing regulatory requirements [19–21]. About 95% of new drugs fail in clinical trials, with failures occurring at each phase of development. The reasons for this range from a lack of transferability of findings from initial animal models to humans, to initially unforeseen safety or toxicity issues, to a lack of efficacy in the intended clinical indication [18]. It is clear that various stages of drug development can benefit from further optimization through innovations that, for example, enable to better predict the behavior of a particular target molecule in phase 0–3 trials than has historically been possible.

Systems medicine is an indispensable paradigm for bridging the translational gap between initial preclinical research ("the bench") and the clinical application of a drug in patients ("the bedside"). In contrast to traditional clinical medicine, which too often perceives diseases as dichotomous "on–off" conditions ("sick" vs. "healthy," "pathological" vs. "physiological") that affect a specific organ, the systems medicine framework approaches disease from the perspective of dysregulations of networked biological systems. Each of these systems functions along a continuous spectrum that is specific to the individual and their interaction is also specific to the individual so that dysregulations in a particular system in one person will not reliably lead to the same consequence in another. A logical consequence of this is that personalized medicine is required to adequately treat disease. This framework recognizes that in a complex system like the human body, nothing occurs in isolation and that (pathological) changes in a particular system or organ will inevitably affect other elements of the system. It intuitively follows that drugs that may rectify an imbalance in one system that is critical for the treatment of a particular disease can easily be associated with (side) effects affecting other organs. Hence, it is imperative to holistically study the function and structure of all major organ systems in relation to the expression of each molecule that is considered a potential drug target as part of the initial prioritization of candidate targets. This approach helps to avoid unforeseen toxicity issues and adverse effects in organs, but also to capture any positive effect in the body (currently still addressed as "drug repurposing") that may not appear to be related at first glance.

One of the biotechnological revolutions that have helped popularize the concept of systems medicine is the advent of modern, high-throughput "omics" technologies that allow for the generation of high-dimensional molecular profiles of individuals. This has allowed diseases to be dissected into molecular processes that constitute them, which operate at on all levels—genetic, epigenetic, transcriptional, translational, post-transcriptional, post-translational, and many more. These technologies have made it considerably easier to capture the molecular complexity and person-specific nuances involved in disease. At the same time, these advances have allowed biomedical research to move away from simplistic, single target-oriented research toward the study of how potential interventions may affect complex systems in their entirety. In preclinical research, this is reflected by an increasing reliance on methods like single cell transcriptomics, in which the reprogramming of entire gene networks can be studied after experimental manipulation at the level of individual cells [22, 23]. In clinical research, examples of this thinking include studies in which the effects of drugs on blood plasma samples are studied based on proteomic analyzes before and after the intervention to evaluate which pathways in the circulation are affected. This has been done for recent cardiovascular blockbuster drugs like sodium glucose co-transporter 2 (SGLT2) inhibitors, used in diabetes mellitus, chronic kidney disease and heart failure [24, 25], and more recently for GLP-1-receptor agonists in individuals with obesity [26]. As these techniques become more prevalent in biomedicine, our understanding of the function of specific genes, proteins and other molecules is also improving. This contributes to better predictions of how a particular therapy may affect systems other than the one it was intended to target. This will eventually make explicit "drug repurposing" trials, in which established drugs are tested for efficacy and safety when administered in another indication, unnecessary.

As already mentioned, another common reason for the failure of drug candidates is the poor generalizability from the initial model system (e.g., animal model) used for preclinical testing to humans. Reasons for this include the fact that the model system does not mimic the human disease, or human biology and life in terms of its length and diversity—and thus complexity—closely enough, that the expression of the target molecule differs between species, or that the effect of the target on the condition in humans is simply less than in other animals. To prevent the risk of target molecules identified and tested in animal models being poorly translated to humans, it is useful to screen and prioritize target candidates based on their (likely) effects on organ function and clinical outcome in humans before designing and conducting the first preclinical experiments. But how can this be done without conducting costly and time-consuming randomized clinical trials? The concept of "digital twins" can be instructive in understanding how the virtualization of clinical research may contribute to more efficient drug target discovery. A digital twin is a high-fidelity digital representation of a real-world entity that can be virtually manipulated to explore outcomes without incurring any real risks or costs. This concept was initially used in manufacturing, for instance by NASA [27], to test the physical behavior of rockets and their components before they were built and launched. Since then, the concept has proliferated into other fields, including biomedicine, where it can be used to test out new surgical techniques, pharmaceutical compounds, and other interventions associated with substantial risk.

While there is no substitute for randomized clinical trials to ascertain the actual causal impact of influencing a particular target with a particular drug formulation, the digital twin approach is useful for gaining initial insights into off-target effects, unintended side effects, and, in general, the likely efficacy and safety of a treatment, thus helping to prioritize targets for drug development. To implement this, well designed and large prospective cohort studies with biomaterial collection are important. To further improve these investigations, a longitudinal biomaterial collection will help to identify likely causal targets and shed light on the influence of molecular dynamics, which can critically influence the effect of specific molecules. In the *curATime* consortium, human multi-omics investigations as just described are taking place as part of the "*biosignATure*" subproject and are complemented by preclinical in vivo and in vitro research in various other subprojects.

One of the technological developments that have taken place in lockstep with the "multi-omics" revolution is that of artificial intelligence (AI). As the volume and complexity of data have exploded in the past two decades so has the need for methodological advances that allow these data to be meaningfully utilized. Accordingly, the largely hypothesis-driven biomedical research of the past (e.g., candidate gene studies) has largely been supplanted by a hypothesis-free, "discovery" science that acknowledges the deep complexity of biology. Rather than being bound by preconceived notions about the structure and behavior of a biological system, this new approach allows unbiased, high-resolution snapshots of diseases to be captured, which only make sense when the initial complexity is passed through a series of filters to isolate the most relevant signal. This is essentially what modern AI methods attempt—to allow for maximum flexibility in the structure and size of the input data and to distil relevant insights from it without the need for human intervention. Examples of such methods are neural networks, i.e., networks of separate regression functions that collectively function as "universal function approximators" [28], meaning that they are able to infer any, even highly complex, relationship between a set of inputs and outputs. Countless "architectures," i.e., specific designs of such networks, have since been proposed, each with unique properties that render them particularly suitable for tasks as wide-ranging as medical image analysis, translating and generating text (e.g., chatGPT), or enabling automated and autonomous navigation in the real world by robots (including cars).

AI methods have become so prevalent in research and society that they are now being scrutinized by legal and ethics experts, who have noted that the use of such methods can have unfortunate side effects. These side effects include biases in data collection deriving from societal prejudices—such as racism and sexism—and are exacerbated by the use of these strongly data-driven methods. This has led to a new movement of AI researchers focusing on building fair and "unbiased" models that can counter such biases by explicitly accounting for their disproportionate presence in the data. Other typical criticisms of AI methods are that they are "black boxes" because their inner workings are so complex that it is not possible to understand how their predictions are derived, and that the algorithms are too data-hungry, in that they only function correctly when fed extremely large datasets that are prohibitively expensive and difficult to collect in the biomedical field. Within the *curATime* consortium, researchers on the cutting edge of AI are actively working on solutions to these problems by focusing on methods that are particularly suitable for small data volumes as well as methods that provide intuitive insight into how they work, known in the community as "explainable AI" (or XAI). These and other innovations will make AI methods increasingly well suited to the task of aiding understanding of disease biology, as well as identifying targets and generating lead compounds for new therapeutics, and developing innovative tools for diagnostics, prognostics, and therapy monitoring instruments.

The final area of technological innovation required to make personalized cardiovascular medicine a reality is pharmaceutical biotechnology. One of the major challenges in the development and application of therapeutics is that these are always associated with at least some side effects. Undesired side effects can be due to the target of the drug having a physiological function in systems other than those intended to be treated ("pleiotropy") but also of a target being modulated too strongly, for too long or too widely in the body. For example, a drug designed to inhibit a pathological process in a single organ, such as the heart, should ideally leave other organs unaffected. Moreover, a systemically administered drug may not only inadvertently enter off-target organs but may even accumulate there to greater extent than in the target organ, diminishing efficacy and exacerbating side effects [29]. To mitigate these issucs, tissue- and cell-specific drug delivery has emerged as the new holy grail of precision medicine, with oncology leading the way in developing and testing experimental targeted treatments [29–34]. Targeted delivery to a specific tissue and cell type first requires careful mapping of the tissue and cell-specifically expressed targets and then the development of a system that comprises a ligand that can bind this site, the cargo to be delivered, and a so-called linker, which is the vehicle used to transport both [29]. Such a system faces many obstacles, including intravascular, endothelial, extracellular, and cellular barriers, all of which must be traversed to deliver the therapeutic cargo to its destination. This is an active field of research in which biophysical and biochemical engineers are playing the leading role.

Amid the societal upheaval wrought by the SARS-CoV-2 pandemic, another rapidly adopted innovation was that of messenger (m)RNA-based therapeutics. Several companies were working on this simultaneously, but the mRNA-based vaccine that ultimately garnered the most commercial success was the Pfizer-BioNTech COVID-19 vaccine. While mRNA-based drugs were initially dismissed as immunogenic (i.e., triggering an unwanted immune response), nucleoside modifications, such as those pioneered by Nobel laureate Katalin Karikó, have since rendered these objections obsolete [35]. Meanwhile, mRNA therapeutics are considered to be a highly beneficial class of drug, characterized by high transfection efficiency and low toxicity, alongside other advantages such as a relatively long duration of action [36]. The lighthouse project "*curAIntervent*" of the *curATime* consortium aims to innovate in the field of locoregionally delivered RNA therapeutics that will be useful for the treatment of cardiovascular diseases. The research conducted in this project combines the expertise of drug developers at BioNTech subsidiary resano GmbH with the preclinical modeling and expertise of academic researchers specializing in nanoparticle development.

In summary, the *curATime* consortium will address several key areas in current research to help realize cardiovascular precision medicine. The consortium's focus on the critical areas of innovation of optimized target identification, AI methods for complex biological data, and tissue- and cell-specific delivery systems for new therapeutics reflects the open research questions in current medical research. The close interaction between academia and biomedical industry at the forefront of biomedical and pharmaceutical research and development will continue to be a key factor for the success of this endeavor.

Declaration of Interests

V.t.C. was funded by a grant from the Federal Ministry of Education and Research, the future cluster curATime (BMBF 03ZU1202AA). Outside the submitted work, VtC has received consulting fees from Bayer AG. P.S.W. is principal investigator in the future cluster curATime (BMBF 03ZU1202AA, 03ZU1202CD, 03ZU1202DB,

03ZU1202JC, 03ZU1202KB, 03ZU1202LB, 03ZU1202MB, and 03ZU1202OA). PSW is funded by the Federal Ministry of Education and Research (BMBF 01EO1503), Ministry of Science and Health of the State of Rhineland Palatinate (MWG RLP 724-0010#2021/0030-1501), and the Federal Institute for Occupational Safety and Health (BAuA, F2447/537727) outside the present work. PSW is principal investigator of the German Center for Cardiovascular Research (DZHK) and principal investigator of the DIASyM research core (BMBF DIASyM research core (BMBF 161L0217A, 031L0217A)). Outside the submitted work, PSW reports grants from Bayer AG, non-financial grants from Philips Medical Systems, grants and consulting fees from Boehringer Ingelheim, grants and consulting fees from Novartis Pharma, grants and consulting fees from Sanofi-Aventis, grants, consulting and lecturing fees from Bayer Health Care, grants and consulting fees from Daiichi Sankyo Europe, lecturing fees from Pfizer Pharma, lecturing fees from Bristol Myers Squibb, consulting fees from Astra Zeneca, consulting fees and non-financial support from Diasorin and non-financial support from IEM.

References

1. Di Cesare M, Bixby H, Gaziano T, Hadeed L, Kabudula C, McGhie DV, Mwangi J, Pervan B, Perel P, Piñeiro D (2023) World heart report 2023: confronting the world's number one killer. World Heart Federation, Geneva
2. Roth GA, Mensah GA, Johnson CO, Addolorato G, Ammirati E, Baddour LM, Barengo NC, Beaton AZ, Benjamin EJ, Benziger CP, Bonny A, Brauer M, Brodmann M, Cahill TJ, Carapetis J, Catapano AL, Chugh SS, Cooper LT, Coresh J, Criqui M, DeCleene N, Eagle KA, Emmons-Bell S, Feigin VL, Fernández-Solà J, Fowkes G, Gakidou E, Grundy SM, He FJ, Howard G, Hu F, Inker L, Karthikeyan G, Kassebaum N, Koroshetz W, Lavie C, Lloyd-Jones D, Lu HS, Mirijello A, Temesgen AM, Mokdad A, Moran AE, Muntner P, Narula J, Neal B, Ntsekhe M, Moraes de Oliveira G, Otto C, Owolabi M, Pratt M, Rajagopalan S, Reitsma M, Ribeiro ALP, Rigotti N, Rodgers A, Sable C, Shakil S, Sliwa-Hahnle K, Stark B, Sundström J, Timpel P, Tleyjeh IM, Valgimigli M, Vos T, Whelton PK, Yacoub M, Zuhlke L, Murray C, Fuster V (2020) Global burden of cardiovascular diseases and risk factors, 1990-2019: update from the GBD 2019 study. J Am Coll Cardiol 76:2982–3021
3. Münzel T, Hahad O, Sørensen M, Lelieveld J, Duerr GD, Nieuwenhuijsen M, Daiber A (2022) Environmental risk factors and cardiovascular diseases: a comprehensive expert review. Cardiovasc Res 118:2880–2902
4. Miller MR, Landrigan PJ, Arora M, Newby DE, Münzel T, Kovacic JC (2024) Water, soil, noise, and light pollution: JACC focus seminar, part 2. J Am Coll Cardiol 83:2308–2323
5. Prüss-Ustün A, van Deventer E, Mudu P, Campbell-Lendrum D, Vickers C, Ivanov I, Forastiere F, Gumy S, Dora C, Adair-Rohani H, Neira M (2019) Environmental risks and non-communicable diseases. BMJ 364:l265
6. Morin L, Johnell K, Laroche ML, Fastbom J, Wastesson JW (2018) The epidemiology of polypharmacy in older adults: register-based prospective cohort study. Clin Epidemiol 10:289–298
7. Eggebrecht L, Nagler M, Göbel S, Lamparter H, Keller K, Wagner B, Panova-Noeva M, Ten Cate V, Bickel C, Lauterbach M, Espinola-Klein C, Hardt R, Münzel T, Prochaska JH, Wild PS (2019) Relevance of polypharmacy for clinical outcome in patients receiving vitamin K antagonists. J Am Geriatr Soc 67:463–470
8. Brown E, Heerspink HJL, Cuthbertson DJ, Wilding JPH (2021) SGLT2 inhibitors and GLP-1 receptor agonists: established and emerging indications. Lancet 398:262–276

9. Rosenstock J, Wysham C, Frías JP, Kaneko S, Lee CJ, Fernández Landó L, Mao H, Cui X, Karanikas CA, Thieu VT (2021) Efficacy and safety of a novel dual GIP and GLP-1 receptor agonist tirzepatide in patients with type 2 diabetes (SURPASS-1): a double-blind, randomised, phase 3 trial. Lancet 398:143–155
10. Sanyal AJ, Kaplan LM, Frias JP, Brouwers B, Wu Q, Thomas MK, Harris C, Schloot NC, Du Y, Mather KJ, Haupt A, Hartman ML (2024) Triple hormone receptor agonist retatrutide for metabolic dysfunction-associated steatotic liver disease: a randomized phase 2a trial. Nat Med 30:2037–2048
11. Ramachandran HJ, Jiang Y, Tam WWS, Yeo TJ, Wang W (2022) Effectiveness of home-based cardiac telerehabilitation as an alternative to Phase 2 cardiac rehabilitation of coronary heart disease: a systematic review and meta-analysis. Eur J Prev Cardiol 29:1017–1043
12. Kuan PX, Chan WK, Fern Ying DK, Rahman MAA, Peariasamy KM, Lai NM, Mills NL, Anand A (2022) Efficacy of telemedicine for the management of cardiovascular disease: a systematic review and meta-analysis. Lancet Digit Health 4:e676–e691
13. Gill SK, Barsky A, Guan X, Bunting KV, Karwath A, Tica O, Stanbury M, Haynes S, Folarin A, Dobson R, Kurps J, Asselbergs FW, Grobbee DE, Camm AJ, Eijkemans MJC, Gkoutos GV, Kotecha D (2024) Consumer wearable devices for evaluation of heart rate control using digoxin versus beta-blockers: the RATE-AF randomized trial. Nat Med 30:2030–2036
14. Li J, Jia H, Zhou J, Huang X, Xu L, Jia S, Gao Z, Yao K, Li D, Zhang B, Liu Y, Huang Y, Hu Y, Zhao G, Xu Z, Li J, Yiu CK, Gao Y, Wu M, Jiao Y, Zhang Q, Tai X, Chan RH, Zhang Y, Ma X, Yu X (2023) Thin, soft, wearable system for continuous wireless monitoring of artery blood pressure. Nat Commun 14:5009
15. Micó-Amigo ME, Bonci T, Paraschiv-Ionescu A, Ullrich M, Kirk C, Soltani A, Küderle A, Gazit E, Salis F, Alcock L, Aminian K, Becker C, Bertuletti S, Brown P, Buckley E, Cantu A, Carsin AE, Caruso M, Caulfield B, Cereatti A, Chiari L, D'Ascanio I, Eskofier B, Fernstad S, Froehlich M, Garcia-Aymerich J, Hansen C, Hausdorff JM, Hiden H, Hume E, Keogh A, Kluge F, Koch S, Maetzler W, Megaritis D, Mueller A, Niessen M, Palmerini L, Schwickert L, Scott K, Sharrack B, Sillén H, Singleton D, Vereijken B, Vogiatzis I, Yarnall AJ, Rochester L, Mazzà C, Del Din S (2023) Assessing real-world gait with digital technology? Validation, insights and recommendations from the Mobilise-D consortium. J Neuroeng Rehabil 20:78
16. Kong G, Chin YH, Chong B, Goh RSJ, Lim OZH, Ng CH, Muthiah M, Foo R, Vernon ST, Loh PH, Chan MY, Chew NWS, Figtree GA (2023) Higher mortality in acute coronary syndrome patients without standard modifiable risk factors: results from a global meta-analysis of 1,285,722 patients. Int J Cardiol 371:432–440
17. Consortium (2023) Tc. CurATime website
18. Sun D, Gao W, Hu H, Zhou S (2022) Why 90% of clinical drug development fails and how to improve it? Acta Pharm Sin B 12:3049–3062
19. Seyhan AA (2019) Lost in translation: the valley of death across preclinical and clinical divide—identification of problems and overcoming obstacles. Transl Med Commun 4:1–19
20. Waring MJ, Arrowsmith J, Leach AR, Leeson PD, Mandrell S, Owen RM, Pairaudeau G, Pennie WD, Pickett SD, Wang J, Wallace O, Weir A (2015) An analysis of the attrition of drug candidates from four major pharmaceutical companies. Nat Rev Drug Discov 14:475–486
21. Cook D, Brown D, Alexander R, March R, Morgan P, Satterthwaite G, Pangalos MN (2014) Lessons learned from the fate of AstraZeneca's drug pipeline: a five-dimensional framework. Nat Rev Drug Discov 13:419–431
22. Ruan Z, Cao G, Qian Y, Fu L, Hu J, Xu T, Wu Y, Lv Y (2023) Single-cell RNA sequencing unveils Lrg1's role in cerebral ischemia–reperfusion injury by modulating various cells. J Neuroinflammation 20:285
23. Wu H, Gonzalez Villalobos R, Yao X, Reilly D, Chen T, Rankin M, Myshkin E, Breyer MD, Humphreys BD (2022) Mapping the single-cell transcriptomic response of murine diabetic kidney disease to therapies. Cell Metab 34:1064–1078.e6

24. Packer M, Ferreira JP, Butler J, Filippatos G, Januzzi JL Jr, González Maldonado S, Panova-Noeva M, Pocock SJ, Prochaska JH, Saadati M, Sattar N, Sumin M, Anker SD, Zannad F (2024) Reaffirmation of mechanistic proteomic signatures accompanying SGLT2 inhibition in patients with heart failure: a validation cohort of the EMPEROR program. J Am Coll Cardiol 84:1979–1994
25. Zannad F, Ferreira JP, Butler J, Filippatos G, Januzzi JL, Sumin M, Zwick M, Saadati M, Pocock SJ, Sattar N, Anker SD, Packer M (2022) Effect of empagliflozin on circulating proteomics in heart failure: mechanistic insights into the EMPEROR programme. Eur Heart J 43:4991–5002
26. Maretty L, Gill D, Simonsen L, Soh K, Zagkos L, Galanakis M, Sibbesen J, Iglesias MT, Secher A, Valkenborg D, Purnell JQ, Knudsen LB, Tahrani AA, Geybels M (2025) Proteomic changes upon treatment with semaglutide in individuals with obesity. Nat Med 31:267–277
27. Allen BD (2021) Digital twins and living models at NASA. In: Digital twin summit
28. Hornik K, Stinchcombe M, White H (1989) Multilayer feedforward networks are universal approximators. Neural Netw 2:359–366
29. Zhao Z, Ukidve A, Kim J, Mitragotri S (2020) Targeting strategies for tissue-specific drug delivery. Cell 181:151–167
30. Kranz LM, Diken M, Haas H, Kreiter S, Loquai C, Reuter KC, Meng M, Fritz D, Vascotto F, Hefesha H, Grunwitz C, Vormehr M, Hüsemann Y, Selmi A, Kuhn AN, Buck J, Derhovanessian E, Rae R, Attig S, Diekmann J, Jabulowsky RA, Heesch S, Hassel J, Langguth P, Grabbe S, Huber C, Türeci Ö, Sahin U (2016) Systemic RNA delivery to dendritic cells exploits antiviral defence for cancer immunotherapy. Nature 534:396–401
31. Krienke C, Kolb L, Diken E, Streuber M, Kirchhoff S, Bukur T, Akilli-Öztürk Ö, Kranz LM, Berger H, Petschenka J, Diken M, Kreiter S, Yogev N, Waisman A, Karikó K, Türeci Ö, Sahin U (2021) A noninflammatory mRNA vaccine for treatment of experimental autoimmune encephalomyelitis. Science 371:145–153
32. Sahin U, Oehm P, Derhovanessian E, Jabulowsky RA, Vormehr M, Gold M, Maurus D, Schwarck-Kokarakis D, Kuhn AN, Omokoko T, Kranz LM, Diken M, Kreiter S, Haas H, Attig S, Rae R, Cuk K, Kemmer-Brück A, Breitkreuz A, Tolliver C, Caspar J, Quinkhardt J, Hebich L, Stein M, Hohberger A, Vogler I, Liebig I, Renken S, Sikorski J, Leierer M, Müller V, Mitzel-Rink H, Miederer M, Huber C, Grabbe S, Utikal J, Pinter A, Kaufmann R, Hassel JC, Loquai C, Türeci Ö (2020) An RNA vaccine drives immunity in checkpoint-inhibitor-treated melanoma. Nature 585:107–112
33. Stadler CR, Bähr-Mahmud H, Celik L, Hebich B, Roth AS, Roth RP, Karikó K, Türeci Ö, Sahin U (2017) Elimination of large tumors in mice by mRNA-encoded bispecific antibodies. Nat Med 23:815–817
34. Rosigkeit S, Meng M, Grunwitz C, Gomes P, Kreft A, Hayduk N, Heck R, Pickert G, Ziegler K, Abassi Y, Röder J, Kaps L, Vascotto F, Beissert T, Witzel S, Kuhn A, Diken M, Schuppan D, Sahin U, Haas H, Bockamp E (2018) Monitoring translation activity of mRNA-loaded nanoparticles in mice. Mol Pharm 15:3909–3919
35. Karikó K (2021) Modified uridines are the key to a successful message. Nat Rev Immunol 21:619
36. Qin C, Yang S, Chu YH, Zhang H, Pang XW, Chen L, Zhou LQ, Chen M, Tian DS, Wang W (2022) Signaling pathways involved in ischemic stroke: molecular mechanisms and therapeutic interventions. Signal Transduct Target Ther 7:215

Part III

New Materials

Challenges in the Implementation of Technologies for the Sustainable Transformation of Chemical Processes into Industrial Practice

9

Robert Franke and Thomas Haas

Contents

R. Franke (✉)
Evonik Oxeno GmbH & Co. KG, Marl, Germany

Lehrstuhl für Theoretische Chemie, Ruhr-Universität Bochum, Bochum, Germany
e-mail: robert.franke@evonik.com; robert.franke@ruhr-uni-bochum.de

T. Haas
Evonik Operations GmbH, Marl, Germany
e-mail: thomas.haas@evonik.com

U. A. K. Betz (ed.), *Science for a Better Tomorrow*,
https://doi.org/10.1007/978-3-031-93623-4_9

Abstract

Industrial research, including idea generation, lab experiments, pilot, and demo plant, needs a convincing profitability calculation. This chapter uses two examples to illustrate the specific challenges of technology developments of a disruptive nature. The Rheticus® project is a new approach of industrial artificial photosynthesis. This project can fulfill many customers' needs, including skin friendliness, biodegradability, viscosity, and application properties. The second example is the MACBETH project, which aims to convert a large-scale carbonylation process from homogeneous catalysis in the liquid phase to heterogeneous catalysis in the gas phase. The specific techno-economic challenges in the development of transformational production processes in the chemical industry are discussed.

Keywords

Rheticus® · MACBETH · Artificial photosynthesis · Syngas fermentation · CO_2 · C1 · Specialty chemicals · Internal rate of return · Hydroformylation · Return on investment · Catalysis

9.1 Introduction

In general, there are two types of research and development. The first is academic research at universities and other public institutions; the second is industrial research. Academic research is fundamental: it is the basis for all applications. The applications themselves are the basis of industrial research. This means industrial research advances the outcomes and knowledge gained from past academic research. Nevertheless, there are substantial differences between these two types of R&D. Academic research is looking for basic knowledge, whereas the purpose of industrial research is always to make profits, often within 3–10 years. The transition of the chemical industry toward defossilized production processes with a partially market-exogenous timeline leads to very high techno-economic challenges for companies. These are illustrated below on the basis of two research projects at Evonik.

9.2 Characteristics of Industrial Research

9.2.1 Preconditions

9.2.1.1 TMRO

The intention to make profit is the most important precondition for industrial research. However, there are also other very important preconditions. Some of these are abbreviated as TMRO readiness levels. TMRO means technology, market, resources, and organization.

The **technology** needed for a research project should be available, ideally within the organization, and ready to be applied. For instance, if the required technology is hydrogenation, the hydrogen supply, reaction technology, and catalysts need to be available.

The project team needs to have **market** access. This includes contact to potential customers. Customer needs are a prerequisite, otherwise there will be nobody to sell the product to. Very often, an unmet customer need is the starting point of industrial research. Resources are necessary to run research projects.

If there are no experienced technicians and scientists available, the research project cannot be properly staffed. Good **resources** include proper lab equipment as well as trained personnel.

Last but not least, the **organization** of the industrial company should support the research project. Engineering, sales, and purchasing functions are needed to push the research project.

9.2.1.2 Cash Flow Calculation

There are several ways of verifying the intention to make a profit. One common method at present is a cash flow calculation.

To perform a preliminary cash flow calculation at a very early stage, certain assumptions must be made. These include an initial R&D phase and a second, successive investment phase. The outcome of the investment phase is the sale of a chemical product. Therefore, the assumptions include a maximum sales volume, a product price, and the cost of production, mainly the total product variables, the time needed for the R&D phase, and the costs incurred in the R&D phase. In addition, an investment volume will be required, along with a time frame for the investment, which is based on the time frame of the market development of the specific chemical products.

The main factors influencing a cash flow calculation are variable costs and product sales. In general, chemical production costs are composed of raw material, capital, and energy costs. Other costs, such as the staff required, interest expense or waste treatment can often be left out of the initial calculation.

Raw material costs are linked to the consumption of specific raw materials, determined by the basic chemical reaction describing the product formation from the raw materials. The consumption of raw materials is linked to the reaction yield. If the production volume of a chemical product is above 5 kt/a, raw material costs often make up the highest portion of the overall cost of production. Therefore, after determining the basic chemical reaction to be used, research is important, because it influences the reaction yield, especially the reaction selectivity, which determines the raw material costs. Consequently, lab experiments to test the chemical reaction, including the catalysts required to obtain maximum selectivity, are crucial.

This means that, in case of high-volume production, running a reaction at elevated temperatures with high selectivity is often better than running a similar reaction at low temperature with low selectivity. As an example, the inlet unit should be heated by the outlet unit using a heat exchanger. In contrast, it is not possible to recover lost raw materials because of poor selectivity.

Raw material prices and, to an even greater extent, the price of utilities, such as electricity and steam, vary significantly from site to site. Therefore, site selection is another important factor in the cost of production.

Energy costs, like electricity or steam costs, have a strong influence on the production of chemicals for two reasons. First, in most cases, chemicals need carbon based raw materials. Their price is linked to the energy price. Second, chemicals are needed in pure form. Very often, the purity is above 99 wt%. Purification often involves separating the product from a solvent and its byproducts. Both need a lot of energy. In many cases, the conversion rate is below 100%, so raw material recycling is necessary. This requires energy, so heat integration must be carefully evaluated. The use of heat integration possibilities is particularly important if energy costs are high, as is the case at present.

Finally, production costs are determined by investment costs. This is the third major factor influencing the cost and profitability of chemical production after raw material and energy costs. Investment costs mainly comprise capital expenditures. Moreover, they determine depreciation charges and fixed costs like maintenance. They also constitute the major risk of the investment. Once the investment has been made, no further influence on this factor is possible. Investment in a chemical plant is generally very specific and the plant cannot normally be used to produce another chemical compound. If subsequent commercialization of the product fails, the investment tends to end up as a stranded asset. Therefore, it is very important to keep these expenditures as low as possible.

Economy of scale is another very important and specific topic in industrial research projects. Scale has a significant influence on production costs and therefore profitability. The capital expenditure is an important cost factor, whereas earnings depend mainly on sales volumes and prices. Economy of scale means that the investment does not increase proportionately to the plant capacity. Doubling the plant capacity does not mean doubling the investment; often the difference is significantly less. Therefore, large-scale plants are usually more profitable than those with smaller capacities, if the selling price of the product remains constant. This is expressed by the degression coefficient x (Fig. 9.1).

The investment does not increase proportionately to capacity. This is expressed by the degression coefficient which is often in the range of 0.6 [1].

Ethylene production is an old example often used to illustrate the impact of plant capacity, see Fig. 9.2.

For instance, if the capacity of an ethylene production plant is raised from 75 to 275 kt/a, the specific production costs will be reduced by 35%.

The cash flow calculation can be used to calculate the internal rate of return (IRR). The IRR is a helpful number to differentiate between potential R&D ideas/projects. Essentially, it approximates the interest an investor can earn by investing in this industrial research if it is successful. The IRR should be above a company's weighted average cost of capital because, as well as competing with bonds, it should also contain a premium to cover the various market and technical risks. The IRR helps to compare different research projects. If the IRR is calculated

Fig. 9.1 Investment and capacity. x = degression coefficient

$$invest_{large} = invest_{small} \cdot \left(\frac{capacity_{large}}{capacity_{small}}\right)^{x}$$

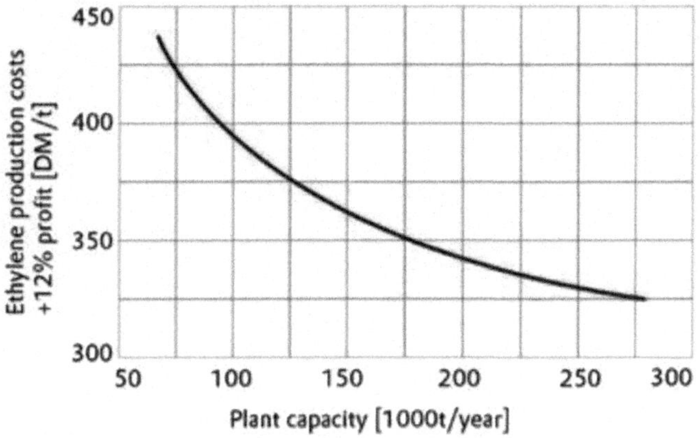

Fig. 9.2 Ethylene production costs including 12% profit for ethylene depending on plant capacity [2]

Table 9.1 Assumptions used to calculate cash flows and the IRR

Construction time 2 years
First sales directly after construction
Fixed costs 5% of the total investment [3]
Overhead costs 20% of the sales
Other costs like interest and taxes are neglected

based on assumptions, as described above, it is extremely important to verify the assumptions in case the project is pursued.

A common set of assumptions is necessary to compare different R&D ideas/projects, see Table 9.1. The cash flows must be calculated using the same time frame. In Table 9.2, the assumptions are a 10 year time frame from the start of the investment, a construction time of 2 years, and initial sales 2 years after the start of the investment. Other assumptions are fixed costs of roughly 5% of the investment costs, while overheads are assumed to be roughly 20% of sales. Interest and taxes are not included in the calculation.

As an example, a virtual industrial research project has been chosen, see Table 9.2. The research project will take 3 years prior to the construction of the first 50 kt/a plant. In this case, it is assumed that 50% of the maximum production volume can be sold in the first year after start of production, 80% in the second year, and 100% in the third year.

The assumed market price of the product is €10 per kg, whereas the sum of all variable costs is assumed to be €4 per kg. The result of the IRR calculation for this project is 14%. Therefore, it should be considered for implementation.

Table 9.2 Cash flow calculation for a virtual industrial R&D project

Business year		2025	2026	2027	2028	2029	2030	2031	2032	2033	2034	2035	2036	2037	$\sum$
Investment (CAPEX) incl. engineering	**T-EUR**	**0**	**0**	**0**	**50.000**	**200.000**	**250.000**	**0**	**0**	**0**	**0**	**0**	**0**	**0**	**500.000**
Sales volume	t	0	0	0	0	0	0	25.000	40.000	50.000	50.000	50.000	50.000	50.000	315.000
Sales price per unit	EUR/t	0	0	0	0	0	0	10.000	10.000	10.000	10.000	10.000	10.000	10.000	
EXTERNAL SALES	**T-EUR**	**0**	**0**	**0**	**0**	**0**	**0**	**250.000**	**400.000**	**500.000**	**500.000**	**500.000**	**500.000**	**500.000**	**3150.000**
Raw, utilities and operating materials	EUR/t	0	0	0	0	0	0	4.000	4.000	4.000	4.000	4.000	4.000	4.000	
$\sum$ VARIABLE COST	**T-EUR**	**0**	**0**	**0**	**0**	**0**	**0**	**100.000**	**160.000**	**200.000**	**200.000**	**200.000**	**200.000**	**200.000**	**1260.000**
Fixed costs excluding depreciation	T-EUR	0	0	0	2.500	12.500	25.000	25.000	25.000	25.000	25.000	25.000	25.000	25.000	215.000
Overhead costs	T-EUR	0	0	0	0	0	0	50.000	80.000	100.000	100.000	100.000	100.000	100.000	630.000
R&D project costs	T-EUR	5.000	5.000	5.000	5.000	5.000									25.000
Cash Flow	**T-EUR**	**−5.000**	**−5.000**	**−5.000**	**−57.500**	**−217.500**	**−275.000**	**75.000**	**135.000**	**175.000**	**175.000**	**175.000**	**175.000**	**175.000**	**520.000**
IRR		**14%**													

9.2.1.3 Other Preconditions

Apart from TMRO, the cash flow and the IRR calculation, the start of industrial research is influenced by further factors such as the industrial state of the art, competitive technology, and the IP situation. A product can often be produced using existing technology or by a combination of existing technologies. It is important that the technology that is the subject of the R&D project is superior to existing technology or that the product is not currently produced, and the new technology would fulfill a customer need.

The IP situation relating to the new technology, especially patents already granted, is also a very important aspect. It is essential to ensure freedom to operate in all major markets and at potential production sites where the new technology might be applied because legal IP issues could hinder production and hold back customer demand, which would have a significant impact on sales. The IP landscape must be viewed from a global perspective because the chemical industry is a global industry. Further criteria are the product carbon footprint (PCF), carbon pricing, life cycle assessments (LCA), and public funding [4].

9.2.2 Project Plan

A project plan, composed of tasks and milestones, is the next step after meeting all the preconditions. All projects, academic and industrial, chemical, automotive, and biological, share this step. It is not specific to industrial chemical research projects.

However, some specific milestones are included in industrial research projects. These specific milestones normally relate to customers or the market. In many cases, representative samples or demonstrators are essential milestones to obtain customer or market feedback, which is important to verify or determine the next milestones.

9.3 Two Practical Examples

9.3.1 Rheticus®

A very important German R&D project in the field of "power to X" is the Kopernikus project [5]. Copernicus, who discovered that the earth is a planet and not the center of the universe, had one student, who promoted his discovery. His name was Rheticus. The Rheticus® project, a much smaller "power to X" project, is being conducted in the context of the Kopernikus project.

Rheticus® is the name of an industrial R&D project examining the conversion of CO_2 and electricity, preferably from wind or solar energy, into useful chemicals. The basic idea of the Rheticus® project is to mimic nature's invention of photosynthesis. It adapts photosynthesis to industrial needs by producing chemicals for our daily life using natural bacteria instead of natural products formed by photosynthesis such as glucose and derivatives, e.g., cellulose, fats, oils, and lignin. Analogously

Table 9.3 Further arguments for the use of C1 chemistry are [6]

Availability of own carbon resources: Countries with huge own carbon resources such as coal, natural gas or biomass may find it beneficial for political or financial reasons to develop their own resources
Innovation: The chances in searching for a competitive advantage are greater where new technology exists. Using crude oil derivatives narrows the opportunities for new processes. Similar technological paths exist for all competitors
Independence: An erratic behaviour of feedstock prices and a shortage in availability makes planning and forcasting very difficult. A reliable, secure feedstock supply and long term contracts at calculable prices are the basis of a prosperous industry. In dependence also embraces concern for national security aimed at using own available feedstocks
Environmental constraints: In our todays world, environmetal considerations have developed into an important factor. Processes, which burden the environment, will be phased out

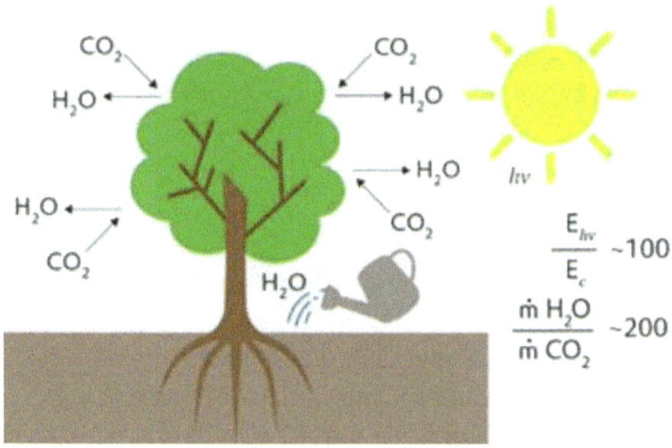

Fig. 9.3 Photosynthesis, energy efficiency, and water utilization

to natural products such as trees and grassland, Rheticus® technology can produce carbon negative products because it consumes carbon as a raw material.

9.3.1.1 C1 Chemistry

C1 chemistry is an old topic of industrial organic chemistry, mainly because methane is much cheaper than other hydrocarbon chemical feedstocks like ethylene or propene (Table 9.3).

9.3.1.2 Artificial Photosynthesis

At first, methane was the main C1 topic. Recently, methane has been replaced by CO_2 as the major C1 compound of interest because of the climate change debate. CO_2 is also the carbon source in natural photosynthesis (Fig. 9.3).

There are limitations of natural photosynthesis if it is applied industrially. First, only 1% of sunlight is utilized to convert CO_2. From the standpoint of the chemical industry, most of the sunlight is reflected, absorbed or otherwise wasted. Second, a lot of water is needed for natural photosynthesis. Roughly 200 kg water is needed to convert 1 kg of CO_2. That is one reason one does not find plants in deserts.

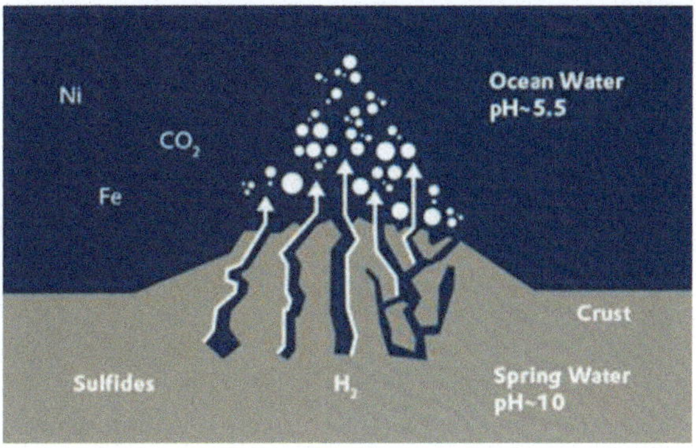

Fig. 9.4 First life on earth. Chemicals from H_2 and CO_2

The Rheticus® project addresses both aspects: the poor energy efficiency and the high water consumption of natural photosynthesis. To simulate natural photosynthesis, photovoltaic cells and electrolyzers are used instead of the natural chlorophyll system in leaves. The energy efficiency of man-made solar panels is already much higher than the natural energy efficiency of the chlorophyll system. Water consumption is improved in the industrial system by recycling water instead of vaporizing it into the air [7].

To achieve high selectivity or carbon efficiency, natural bacteria are used as they can metabolize carbon dioxide and hydrogen. Today, most bacteria digest glucose, mainly to use the energy to convert glucose and oxygen into carbon dioxide. There are a few bacteria whose ability to metabolize hydrogen goes back to ancient times (Fig. 9.4).

Both carbon dioxide and hydrogen were already present in ancient times. The first life was able to gain energy by converting carbon dioxide and hydrogen, mainly to produce acetic acid. This natural know how was stored in the DNA of the organisms, mainly bacteria.

These bacteria are used in the Rheticus® project to convert carbon dioxide and hydrogen into chemicals. Different bacteria produce different chemicals. Hexanoic acid was chosen as the target chemical for the Rheticus® project.

Water is used as a solvent in nearly all biotechnological processes. At the same time, it has the highest vaporization enthalpy compared with almost all solvents. Therefore, vaporizing water to isolate hexanoic acid is not an option. By contrast, the extraction of hexanoic acid is an efficient method. It is particularly advantageous if the extractant has a higher boiling point than the product, because much less energy is needed to evaporate the product than the extractant. An extractant is chosen for its high product solubility and insolubility in water.

The Rheticus® process, see Figs. 9.5 and 9.6, combines fermentation using bacteria to metabolize hydrogen and CO_2, extraction of the chemical product, and distillation to separate the product from the extractant with the high boiling point. The water used in the fermentation process and the extractant are recycled.

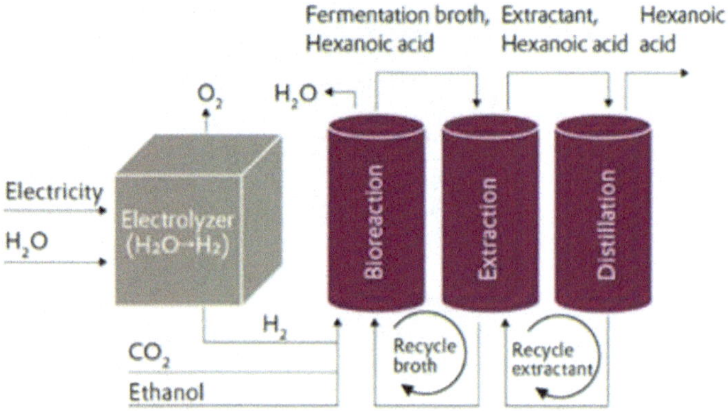

Fig. 9.5 Schematic Rheticus® process

Fig. 9.6 Rheticus® pilot plant, Evonik Industries, 2023

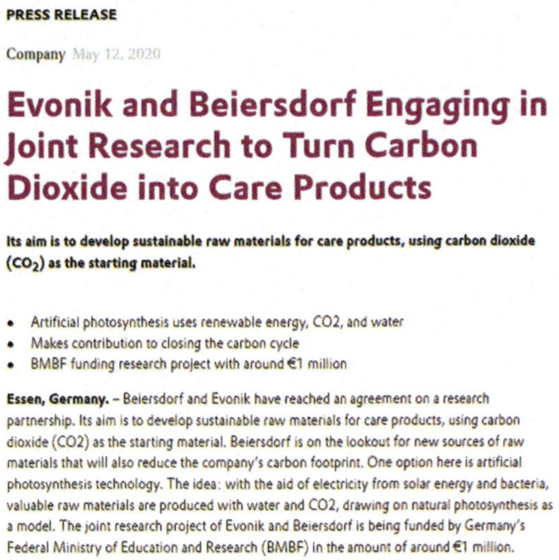

PRESS RELEASE

Company May 12, 2020

Evonik and Beiersdorf Engaging in Joint Research to Turn Carbon Dioxide into Care Products

Its aim is to develop sustainable raw materials for care products, using carbon dioxide (CO_2) as the starting material.

- Artificial photosynthesis uses renewable energy, CO2, and water
- Makes contribution to closing the carbon cycle
- BMBF funding research project with around €1 million

Essen, Germany. – Beiersdorf and Evonik have reached an agreement on a research partnership. Its aim is to develop sustainable raw materials for care products, using carbon dioxide (CO2) as the starting material. Beiersdorf is on the lookout for new sources of raw materials that will also reduce the company's carbon footprint. One option here is artificial photosynthesis technology. The idea: with the aid of electricity from solar energy and bacteria, valuable raw materials are produced with water and CO2, drawing on natural photosynthesis as a model. The joint research project of Evonik and Beiersdorf is being funded by Germany's Federal Ministry of Education and Research (BMBF) in the amount of around €1 million.

Fig. 9.7 Press release, Evonik and Beiersdorf, 2020

Hydrogen is produced indirectly from solar energy. Solar panels produce solar electricity, which is used to convert water to hydrogen by an electrolyzer.

9.3.1.3 Applications

One of the main applications of the Rheticus® technology is personal care products. Personal care customers are not simply concerned about the performance of the products being applied to the skin; they also care about the origin of the raw materials. For example, a joint project was conducted by Beiersdorf and Evonik to maximize the performance of personal care products produced from CO_2 using the Rheticus® technology (Fig. 9.7).

Another important application of Rheticus® products is CO_2 based surfactants. A new nonionic surfactant has been developed together with a leading detergent manufacturer. This CO_2 based surfactant is used as a substitute for tropical oil based products.

Apart from personal and home care applications, there are many more applications where customers care a lot about carbon dioxide and bio-based properties such as biodegradability. These include lubricants and coolants, which are often esters produced from multifunctional alcohols and fatty acids (Fig. 9.8).

At present, fatty acids are often derived from tropical oils. Rheticus® opens a new way to produce these liquids based on CO_2 instead of tropical oils, while ensuring equally good biodegradability.

There are many more potential applications of Rheticus® products, e.g., animal nutrition.

Fats in animal and human nutrition are a common subject of research. These studies most often pay attention to particular fat groups (saturated, unsaturated, polyunsaturated fats or fats grouped by the length of their fatty acid chains into short, medium or long chain fatty acids). Medium chain fatty acids (MCFAs) have two main sources: milk and coconut

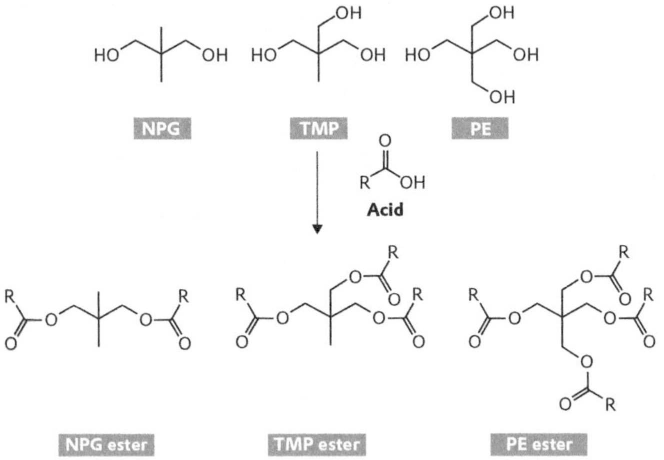

Fig. 9.8 Catalytic esterification of acid and polyol to polyol ester [8]

oil. To date, research has shown these acids have positive effects on health, production, feed digestibility, and lower body and muscle fats in broilers and swine. MCFAs possess antibacterial, anticoccidial and antiviral effects. Also, it has been proven that these acids act synergistically if they are used together with organic acids, essential oils, or probiotics. Nowadays, commercial MCFA products are available for use in animal nutrition as feed additives [9].

9.3.2 MACBETH (Membranes and Catalysts Beyond Economical and Technological Hurdles)

9.3.2.1 Catalysts

If one tries to present a general and generic conceptual definition of a catalyst, one could formulate it as follows: A catalyst is a material that makes it possible to carry out a chemical reaction at a higher rate or under different conditions, for instance at lower temperatures, than would be possible without a catalyst.

The importance of catalysts in production processes in the chemical industry can be demonstrated without further explanation by the following figures: 90% of all industrial chemical processes use catalysts [10], which means that around 15–20% of the total economic output of the developed Western world is directly or indirectly linked to the use of catalysts. To illustrate the economic leverage of the use of catalysis, the following rough facts are given: €1 invested in a catalyst generates €250 on the chemical product side [11]. Since catalysis always results in fewer by-products (i.e., waste from the perspective of a specific production target) and energy savings compared to the uncatalyzed process, catalysis is one important, perhaps even the driver of sustainability.

A fundamental distinction that is usually made in catalysis is between homogeneous and heterogeneous catalysis. In the former, reactants, products and catalyst are present in the same (fluid) phase. This is not the case in heterogeneous catalysis. Here, reactants and products are gaseous or liquid, while the catalysts are usually solids. The relative importance of both classes can be clearly assessed: Approximately, 80% of the catalysts used in industrial practice are heterogeneous [12]. To put it in simplified terms, the main reason for this is the fundamental advantage of heterogeneous catalysts that it is not necessary to separate the catalyst from the reactants and products, as this is a trivial result of the assignment to the fluid phase and solid phase. In detail, other considerations such as the catalyst's input factor and other aspects also play a role, but these will not be discussed further here. On the other hand, there are a number of advantages to using homogeneous catalysts, such as the often higher activity for a particular reaction and the associated lower temperatures that can be used, or the higher selectivity, which means less waste and lower energy consumption in separation steps. The research goal of combining the advantages of both types of catalysis, i.e., developing the heterogenization of homogeneous catalysts, is obvious.

9.3.2.2 Heterogenization of Homogeneous Catalytic Processes Exemplified by Hydroformylation

Although there is a clear dominance of heterogeneous catalysis in large-scale processes for the reasons mentioned above, there are homogeneously catalyzed processes for certain processes in industrial organic chemistry. Examples include hydrocyanation and hydroformylation. In the latter case, there are production plants with capacities >400,000 mt/a. To the best of our knowledge, only homogeneously catalyzed production processes exist for this reaction on this scale.

Hydroformylation is the conversion of olefins to aldehydes by reacting them with synthesis gas in the presence of a catalyst [13]. In view of the application example discussed below, the reaction for an isomeric mixture of *n*-butenes is given in Scheme 9.1.

The aldehydes are important intermediate products for a wide range of chemicals, for which the name "oxo products" has been adopted, or "oxo alcohols" in relation to the most important subset of these products. The total market comprises more than ten million tons per year. The reaction only occurs in the presence of a catalyst, which typically consists of a transition metal complex. In the case of hydroformylation, this is now almost always a rhodium complex with a phosphine or phosphite ligand. Although research into heterogeneously catalyzed variants of hydroformylation has been ongoing since Otto Roelen discovered hydroformylation in 1938, there is still no large-scale process variant that is not catalyzed homogeneously. One variant of heterogeneous catalysis for hydroformylation, which is important for the following accounts, is the class of film coating materials approaches [14]. In this variant, a homogeneous catalyst that is highly optimized and has proven itself in large-scale processes is dissolved in a solvent with a very low vapor pressure. This solution is then dispersed on a porous, solid support with a high surface area. The material produced in this way can be used as a heterogeneous catalyst and shows the same selectivities as the metal complex

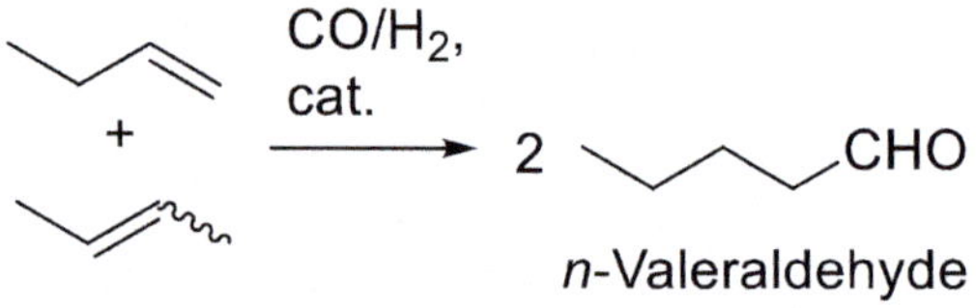

Scheme 9.1 The hydroformylation of *n*-butenes

Fig. 9.9 Film coating material for use as a heterogeneous hydroformylation catalyst

used as a homogeneous catalyst. Figure 9.9 shows a laboratory sample of such a heterogeneous hydroformylation catalyst.

9.3.2.3 Heterogeneous Hydroformylation Catalysts in Integrated Membrane Reactors: The MACBETH Approach

Heterogeneous catalysts often allow for particularly efficient reaction control in reactors that have been optimized specifically for these materials. They are therefore very good starting points for the so-called process intensifications, which make it possible to improve chemical processes in a disruptive way, so that they show significant improvements in terms of energy efficiency, process costs or other features [15]. One possibility that is particularly suitable for the process intensification of continuous chemical production processes is the use of integrated membrane reactors, in which both the reaction and one or more separation steps (e.g., the separation of a product from the reaction mass) take place in a single apparatus. The development of processes based on such reactor types is a major challenge, involving the development of both novel materials (e.g., catalysts and membranes as well as composite materials made from the two) and new apparatuses. Furthermore, questions regarding technical upscaling to very large dimensions and the large-scale production employing the novel apparatuses must be addressed. The technical suitability of the new technology, particularly with regard to long-term stable behavior over periods of years, must be demonstrated. Thus, such developments lead to comprehensive research and development programs, which ultimately require major research collaborations involving industrial companies from different sectors and of different sizes, as well as academic research institutions from a wide range of disciplines. One of these initiatives, known as MACBETH (Membranes and Catalysts Beyond Economical and Technological Hurdles), which is receiving considerable funding from the European Union, had, among other targets, the goal of demonstrating a heterogeneously catalyzed hydroformylation in an integrated membrane reactor at Technology Readiness Level 7 (TRL7) [16]. A schematic representation of the process-intensifying idea of the MACBETH project, which, in addition to hydroformylation, was also investigated for three other key application examples from process chemistry (hydrogen production, propane dehydrogenation, biocatalytic oil cleavage), can be found in Fig. 9.10.

Even the subproject of MACBETH that deals with hydroformylation can only be referred to in part in the context of this chapter. In order to specifically illustrate the process-intensifying aspects of the approach, the focus here is on an investigated variant that enables the heterogeneous hydroformylation of gaseous feeds, notably linear butenes, in a monolith. The monolith itself is a composite material consisting of a modified, highly porous silicon carbide support impregnated with a solution containing a ligand used in homogeneous hydroformylation (BiPhePhos), a liquid that is difficult to vaporize, and a rhodium precursor. A membrane is attached to the outside of the monolith, which allows the aldehyde produced during hydroformylation to be separated (Fig. 9.11).

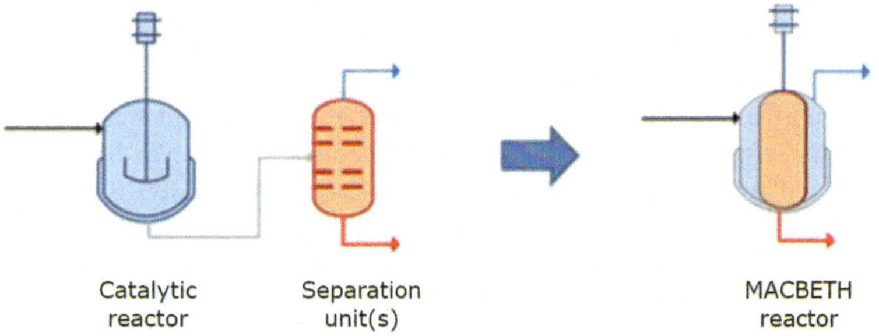

Fig. 9.10 MACBETH in a nutshell: a catalytic reactor and a membrane separation unit are combined in a single apparatus

Fig. 9.11 Monolith with membrane attached, as used in MACBETH

An important part of the research was concerned with the synthesis and characterization of the catalytically active material. In the latter case, it is of course especially interesting to examine the nature of the various materials after a catalytic run. In addition to the degradation products of the catalytically active components, the spectrum of by-products also plays an important role. This is because the reactive products of hydroformylation, the aldehydes, tend to undergo complex consecutive reactions, in which not only the product itself is lost, but the resulting long-chain and low-volatility by-products in particular have a negative impact on the lifespan of the catalyst and the efficiency of the process [17]. Experiments employing mini plants have shown that such catalytic materials can be operated stably and effectively in integrated membrane reactors for over 2000 h.

9.3.2.4 From the Miniplant to the Demonstration Plant

The development of large-scale, continuous chemical processes always involves a crucial step: upscaling. This step is the last before a project study and basic engineering can be carried out, which then lead to the launch of a specific investment project, followed by mechanical completion and production in a new plant. The costs of this development phase far outweigh the costs of the preceding activities. In view of this, intensive discussions with the management of a company are necessary, which, in addition to evaluating the technical aspects, also include a quantitative assessment of the risks, as well as market aspects and general strategic considerations. Despite enormous progress in the simulation of chemical processes, upscaling is carried out using a demonstration plant, especially in the case of new, disruptive technologies that are characterized by the fact that the materials and reactors used, or even process interconnections, have never before been the subject of practical testing either in-house or, not untypically, worldwide. The demonstration plant provides a wealth of insights. In addition to addressing reaction engineering topics related to material and heat transport, for example, it can be used to test the availability of materials and equipment, evaluate mechanical or chemical boundary conditions in practice, and, above all, experience the process capability under the real conditions of a production process embedded in a large-scale plant. Ideally, the operation of the demonstration plant is fully integrated into the process control system of the large-scale production plant and is operated entirely by the operating personnel of the plant. The enormous effort associated with

Fig. 9.12 Impressions of the integrated membrane reactor of the MACBETH demonstration plant. The monoliths shown in Fig. 9.11 are inserted into the reactor that was specially developed for this technology

Fig. 9.13 Impressions of the installation of the modules of the MACBETH demonstration plant in the oxo production plant in the Chemical Park in Marl and a complete view of the demonstration plant

the demonstration plant is ultimately aimed at derisking with regard to unknown, potentially economically harmful aspects of new technologies for a company.

For the hydroformylation using the MACBETH technology, a corresponding demonstration plant was implemented and successfully operated in the oxo production plant in the Chemical Park in Marl, Germany. Figure 9.12 shows the integrated membrane reactor, which is a completely new development. The concept of upscaling is a characteristic of this part of the MACBETH technology. The increase in production capacity is achieved by what is known as numbering-up, which has the advantage that decisive technical reaction parameters used in the reactor in the demonstration plant are identical to those of the later large-scale plant. Furthermore, during the development of the technology, emphasis was placed on ensuring that it largely conforms to modular design principles. Two important advantages associated with this are the flexible expansion of production capacity and the possibility of installing the demonstration plant in facilities at other chemical sites (Fig. 9.13).

The demonstration plant will continue to operate even after the 5-year MACBETH research project funded by the European Union has come to an end. The main objective here is to test the technology for long-term feasibility under real production conditions. To do this, the plant must run successfully for several years.

9.4 The Tense Field of the First Production Plant

9.4.1 The Defossilizing Transformation First and Foremost Requires Process Innovations

The chemical industry is currently confronted with a wide range of disruptive influences that are related to both a sharp increase in ecological sustainability requirements and a wide range of new technological developments, some of which are exogenous to chemistry, such as the rapid developments in the field of artificial intelligence or the greatly accelerated digitalization in all areas of the economy, but which are also strongly influenced by chemical innovations, e.g., in the field of new materials, which is a prerequisite for the further development of energy storage systems, for example. The research and development fields of the chemical industry are therefore both exogenously driven and intrinsically driven. The challenges of climate change are accelerating the tendency of governments to roll out carbon dioxide-reducing regulations for all industries. The chemical industry is therefore confronted with the challenge of reducing emissions by around 60% by 2030. In addition to this framework condition, since around 2020 there have been sharp increases in inflation and very significant disturbances in the supply chains of almost all chemical sectors and, beyond that, further areas of industry that are of indirect signficance for the chemical industry. To illustrate the enormous challenge that the transformation of the chemical process industry comprises, some figures for the year 2013 are given here [18]: 513 million metric tons of fossil fuels and 163 million metric tons of olefins and aromatics were used, together with 1000 million metric tons of other reactants such as water, oxygen, nitrogen, etc., to produce 820 million metric tons of chemical products in complex chemical production process chains. Among other substances, 140 million metric tons of methane and 287 million metric tons of carbon dioxide were produced.

It therefore follows from the logic of the chemical industry that a thorough revision of existing chemical processes and the development of new processes will be necessary, both for existing and for new products, with particular attention being paid to process-intensifying aspects and those related todigitalization and artificial intelligence. The development of new catalysts will also be a very important aspect here. This will be crucial for one of the greatest challenges in process development: the processing of CO_2 as a new raw material to replace fossil sources.

Process innovations are almost always linked to investment projects. Depending on their degree of novelty, they inevitably involve high risks at times. In addition, it is highly likely that for a number of elements of the new production process chains, for example for the use of CO_2 as a C1 building block, it will be technologically unavoidable that the overall economic efficiency will be worse or even considerably worse over a very long period of time than the conventional technology based on fossil raw materials. For economic reasons, there is naturally a strong countermovement to such projects. However, it is foreseeable that ultimately even very large investments will be unavoidable and that entrepreneurial activity in the chemical industry also involves betting on such investments.

9.4.2 A Simple Case Study of Investment Risks

The real challenge in investment planning and thus in the methodology of investment appraisal is that the returns on an investment are generally uncertain because the incoming cash flows are subject to risk. It is therefore essential to identify the opportunities and risks associated with an investment and to take them into account appropriately in the decision-making process. In this context, it is crucial that data on the expected cash flows are available. If this is not the case, the discounted cash flow (DCF) method, for example, cannot be used. Monte Carlo simulation provides a comparatively systematic and well-defined method for quantifying the expected cash flows and the risk-related discount rate for investment calculations. The determinants of cash flows are treated as stochastic variables. All scenarios defined in the space of the distribution functions (which can routinely be several million scenarios if it seems sensible) are captured quantitatively.

To avoid the rather tedious task of introducing the usual calculation rules for investment calculations, which must precede a presentation of the results of such simulation calculations, a somewhat unconventional approach to the quantitative presentation of a risk analysis for a typical investment project in a chemical production plant is presented below. The profit per year generated by an investment could be expressed as:

$$P = SV \cdot PP - [FC + VC \cdot SV] \tag{9.1}$$

where annual profit = P, sales volume = SV, product price per sales unit = PP, fixed costs = FC and variable costs = VC. Let I be the investment with a straight-line depreciation D at the residual value of $DR = 10\%$ within 10 years and annual maintenance costs MC of $MCR = 7\%$. Together with the labor costs LC, the fixed costs are then: $FC = LC + D + MC$. It is assumed that probability distributions exist for all the values considered and that these are all normally distributed. The investment risk is considered by standard deviations, as for example in the Six Sigma methodology. There is an approximately 68% probability that the expected annual profit P' falls within the interval $P \pm \sigma(P)$, with the standard deviation of the profit being $\sigma(P)$. We will now estimate the standard deviation of profit P using an approach analogous to Gauss' law of error propagation, whereby we assume that the input variables for P in Formula (9.1) are uncorrelated. In terms of product price and sales volume, this is only an approximation. Mathematically, it would be easy to go beyond the simplification, but then assumptions would have to be made about

Table 9.4 Quantities of the investment project

Quantity	Base value	Optimistic estimate	Pessimistic estimate
PP [€/mt]	1000	1200	600
SV [mt/a]	100,000	100,000	80,000
I [€ million]	100	80	120
VC [€/mt]	650		
LC [€ million/a]	2		

the correlations and the explanations would become lengthy without this illustrative example gaining in value. Therefore, it is omitted here. After carrying out the partial differentiations, the result for the variance (i.e., the square of the standard deviation) of the annual profit is:

$$\sigma^2(P) = (PP - VC)^2\sigma^2\,(SV) + SV^2\sigma^2\,(PP) + \frac{(DR + MCR)^2}{10,000}\sigma^2(I) \qquad (9.2)$$

We utilize a project management approach to estimate the standard deviations of sales volume, product price, and investment. They are estimated by the value of the difference between an optimistic estimate and a pessimistic estimate divided by 6. As a case study, we consider the investment project for a chemical production plant with a capacity of 100,000 mt/a, which is defined in Table 9.4.

With a probability of about 68%, the annual profit of this investment project is calculated in the range from €4.5 million to €25 million for this case, using the approximation described above, which corresponds to a return on investment in the range from 4.5% to 25%. Such a situation is by no means unusual. It illustrates that under these assumptions for the variations of the input variables, which are in a typical range for products manufactured on this production scale, there is a not insignificant risk of not achieving economic targets. Although it can be assumed with a probability of 68% that no real losses are to be expected, the uncertainty is considerable and the range of variation in the return on investment is substantial.

In the course of the defossilization of chemical production processes, it will often occur that a technology developed for the production of an existing product in an existing market with an established price structure requires a significantly higher investment than the conventional technology. For disruptive technologies, there is naturally no exemplary project for the first investment project of its kind that provides a well-defined starting point for extrapolation with regard to the uncertainties in the scale of investment. This can lead to a situation in which a certain new production process has been tested in a demonstration plant, but the step to a production plant is not taken. This does not necessarily mean that the developed technology has failed per se. In a positive case, it may be implemented later, possibly after, for example, licensing by other companies. To illustrate this tense field, Fig. 9.14 shows the return on investment as a function of the investment for our case. The optimistic estimate assumes that only half the investment sum needs to be used,

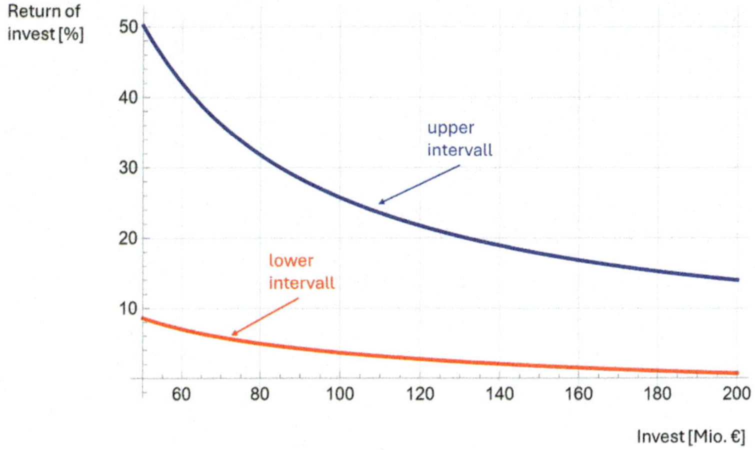

Fig. 9.14 Upper and lower interval limits of the return on investment that will occur with 68% probability for the investment project defined in Table 9.4, with variation of the investment amount with a deviating optimistic estimate of half the investment requirement and a pessimistically estimated investment requirement of twice the amount

while the pessimistic estimate reflects the case in which double the investment is required. This may seem drastically exaggerated at first glance, but it does reflect cases in which materials or apparatus that have never been used on a large scale before are the subject of the investment. We can see the significant reduction of the interval that occurs, together with increasingly poorer economic attractiveness, as the investment in the scenario increases. This behavior of economic potential contradicts the entrepreneurial expectation that a new technology, which initially entails a significantly increased entrepreneurial risk, will ultimately be introduced essentially to achieve significantly improved profitability compared to the existing technology.

9.4.3 Some Considerations on the Issue of Initial Investments in a Disruptive Production Plant

The methodology used to control industrial research projects is congruent in many respects with the methodology of the technological readiness level (TRL). This cannot be discussed here and reference is made to [19]. What is crucial for the purposes of this study is that research expenditure and the state of technological knowledge increase as the various TRL levels are passed through. Conversely, the techno-economic risks naturally decrease steadily as the lack of knowledge decreases. The project risks are at their lowest in the phases TRL 6 and TRL 7. In this phase, a demonstration plant is usually constructed and operated, the purpose of which has been explained in Sect. 9.3.2.4 using the practical example

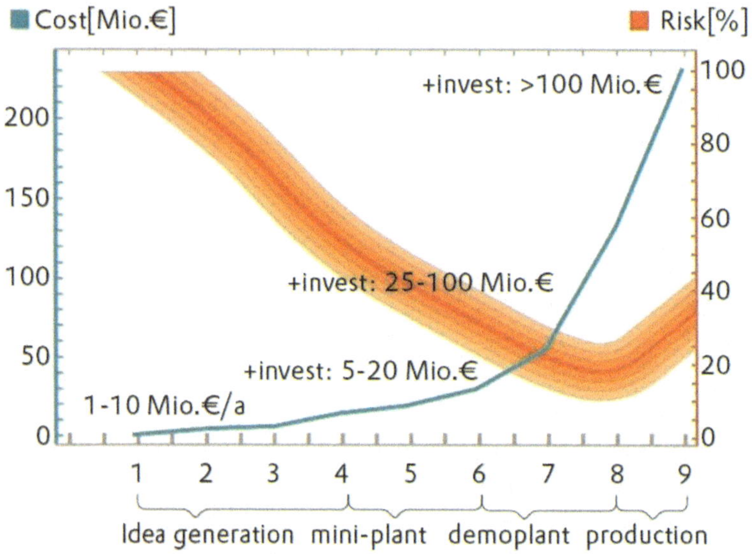

Fig. 9.15 Typical financial expenditures and risk behavior depending on the TRL degree

of the MACBETH technology. Figure 9.15 illustrates the financial expenses and development risks associated with the various TRL levels. The expenditures are representative for typical projects in the chemical industry that aim to capitalize on a process innovation. There are, of course, projects with even higher expenditures, but these are certainly not part of the mainstream. As already mentioned in Sect. 9.3.2.4, entry into the TRL 6–7 phase is a critical issue from a decision-maker's perspective, where it must be weighed up whether to incur considerable costs without it yet being clear whether there will ever be any revenue.

However, the construction and operation of the demonstration plant then represents the phase in which risk avoidance in relation to process innovation is at its maximum and a very high level of information is gained with regard to large-scale implementation. The next step, the implementation in large-scale production and thus the ultimate goal of a research project, the sustainable achievement of added value, is, however, associated with a significant increase in entrepreneurial risk in the case of disruptive and, in a best case scenario, completely new global technology. This phenomenon, which may seem counterintuitive at first glance, will be explored in some more detail in the following.

One very important aspect was already explained at the end of the last section with the help of the economic case study and is rooted in the relatively higher capital expenditure of a disruptive technology. Another problem area is the supply situation with regard to the specific assets with a consumption character that are necessary for the disruptive technology, which will be required in the absolute majority of such technologies. Examples include catalysts, membranes, special excipients, and many more. If there are problems in the value chains here, such production processes can be threatened by serious production losses. The risk here is also particularly high because, in contrast to conventional technology, there is no experience to fall back on. In addition, there are also serious market risks that are often not well recognized outside the chemical industry. Even the manufacture of an existing product in an existing market using a completely new, disruptive production technology carries commercial risk. This is due to the very strict specifications required for products, which result in fundamental reservations with regard to new production methods, and are associated with a phase of establishment that can sometimes last for years, during which both volumes and prices are significantly below market expectations.

The socially required transformation of the chemical process industry with the aim of defossilization and CO_2 neutrality, which requires disruptive technologies that may be characterized by a lack of marketability in the initial phase, will only happen rapidly if the risks in this initial phase are significantly reduced by public funding. For this purpose, financing instruments must be developed to support such capital-intensive investment projects.

9.5 Summary and Outlook

Industrial research is significantly different from academic research. To start an industrial research project, the intention to earn money from the outcome of the research project is an important precondition. The outcome can be measured by an internal rate of return. This return has to be significantly higher than present market interest rates to compensate for inherent technological and market risks. The Rheticus® project is an example. The outcomes of this project are new high-value specialty chemicals produced from CO_2 and regenerative energy. The IRR of this project exceeds present market interest rates but also needs high capital expenditure. It fulfills many customers' needs for new, sustainable, biodegradable products with superior properties. Disruptive technologies based on process intensification are of crucial importance for the defossilization of chemical process chains. A typical example here is the heterogenization of a large-scale homogeneous catalytic production process such as hydroformylation. The approach presented here is the MACBETH technology. These technologies are usually technological world premieres and naturally entail higher risks during implementation. For economic reasons, there is usually a countermovement here, since a correspondingly significant increase in surplus is either uncertain or—as is particularly the case in the context of defossilization—almost impossible. A market-exogenous timeline can therefore only be accompanied in a way that adds value if market-exogenous incentives are also provided.

References

1. Tribe M, Alpine R (1986) Scale economies and the 0.6 rule. Eng Costs Prod Econ 10:271–278
2. Fitzer E, Fritz W (1989) Technische Chemie, Einführung in die Chemische Reaktionstechnik, Springer-Lehrbuch, 3rd edn. Springer-Verlag, Berlin, p 16
3. Ramanathan A et al (2022) A thermo-economic approach to energy from waste. Elsevier, Amsterdam, pp 171–193
4. Sueltemeyer J, Xu L, Avient (2023) Product carbon footprint explained webinar, p 26
5. Kopernikus-Projekte: Größte Forschungsinitiative zur Energiewende gestartet, Projektträger Jülich (2016)
6. Keim W (1986) C1 chemistry: potential and developments. Pure Appl Chem 58(6):825–832
7. Haas T et al (2018) Technical photosynthesis involving CO2 electrolysis and fermentation. Nat Catal 1:32–39
8. Cheryl-Low YL, Lee HV, Abd Hamid SB (2017) Catalytic upgrading of bio oil model compound into polyol ester via green alginate catalyzed esterification reaction. Process Saf Environ Prot 111:160–169
9. Baltić T et al (2017) Changes in total viable count and TVB-N content in marinated chicken breast fillets during storage. IOP Conf Ser Earth Environ Sci 85:012073
10. Hegedus LL (1987) Introduction. In: Hegedus LL, Rutherford A (eds) Catalyst design: progress and perspectives. Wiley, New York, pp 1–10
11. Centi G, Perathoner S (2009) Methods and tools of sustainable industrial chemistry: catalysis. In: Cavani F, Centi G, Perathoner S, Trifiró F (eds) Sustainable industrial chemistry. Wiley-VCH, Weinheim, pp 73–198
12. De Jong KP (2009) General aspects. In: De Jong KP (ed) Synthesis of solid catalysts. Wiley-VCH, Weinheim, pp 3–10
13. Franke R, Selent D, Börner A (2012) Applied hydroformylation. Chem Rev 112(11):5675–5732
14. Zhao F, Fujita S, Arai M (2006) Developments and applications of supported liquid phase catalysts. Curr Org Chem 10:1681
15. Stankiewicz AI, Moulijn JA (2000) Process intensification: transforming chemical engineering. Chem Eng Prog 96:22–34
16. Stenger F, Franke R, Galluci F, Liese D, Fabio A, Cosentino V (2023) MACBETH: a revolution in catalytic reaction technology: insights, current status and future perspectives. Johnson Matthey Technol Rev 67:213–218
17. Madani M, Schill L, Zahrtmann N, Portela R, Arsenjuk L, Franke R, Fehrmann R, Riisager A (2023) Influence of support structure on catalytic performance of supported liquid-phase (SLP) catalysts in hydroformylation of 1-butene. Top Catal 66:1440–1450
18. Levi PG, Cullen JM (2018) Mapping global flows of chemicals: from fossil fuel feedstocks to chemical products. Environ Sci Technol 52(4):1725–1734
19. Buchner GA, Stepputat KJ, Zimmermann AW, Schomäcker R (2019) Specifying technology readiness levels for the chemical industry. Ind Eng Chem Res 58:6957–6969

Polymers: The Hidden Heroes of Life

10

D. Dormann, R. Fröhlich, T. Dias, N. Belo do Couto, N. Heinss, S. Hutten, K. Küssner, A. Ceron-Noriega, L. Chargui, M. Gelléri, A. Gopalan, K. Hertäg, R. Piccinno, C. Rickert, A. Schaaf, E. Schumbera, S. Shoup, F. Stiehler, P. Turunen, M. A. Andrade-Navarro, P. Beli, P. Besenius, M. Bonn, R. Dahm, S. Dhiman, S. Gerber, M. Girard, B. Hülsmann, R. Ketting, E. Kim, J. König, K. Kremer, O. Kukharenko, K. Landfester, K. Luck, A. Methner, J. Michels, M. M. Möckel, S. Morsbach, S. Ritz, S. Schick-Nickolaus, F. Schmid, C. Schmidt, D. Schneider, S. Schweiger, T. Speck, L. Stelzl, M. te Vrugt, H. D. Ulrich, P. Virnau, A. Wachter, A. Walther, T. Weil, S. Wittmann, and E. A. Lemke

D. Dormann (✉) · A. Gopalan · P. Beli · L. Stelzl · E. A. Lemke (✉)
Johannes Gutenberg University Mainz, Mainz, Germany

Institute of Molecular Biology, gGmbH, Mainz, Germany
e-mail: ddormann@uni-mainz.de; edlemke@uni-mainz.de

R. Fröhlich · T. Dias · N. Belo do Couto · N. Heinss · S. Hutten · K. Küssner · C. Rickert · E. Schumbera · M. A. Andrade-Navarro · P. Besenius · S. Dhiman · B. Hülsmann · F. Schmid · C. Schmidt · D. Schneider · M. te Vrugt · P. Virnau · A. Wachter · A. Walther
Johannes Gutenberg University Mainz, Mainz, Germany

M. Gelléri · R. Piccinno · A. Schaaf · S. Shoup · F. Stiehler · P. Turunen · R. Dahm · R. Ketting · J. König · K. Luck · M. M. Möckel · S. Ritz · S. Schick-Nickolaus · H. D. Ulrich · S. Wittmann
Institute of Molecular Biology, gGmbH, Mainz, Germany

A. Ceron-Noriega · S. Gerber · A. Methner · S. Schweiger
University Medical Center, Johannes Gutenberg University Mainz, Mainz, Germany

L. Chargui · M. Bonn · M. Girard · K. Kremer · O. Kukharenko · K. Landfester · J. Michels · S. Morsbach · T. Weil
Max Planck Institute for Polymer Research, Mainz, Germany

K. Hertäg · T. Speck
University of Stuttgart, Stuttgart, Germany

E. Kim
Max Planck Institute of Biophysics, Structure and Dynamics of Chromosomes, Frankfurt am Main, Germany

U. A. K. Betz (ed.), *Science for a Better Tomorrow*,
https://doi.org/10.1007/978-3-031-93623-4_10

Contents

Abstract

Polymers are incredibly versatile and integral to daily life, industry, and advanced technologies. Nylon, polyester, adhesives, plastic bags and bottles would not exist without them. However, polymers are the invisible architects of life itself, and we, humans, would not exist without them either. They store genetic information, orchestrate biochemical processes, and structure the cellular environment—and yet science is only beginning to understand their true potential as functional entities of life. The Collaborative Research Center (CRC) 1551 on "Polymer Concepts in Cellular Function" is dedicated to researching these dynamic molecules. Using an interdisciplinary approach, the scientists of this research consortium combine findings from biology, chemistry, physics, and computer science to decipher how apparent molecular flexibility and disorder enables highly precise biological functions.

This article introduces the world of polymer research, showcases the interdisciplinary approaches of CRC1551, and highlights how new research findings on biopolymers could lead to groundbreaking applications in basic biology, biotechnology, medicine, and materials science. The possibilities are limitless, from the targeted manipulation of cellular processes with optogenetic tools to the development of sustainable polymer-based materials. At the same time, scientific discourse is not limited to laboratories: the CRC1551 is actively involved in science communication to promote the dialogue between science and society and ensure the responsible use of new technologies.

Research at the CRC1551 shows that polymers are more than molecular building blocks—they are the hidden heroes of life and the key to an innovative and sustainable future.

Keywords

Biopolymer · DNA · RNA · Intrinsically disordered protein (IDP) · Life sciences · Polymer sciences · Phase separation · Multiscale modeling · Aging

10.1 Introduction

Nylon is a synthetic polymer known for its strength, elasticity, and durability. It was first developed by Wallace Carothers and his team at DuPont in 1935 and became commercially available in 1938. Nylon was originally introduced as a replacement for silk in products like stockings but quickly found applications across various industries. The smallest nylon unit contains an amide, a special chemical entity that contains carbon and nitrogen. Through a chemical reaction, a "polymerization," multiple such amide-containing monomer units form a polymer which can have very different properties than the individual monomer unit; for example, in the case of nylon, long polymers can be woven into threads. We encounter other polymers in many plastic bags or bottles, such as polypropylene or polyethene. We know hard plastics, soft plastics, plastics that appear more durable than others, and some that can be stretched. By tuning the length of the polymer, its connectivity with other polymers, and its chemical identity, an unbelievable number of different material states can be achieved, from simple children's toys to special parts of a spacecraft.

In fact, many of us might think of polymers as something man-made that in worst case end up in the ocean polluting our environment and hurting our planet. However, essential polymers appeared on Earth billions of years before the plastic bag, and these will be the focus of this article.

In every cell of our body, there are silent heroes who rarely receive the attention they deserve. These heroes are also polymers—long chain-like molecules that perform a breathtaking variety of functions. For example, they store genetic information in the form of DNA, translate it into proteins, and form the molecular machines that make life possible, to just name a few. But their importance goes far beyond this classic view. Polymers are not only the building blocks of life, they act as dynamic architects of life: they can spontaneously organize themselves into supramolecular structures that regulate biological reactions and create order in the apparent chaos of a cell. This is even more remarkable since polymers are typically flexible and tumble/wobble around, like well-cooked spaghetti in hot water, and are thus considered disordered molecules.

This fascinating topic is at the heart of the Collaborative Research Center (CRC) 1551 "Polymer Concepts in Cellular Function." Using a unique interdisciplinary approach, researchers of this consortium are exploring the complex roles of polymers in cellular processes. They focus on one fundamental question: How can biopolymers that appear so disordered and flexible perform such precise and vital tasks? Scientists in the CRC1551 combine the latest approaches from biology, physics, chemistry, and computer science to find answers. By combining experiments and theoretical approaches, they try to gain a comprehensive picture of how polymers contribute to the organization and function of a living cell. Not only are the fundamental biophysical principles of polymers being investigated, but also their role as "dynamic building blocks" of life. However, the CRC1551 is not just a scientific project but also a vision for the future that could profoundly affect basic science, medicine, materials design and biotechnology.

This article takes you on a journey through the fascinating world of polymers. We shed light on how these "hidden heroes" organize our cells, influence diseases, and could form the basis for future innovations. However, the focus of this article is not only on the science but also on the people and visions that drive the CRC network. Let us delve into the molecular world together—a world that seems chaotic at first glance but is characterized by order, functionality and impressive precision.

10.1.1 Polymers as Functional Units of Life

Polymers are central to every key function of life, as evident from what has been termed the "central dogma of molecular biology": Long polymeric deoxyribonucleic acid (DNA) is transcribed into long polymeric ribonucleic acid (RNA), which is further translated into polypeptides or proteins [1] (Fig. 10.1). The monomer unit in DNA is 1 of 4 chemically different nucleotides consisting of a base, a sugar and a phosphate group that polymerize via a sugar-phosphate linkage. The bases of two separate polynucleotide strands pair up to form double-stranded DNA that has a characteristic helical structure [2] and, in human cells, can be 2 m long if stretched out. DNA can be transcribed to RNA through an enzymatic process called transcription, where the monomer unit is one of four RNA nucleotides that can polymerize into single strands, typically between 20 and a few thousand units. If stretched out, a long RNA molecule could approach the micrometer regime, and thus be roughly three orders of magnitude smaller than a genomic DNA molecule.

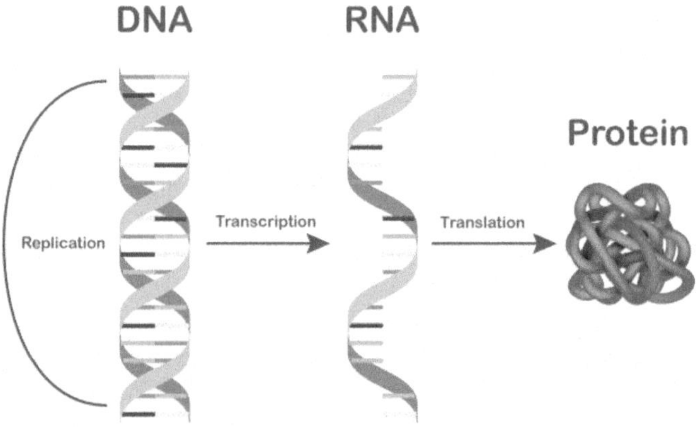

Fig. 10.1 Central dogma of molecular biology

Three consecutive monomer units (i.e., RNA nucleotide units) form a codon, and in a specialized process called translation, the RNA code is decoded into proteins, which are polymers mostly comprised of 20 naturally occurring amino acids. These amino acids are concatenated by peptide bonds (formed between the amino group of one amino acid and the carboxy group of the next), thus also giving rise to the term (poly)peptide. Peptides and proteins are thus, in principle, interchangeable terms. In essence, proteins are the molecular machines or tools inside our cells. Many proteins fold into a specific three-dimensional structure that enables the proteins to carry out a specific function [3], and they typically are a few nanometers in diameter. Proteins are the nano-engines of cells, each crafted to perfection to perform a particular task in a cell or tissue.

Together, the three different classes of biopolymers (DNA, RNA, and proteins) orchestrate a network of cellular functions that ultimately give rise to life. These cellular processes are so complex that we are far from a complete understanding despite intensive research in the last decades.

However, why should we care about biopolymers and their cellular functions? The answer is that the importance of function also comes with the risk of malfunction and disease. We want to showcase this in the next section by highlighting a class of biopolymers that is particularly important for cellular function but also makes cells prone to malfunction, disease, and aging.

10.1.2 Intrinsically Disordered Proteins (IDPs) as Polymers with Particular Importance for Cellular Function and Malfunction in Disease and Aging

While structured proteins have been intensely investigated for decades, the so-called intrinsically disordered proteins (IDPs) have now moved into the spotlight. This class of proteins seems to defy the traditional understanding of form and function. IDPs got enriched during the evolution of more complex species and account for around 30% of the total proteome in a eukaryotic cell. In contrast to folded proteins, IDPs lack a stable secondary and tertiary structure if they are on their own, i.e., in their native state in isolation [4, 5]. Their intrinsic flexibility and dynamics in the disordered state make them invisible to conventional structural biology approaches developed for studying folded proteins, such as X-ray crystallography or cryo-electron microscopy. Consequently, our structural and molecular knowledge of IDPs is comparatively underdeveloped, and their properties are much harder to grasp. We can picture them as a dynamic ensemble of protein states (e.g., billions of different states that can interconvert into each other within nanoseconds). Instead of adopting a defined structure, they constantly fluctuate—and it is precisely this disorder and dynamics that gives them extraordinary abilities [4, 6]. An example are intrinsically disordered nucleoporin molecules that form a selective permeability barrier in nuclear pore complexes, which are basically guardians of our genome, as they regulate access to our genome. Flexible, multivalent IDP polymers have been shown to enable much faster interaction networks with other biomolecules, which can facilitate highly specific, yet ultrafast transport in and out of the nucleus [7, 8]. Another specific ability that many IDPs have is to phase separate and form droplet-like microcompartments (discussed in more details below) [9], which are likely essential to many cellular processes, such as transcription, DNA damage repair, protein and RNA transport, and cell signaling [10, 11].

However, in contrast to folded proteins, the structural flexibility of the disordered state also makes IDPs particularly prone to misfolding into amyloid structures [12]. Amyloids are highly stable, β-sheet-rich structures where protein strands align side-by-side, forming very stable protein fibers [13, 14]. The formation of amyloid fibers follows a nucleation-dependent polymerization process, which involves: (1) primary nucleation: formation of initial β-sheet structures, (2) elongation: rapid fiber growth through addition of monomers, (3) secondary nucleation: new amyloid fibers form on the surface of existing ones or through fiber fragmentation [15]. Amyloids are frequently involved in detrimental human diseases, such as neurodegenerative diseases (e.g., Alzheimer's and Parkinson's disease) [14], systemic amyloidosis, and metabolic disorders (e.g., type 2 diabetes) [16].

Amyloid structures are incredibly stable and persist in cells, causing irreversible damage, in particular in long-lived cells. Thus, over the lifetime of an organism, amyloids accumulate in certain tissues, contributing to cellular dysfunction and age-related diseases. Cells use a network of defense mechanisms known as proteostasis to manage IDPs and prevent them from forming harmful amyloids [17]. Each amyloid formation event that escapes cellular defense mechanisms can be thought of as a "tick" on the cellular clock of aging (Fig. 10.2) [18]. Aging is an inevitable part of life, but the molecular mechanisms underlying this process are still not fully understood.

The CRC1551 is intensively devoted to understanding the role of IDPs in aging processes. For example, many IDPs play a crucial role in forming or stabilizing phase-separated droplet-like organelles essential for cell function. However, with increasing age, disturbances in phase separation processes can occur, leading to the formation of amyloids and, ultimately, cellular malfunction [19]. One of the goals of the CRC is to understand the principles of how these molecular "aging clocks" work and how this, in the future, can help design targeted interventions to influence them. Current research shows that IDPs act as a kind of molecular sensor that responds to cellular stress and can either activate protective mechanisms or accelerate cell aging [19, 20]. This knowledge could be used to develop new approaches for the treatment of age-related diseases such as Alzheimer's or Parkinson's disease.

10.1.3 DNA as an Example of How Different Life Scientists and Polymer Scientists Can Think About the Same Biomolecule

DNA is widely known as the storage device for the genetic code of life. For many biologists, studying and understanding its sequence is thus of primary importance and sequencing technologies—known to many as a primary crime scene

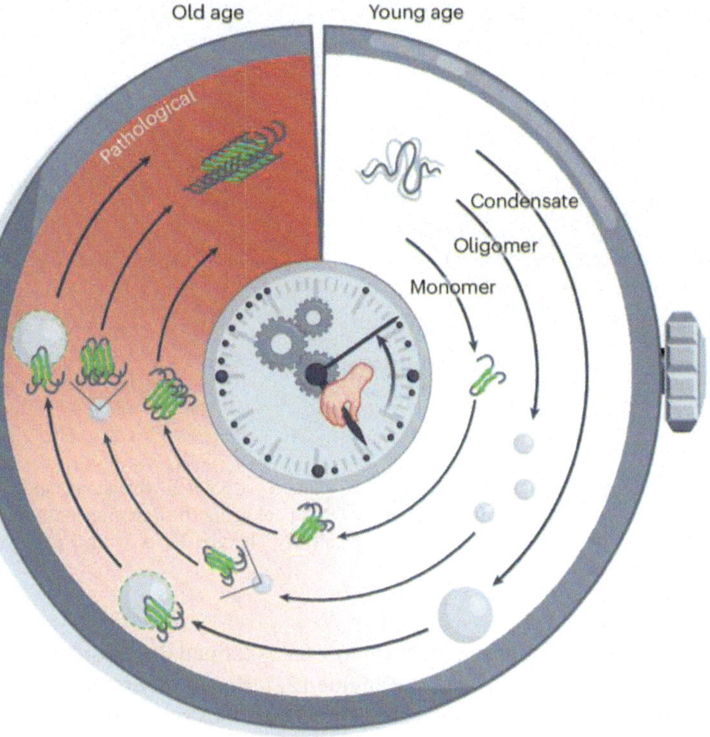

Fig. 10.2 The IDP clock: different pathways of amyloid formation. Monomer pathway: single misfolded proteins trigger a chain reaction, similar to prion diseases. Oligomer pathway: small protein clusters increase the chance of β-sheet formation, eventually elongating into fibers. Condensation pathway: IDPs form liquid-like droplets (condensates) that may give rise to amyloid fibers over time. These pathways are not mutually exclusive and may overlap. For example, liquid droplets can transition into solid-like amyloids in vitro and in vivo, as seen in stress granules. Different cellular defense mechanisms (proteostasis) exist in healthy cells that can turn back the clock (indicated by the hand pulling on the pointer, reproduced with permission from Dormann & Lemke, Nature Cell Biology 2024)

investigation tool—are even drivers of basic science and biotechnology nowadays. In fact, sequencing devices have not just grown in performance but also shrunk in size from freezer-sized machines to USB stick-level devices, making sequencing a widely available and mobile technology.

A fascinating problem is how DNA replicates, a process essential to the birth of new life, as the DNA needs to be replicated before it can be shared between mother and daughter cells [21]. For humans, this means that the 2 m-long DNA needs to be copied—ideally without errors—and then stored again in a compact micrometer-sized cell nucleus. A biological approach inherently involved searching for proteins and enzymes that perform this packaging process. Let us start with a simple case: DNA packaging into a viral capsid. Here, a motor protein at the mouth

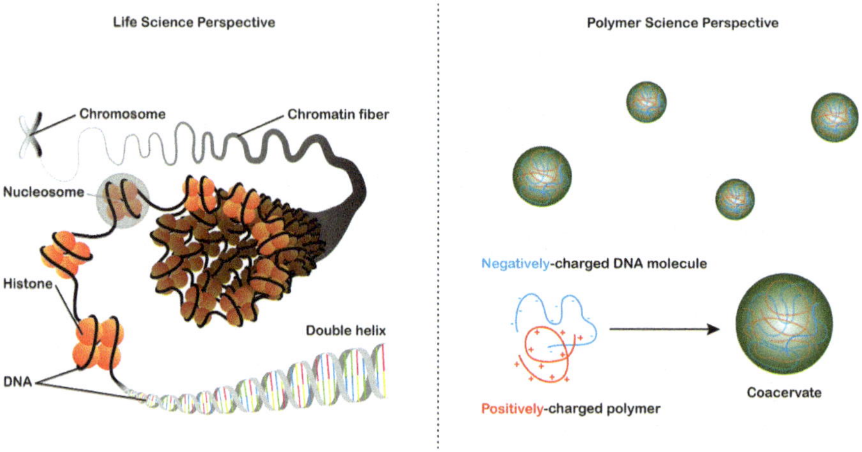

Fig. 10.3 DNA packaging viewed from a life science vs. polymer science perspective. The left panel illustrates different hierarchical levels of DNA packaging (histones/nucleosomes in orange). The right panel illustrates coacervation of negatively charged DNA with a positively charged polymer

of a spherical shell compresses DNA into what is termed the capsid, reaching power densities twice that of a typical car engine [22]. While in virus-like structures, the DNA stays compressed until it infects a host, the DNA in a human nucleus needs to be accessible at all times and thus cannot be compressed the same way. Instead, a hierarchy of assembly machinery was identified, including negatively charged DNA wrapping around positively charged histones, ultimately yielding chromosomes [23] (Fig. 10.3, left). However, DNA packaging is still an area of contemporary research, and motor domains that actively compact DNA by loop formation have only been known for a few years [22, 24].

Despite this remarkable progress in biosciences, only about a decade ago did phase separation of DNA as a mechanism for DNA compaction move into the focus of life scientists [25]. Phase separation is a physicochemical phenomenon where a mixture of two species separates into a dense and soluble phase [26]. An example from everyday life is salad dressing, where oil and vinegar demix into oil droplets that float in the vinegar.

Every polymer mixed with a solute (like water) has a phase diagram that describes under which conditions (temperature, pressure, concentration, etc.) phase separation occurs. It is something so basic that it is usually taught in Physical Chemistry 101. Thus, everyone with the most basic training in polymer science knows this phenomenon by heart. In fact, DNA phase separation, also called coacervation, has been studied since the 1920s [27, 28]. In the 1970s, experimental validation showed that DNA can nicely compact when this negatively charged polymer is mixed with positively charged cations [29] (Fig. 10.3, right). In fact, many polymer scientists lost interest in this topic because it became so well understood due to progress in that field.

However, life scientists only show dramatically increased interest in phase separation processes since seminal work by Hyman, Jülicher, and Rosen [30, 31] demonstrated that biomolecules readily phase separate into droplet-like organelles to mediate crucial cellular functions. Before 2009, the phenomenon of phase separation was hardly studied in biology.

10.1.4 What Life Scientists and Polymer Scientists Can Learn from Each Other

The example described under 10.1.3 teaches us that DNA coacervation as a relevant packaging mechanism in human cells could already have been included in mainstream biological thinking in the 1970s. In the 1970s also hardly anybody believed in the existence of intrinsically disordered proteins (IDPs), and thus, the recognition of their functional importance described in 10.1.2 had to wait for science to develop. Despite IDP malfunction being at the root of most neurodegenerative diseases, mainly amyloid formation by templated misfolding was considered an amyloid formation mechanism until about a decade ago. In the meantime, we know that IDPs, including many proteins involved in neurodegenerative diseases, can phase separate into droplets [19], and this now even has become a basic science experiment that can be easily demonstrated with egg white at an open school day. Including phase separation properties into thinking about IDPs and their tendency to form amyloids in neurodegenerative diseases is something that, in principle, could have already happened decades earlier if polymer sciences and life sciences had not followed parallel paths, with insufficient cross-nurturing between the two disciplines.

The above examples might give the impression that only life scientists can learn from polymer scientists. However, this is not the case. In fact, while polymer scientists have developed a very good understanding and theories for simple polymer mixtures, the complexity of life science scenarios far exceeds what is state of the art in the polymer sciences. Biological systems are much more complex with respect to composition and thus represent excellent systems to understand hetergeneous multicomponent problems (Fig. 10.4). Moreover, the fact that living systems are not at equilibrium but constantly consume energy presents another major challenge and interesting research problem to the polymer science community (Fig. 10.4).

10.2 The Scientific Approach and Focus of the CRC1551

The Collaborative Research Center 1551 on "Polymer Concepts in Cellular Function" aims to decipher the molecular secrets of polymers—in particular, their ability to transform apparent chaos into biological precision. One central idea is

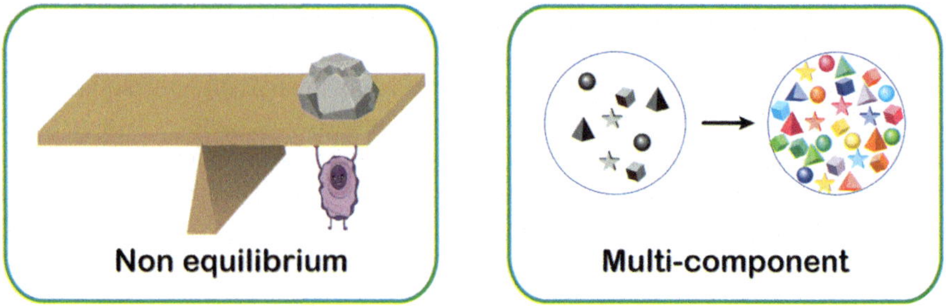

Fig. 10.4 Two important polymer concepts studied in CRC1551. These scenarios are commonly found in the life sciences and pose interesting challenges and research areas to the polymer science community

that polymers are not just carriers of information or cellular building blocks but dynamic actors that actively contribute to the organization and regulation of cellular processes. This insight forms the basis for diverse research projects carried out by an interdisciplinary team. A CRC1551 team always consists of at least one polymer scientist who approaches the research problem from a polymer point of view and a life scientist having a mechanistic view of molecular and cellular processes. The traditional disciplines of physics, chemistry and biology blur at this stage; hence, it is challenging to categorize CRC1551 research into disciplines and terms like Synthetic Biology, Bioengineering, Chemical Biology, Biotechnology or Supramolecular Chemistry, and multiscale modeling might be better, yet not sufficient descriptors of the transdisciplinarity needed in this field of science. Thus, we like to talk about training the next generation of "polymer-life-scientists."

10.2.1 Focus on DNA, RNA, IDPs and Synthetic Polymers as Biological Tools

CRC1551 studies all major classes of biopolymers, in particular the three main classes introduced above (DNA, RNA, and IDPs). Of course, other biopolymers exist in the cell, such as the cytoskeleton. Cytoskeletal assemblies typically give the cell stiffness or structure, and they form through a more ordered assembly, often under energy consumption. To keep focus, cytoskeletal elements are not among the core research topics of the CRC1551, even though all polymers share similarities, and there can always be something learned from each of them. The concept of how a polymer behaves in solution versus on a surface is crucial for understanding biology; thus, interfaces and surfaces are part of CRC1551 research. In general, every polymer where the function might originate from the polymer properties and/or the apparent disorder of the molecule is of central interest to the CRC1551. We are convinced that studying polymers with different chemical identities and different length scales and discerning their similarities and differences in function can lead to the discovery of general concepts.

While the CRC focuses on natural biopolymers, synthetic polymers also play an important role. These artificially produced molecules make it possible to imitate or even extend biochemical processes. CRC1551 uses synthetic polymers to develop new materials that can function as molecular tools in the cell or as bottom-up mimics to step-by-step design and understand complex processes that are difficult to study directly. A particularly innovative approach is the development of "intelligent polymers" that react to external stimuli or signals in the cell and then respond intelligently and show interactive behavior.

10.2.2 Simulation and Theory: From Molecules to Models

The CRC combines experimental approaches with theoretical models to understand the complexity of polymer interactions. Computer simulations can, for example, help analyze the dynamics of IDPs, their interaction with RNA, or the formation of phase-separated droplets at the molecular level. Such models consider the physicochemical properties of molecules and their interactions with other components. Such work has shown that even small changes in the droplets' composition can significantly affect their function [32]. Computer simulations and theoretical models are tools for data analysis and guideposts for future experiments. They allow testing hypotheses and a better understanding of the physical principles underlying a certain molecular process. With growing computational power, phase separation can be simulated in a realistic cellular environment.

The CRC1551 combines the above-mentioned research areas into a holistic vision. The research carried out in this network not only advances our basic understanding of biology but also provides a platform where scientists from different disciplines can work together to shape the future and use knowledge from polymer research and life science for biomedical and technological innovations.

10.2.3 CRC1551: A Network for Exchange, Promoting Young Talent, International Cooperation and Science Communication

The CRC1551 is divided into several subprojects, each addressing a specific question—these range from investigating individual IDPs, DNA, RNA to developing synthetic materials. In addition, central platforms provide the latest technologies in bioimaging, biopolymer synthesis and analytics. Regular workshops and conferences promote exchange between the subprojects and with international partners. These events are a place for scientific discussion and an opportunity to initiate new collaborations.

10.2.3.1 Promoting Young Talent: Science of Tomorrow

A central mission of the CRC1551 is to promote the science of tomorrow by training young researchers. Young scientists can work at the forefront of biopolymer research and learn from experienced mentors. Specifically, the CRC1551 offers a structured graduate education that imparts both scientific and general professional skills. The network aims to train a new generation of blue-sky-thinking polymer-life scientists to approach a research problem with a true hybrid perspective that emerges from combining a polymer-based view with a biology-based view.

PhD students, postdocs, and even principal investigators have the opportunity to move smoothly between different projects, expand their knowledge, and gain new insights through cross-collaborations and laboratory exchanges. Workshops on career opportunities, both within and outside of science, are an important part of the CRC1551 training program. The aim is to prepare young researchers for a career in research and beyond.

10.2.3.2 International Cooperation: Overcoming Borders

The challenges of polymer research do not end at national borders. CRC1551 scientists collaborate with scientists worldwide to gain access to the best technologies and minds. Moreover, they regularly participate in visible international meetings, such as the "Curious 2024 Future Insights" and the Biophysical Society Meeting in the USA. Moreover, the CRC also organizes its own meetings to actively bring scientists together, and initiates subgroup sessions on "Polymer Concepts in Cellular Function," e.g., at the Annual Meeting of the American Society of Cell Biology. Many of the CRC's groundbreaking results are generated by international teams and published in leading journals. These publications not only increase the visibility of the network but also strengthen global collaboration.

10.2.3.3 The Social Dimension of Polymer Research: Sharing Knowledge Through Science Communication

Research on biomolecules raises not only scientific but also ethical questions. The targeted manipulation of molecular processes entails both opportunities and risks. Thus, the CRC1551 is actively involved in science communication, involving the public in the discourse and showcasing that research is being carried out in a responsible way that will ultimately benefit society. Science communication plays an essential role in conveying the relevance of polymer research and inspiring society for science. CRC1551 scientists have been involved in high school projects, developing teaching materials and experimentation kits to introduce young students to the world of polymer research. Events such as open days, podcasts, interactive exhibitions, and even participation in the local carnival parade, make the work of the CRC1551 accessible to a broader public.

10.3 Conclusion and Outlook: Science as the Key to the Future

Research on biopolymers opens the door to a world that leads deep into the molecular mechanisms of life. The Collaborative Research Center 1551 (www.crc1551.com) has set itself the goal to decipher these mechanisms and understand fundamental concepts of polymer function in cells. A central goal of the CRC1551 is to redefine the fundamentals of molecular biology. The study of

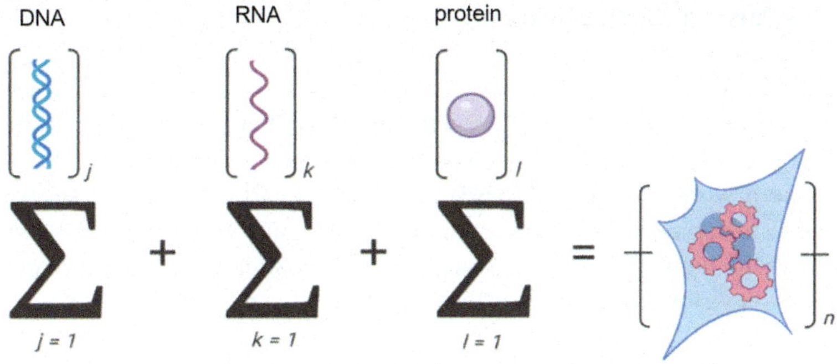

Fig. 10.5 A central goal of CRC1551. Understanding how the sum of biopolymers and their dynamics lead to the emergence of complex cellular function

the polymeric nature of DNA, RNA, and IDPs and their polymeric interactions will revolutionize our view of cellular processes (Fig. 10.5) and offer a deeper understanding of biological malfunctions and disease.

The strength of polymer research lies in its comprehensive approach: it connects the nano- and microworld of molecules with the macro-processes of life, creating a coherent picture that explains both the precision of biological processes and the apparent disorder. These insights are the basis for innovations that go far beyond basic research and could one day provide answers to some of the greatest challenges of our time—be it in medicine, materials science or energy generation.

Research at the CRC1551 produces concrete approaches that could lead to groundbreaking innovations in the near future. Molecular tools offer the possibility of manipulating cellular processes in real-time. These technologies could enable pioneering advances in personalized medicine, for example, in treating neurodegenerative diseases or in developing targeted cancer therapies. Biotechnology also benefits from polymer research. The ability to control biological processes using tailor-made polymers could revolutionize the production of biomaterials and lay the foundation for sustainable, environmentally friendly technologies.

Science not only has the task of gaining new knowledge but also of considering it in a social context, thus involving the public in the scientific discourse. The developments in polymer research have the potential to change our lives fundamentally—but they also require responsible action. Through school programs, public lectures and citizen science, the CRC1551 aims to promote understanding of science while raising awareness of research's ethical and societal dimensions. This dialogue is crucial to promote the acceptance of new technologies and ensure sustainable use of scientific knowledge.

10.3.1 A Vision for the Future

Research into biopolymers has only begun to exploit the potential of these fascinating molecules. The CRC1551 has established itself as a pioneer in this field by combining interdisciplinary approaches with innovative technologies.

Imagine a world in which diseases are detected early and treated specifically, sustainable materials reduce environmental impact, and cellular processes can be controlled with unimaginable precision. This is the vision that drives research at the CRC1551—a vision that connects science and society and paves the way to a better future.

The work of the CRC1551 is a journey into the depths of the molecular world. It is characterized by curiosity, creativity and the will to expand the boundaries of knowledge. But this journey does not end with the solution of scientific questions. It begins where science meets application and where innovations arise that can improve the lives of all people. Polymers—the hidden heroes of life—show us that even in the apparent disorder, there is an order that has the potential to transform our world.

Acknowledgement We acknowledge funding from the Deutsche Forschungsgemeinschaft (DFG), CRC1551 "Polymer concepts in cellular function" (project number 464588647).

References

1. Crick F (1970) Central dogma of molecular biology. Nature 227:561–563
2. Watson JD, Crick FH (1953) Molecular structure of nucleic acids; a structure for deoxyribose nucleic acid. Nature 171:737–738
3. Anfinsen CB (1972) The formation and stabilization of protein structure. Biochem J 128:737–749
4. Dyson HJ, Wright PE (2005) Intrinsically unstructured proteins and their functions. Nat Rev Mol Cell Biol 6:197–208
5. Dunker AK, Silman I, Uversky VN, Sussman JL (2008) Function and structure of inherently disordered proteins. Curr Opin Struct Biol 18:756–764
6. Wright PE, Dyson HJ (2015) Intrinsically disordered proteins in cellular signalling and regulation. Nat Rev Mol Cell Biol 16:18–29
7. Milles S et al (2015) Plasticity of an ultrafast interaction between nucleoporins and nuclear transport receptors. Cell 163:734–745
8. Aramburu IV, Lemke EA (2017) Floppy but not sloppy: interaction mechanism of FG-nucleoporins and nuclear transport receptors. Semin Cell Dev Biol 68:34–41
9. Borcherds W, Bremer A, Borgia MB, Mittag T (2021) How do intrinsically disordered protein regions encode a driving force for liquid-liquid phase separation? Curr Opin Struct Biol 67:41–50
10. Boeynaems S et al (2018) Protein phase separation: a new phase in cell biology. Trends Cell Biol 28:420–435
11. Hirose T, Ninomiya K, Nakagawa S, Yamazaki T (2023) A guide to membraneless organelles and their various roles in gene regulation. Nat Rev Mol Cell Biol 24:288–304
12. Mukherjee S, Poudyal M, Dave K, Kadu P, Maji SK (2024) Protein misfolding and amyloid nucleation through liquid-liquid phase separation. Chem Soc Rev 53:4976–5013

13. Dobson CM, Knowles TPJ, Vendruscolo M (2020) The amyloid phenomenon and its significance in biology and medicine. Cold Spring Harb Perspect Biol 12:a033878
14. Scheres SHW, Ryskeldi-Falcon B, Goedert M (2023) Molecular pathology of neurodegenerative diseases by cryo-EM of amyloids. Nature 621:701–710
15. Arosio P, Knowles TPJ, Linse S (2015) On the lag phase in amyloid fibril formation. Phys Chem Chem Phys 17:7606–7618
16. Iadanza MG, Jackson MP, Hewitt EW, Ranson NA, Radford SE (2018) A new era for understanding amyloid structures and disease. Nat Rev Mol Cell Biol 19:755–773
17. Labbadia J, Morimoto RI (2015) The biology of proteostasis in aging and disease. Annu Rev Biochem 84:435–464
18. Dormann D, Lemke EA (2024) Adding intrinsically disordered proteins to biological ageing clocks. Nat Cell Biol 26:851–858
19. Alberti S, Hyman AA (2021) Biomolecular condensates at the nexus of cellular stress, protein aggregation disease and ageing. Nat Rev Mol Cell Biol 22:196–213
20. Franzmann TM, Alberti S (2019) Protein phase separation as a stress survival strategy. Cold Spring Harb Perspect Biol 11:a034058
21. Chagin VO, Stear JH, Cardoso MC (2010) Organization of DNA replication. Cold Spring Harb Perspect Biol 2:a000737
22. Mo Y et al (2020) Function of a viral genome packaging motor from bacteriophage T4 is insensitive to DNA sequence. Nucleic Acids Res 48:11602–11614
23. Maeshima K, Ide S, Babokhov M (2019) Dynamic chromatin organization without the 30-nm fiber. Curr Opin Cell Biol 58:95–104
24. Ganji M et al (2018) Real-time imaging of DNA loop extrusion by condensin. Science 360:102–105
25. Larson AG et al (2017) Liquid droplet formation by HP1α suggests a role for phase separation in heterochromatin. Nature 547:236–240
26. Choi J-M, Holehouse AS, Pappu RV (2020) Physical principles underlying the complex biology of intracellular phase transitions. Annu Rev Biophys 49:107–133
27. Bungenberg de Jong HG, Kruyt HR (1929) Coacervation (partial miscibility in colloid systems). Proc K Ned Akad Wetensch 32:849–856
28. Smith AE, Bellware FT, Silver JJ (1967) Formation of nucleic acid coacervates by dehydration and rehydration. Nature 214:1038–1040
29. Lerman LS (1971) A transition to a compact form of DNA in polymer solutions. Proc Natl Acad Sci U S A 68:1886–1890
30. Brangwynne CP et al (2009) Germline P granules are liquid droplets that localize by controlled dissolution/condensation. Science 324:1729–1732
31. Li P et al (2012) Phase transitions in the assembly of multivalent signalling proteins. Nature 483:336–340
32. Gruijs da Silva LA et al (2022) Disease-linked TDP-43 hyperphosphorylation suppresses TDP-43 condensation and aggregation. EMBO J 41:e108443

Part IV

The AI Revolution

The Future of Drug Discovery: An AI-Driven Perspective

11

Muhammad Nabeel Asim, Michael Böhm, and Andreas Dengel

Contents

M. N. Asim (✉)
German Research Center for Artificial Intelligence (DFKI), Kaiserslautern, Germany

Intelligentx GmbH (intelligentx.com), Kaiserslautern, Germany
e-mail: muhammad_nabeel.asim@dfki.de

M. Böhm
Saarland University, Cardiology, angiology and internal intensive car, Saarbrücken, Germany

A. Dengel
German Research Center for Artificial Intelligence (DFKI), Kaiserslautern, Germany

Department of Computer Science, Rheinland Pfälzische Technische Universität, Kaiserslautern, Germany

Intelligentx GmbH (intelligentx.com), Kaiserslautern, Germany

U. A. K. Betz (ed.), *Science for a Better Tomorrow*,
https://doi.org/10.1007/978-3-031-93623-4_11

Abstract

Conventional drug development landscape is a complex and time-intensive process, encompassing multiple stages from target identification to clinical trials and regulatory approval. Artificial intelligence is a revolutionizing drug development process through advanced computational methodologies and predictive analytics. Specifically, the combination of AI-powered data-driven forecasting and wet-lab experiments is enabling rapid identification of promising targets, precise design of novel molecules, and strategic optimization of clinical trials. Although the full potential of AI in drug development is yet to be realized, its integration has already demonstrated remarkable impact by reducing development timelines from 10–15 years to 5–7 years and costs from 2–3 billion dollars to 1–1.5 billion dollars. To unlock AI potential and explore new opportunities for development of innovative drugs and therapies, established pharmaceutical companies are collaborating with AI companies. Primary objective of this chapter is to present high-level overview of drug development phases and possible AI applications at each phase. Moreover, it presents promising AI-focused start-ups and their work on revolutionary AI-driven drugs.

Keywords

Artificial intelligence · Drug design · Machine learning · Deep learning · Drug discovery pipeline · Pharmacodynamics · Pharmacokinetics · Computational drug design · Drug development · Drug development tasks

11.1 Introduction

Drug development is a multidisciplinary field that acquires expertise from biology, chemistry, pharmacology, and clinical medicine disciplines [1]. Despite remarkable advancements in these fields, clinical development process, the path from a promising idea to a life-saving medication, typically requires 10–15 years [2]. The prolonged timeline requires extraordinary patience and persistence from researchers and stakeholders. Specifically, development of a single drug usually costs between 2 and 3 billion dollars [2]. Primarily both prolonged timeline and high cost are driven by the intensive wet-lab research required for target identification and the extensive multiphase clinical trials that are essential to ensure safety and efficacy [2]. Another challenge is the high attrition rate, where numerous costly and time-consuming failed attempts precede the successful development of a viable drug candidate [3]. Generally more than 90% drug candidates remain fail to secure market approval [3]. The high failure rate is driven by multifarious factors including inadequate efficacy, where drugs do not work well as expected [3], and unexpected toxicity, where the drug exhibits harmful side effects that were not anticipated during preclinical studies [4]. Unfavorable pharmacokinetics, such as poor absorption, rapid metabolism, or inefficient distribution, can hinder the ability of a drug to reach its target site and exert its desired effect [4].

Significant challenges in drug development landscape require the integration of AI into the drug development life cycle [5]. This emergence can drive a transformative shift toward personalized medicine, tailored treatment strategies, and fast paced drug discovery processes [5]. AI algorithms leverage the vast amounts of genomics, proteomics, omics, compounds, and clinical data to identify promising drug targets and design novel compounds with precisely tailored properties [6]. In addition, AI potential is being utilized to optimize clinical trial designs, which is a critical phase in the drug development process [6]. The AI solutions are not only streamlining traditional drug discovery pipeline but also opening up new possibilities for identification of new therapeutic areas for already approved drugs [6]. Specifically, rapid development of large-scale biological knowledge graphs enabled AI models to simultaneously analyze existing drugs across multiple disease pathways, integrating diverse data sources—from biological information and clinical trials to patient outcomes—to identify promising drug candidates through sophisticated network analysis [6].

The multidisciplinary collaboration of pharmacology, clinical medicine, and AI disciplines has spawned over 200 AI-focused start-ups, which have forged strategic partnerships with established pharmaceutical giants [7]. These collaborative ventures are leveraging cutting-edge AI technologies to optimize and accelerate various stages of the drug development process [7]. The synergy between AI innovation and pharmaceutical expertise has already yielded numerous AI-driven drug candidates [7]. To further accelerate and expedite research in AI-driven drug development, a comprehensive literature review is needed to establish a solid foundation. The review should provide insights from diverse perspectives, including what phases of drug discovery pipeline have witnessed the implementation of AI models? How effectively have AI models been employed across different phases of the drug discovery pipeline? What are the key technical and practical challenges AI models face in drug discovery pipeline? What limitations must be overcome to achieve more comprehensive AI integration throughout the entire drug discovery pipeline?

The focus of this chapter revolves around aforementioned questions with the following contributions: (1) It provides a generalized drug development life cycle. This is primarily done to establish a clear path for understanding how AI technologies can be effectively integrated into traditional drug discovery pipeline phases. (2) It systematically explores AI applications across different phases of drug discovery pipeline and elucidates specific tasks where AI models have demonstrated particular efficacy. (3) It presents detailed case studies of AI-designed drug candidates that have successfully progressed to clinical trials. These contributions collectively provide a realistic assessment of AI's current capabilities and limitations in drug discovery while pointing toward future opportunities for more comprehensive integration of AI technologies across drug discovery pipeline. As these technologies continue to mature and demonstrate their value, they are likely to play an increasingly central role in pharmaceutical research and give birth to fully AI-enabled drug discovery success story.

11.2 AI Applications in Different Stages of Drug Development Pipeline

Considering the availability of huge Electronic Health Records (EHRs), medical literature, patient demographics, unstructured clinical notes, and multi-omics data, AI models are being increasingly applied to various phases of drug discovery pipeline. Figure 11.1 graphically illustrates different phases of drug discovery pipeline and their associated AI-driven subtasks. A detailed overview of each phase and its corresponding AI-driven subtasks is provided in the following subsections.

11.2.1 Target Identification and Validation

Target identification involves the identification of genes, proteins, metabolites, and key pathways that can be modulated by drugs to treat a disease [8]. Targets are selected based on two key criteria: their causal role in disease mechanisms and their suitability for drug development [8]. To find the actual cause of a disease, scientists usually look for chain of events that leads to the disease, like a genetic mutation might trigger specific protein [9], which causes specific cell process, that finally leads to complex disease [10]. Understanding this chain helps the scientists to identify the right targets responsible for the development of a disease. The second criterion, druggability, indicates the ability of a target to be modulated by a particular drug. Unique appearances, localization, and structural characteristics of targets affect their druggability [11]. This is because some drugs are suitable for

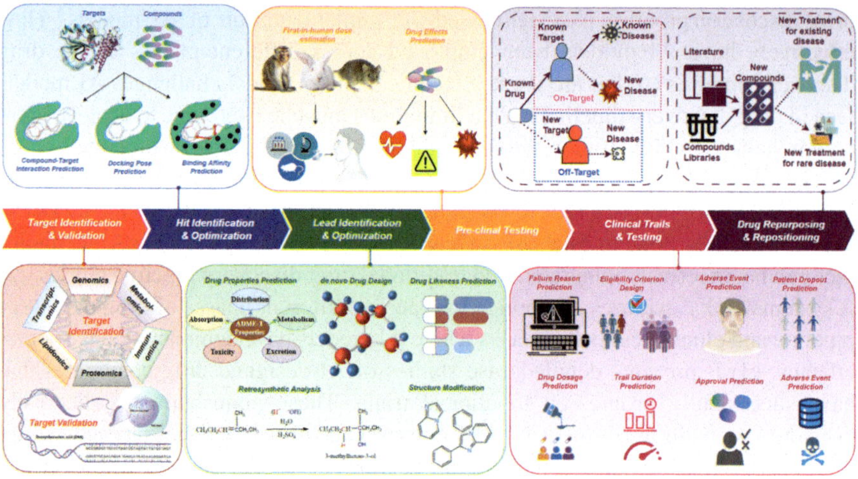

Fig. 11.1 AI-Driven Drug Discovery Pipeline: a visual representation of the key phases in drug discovery and the AI-driven tasks that can accelerate each stage

targets having well-defined binding pockets, while other drugs are more appropriate for large surface targets.

In the context of drug development, first of all targets are identified through various analyses including mutations identification [12], genomic variants identification [13], genotype-phenotype correlations [14], biomarker identification [15], and variant calling [16].

Another important task is the analysis of biological pathways that regulate essential cellular processes, namely metabolism, signaling, and gene expression [17, 18]. These pathways are interconnected networks of molecular entities, and their proper functioning is essential to maintain normal cellular behavior [17, 19, 20]. However, disruptions or abnormalities in pathways can contribute to disease progression [21–23]. Aforementioned diverse types of analyses through wet-lab experiments are expensive and time consuming. In order to perform these analyses at large scale, AI potential is being harnessed [24, 25]. For instance, knowledge graph embedding and AI methods-based applications have been developed to find interactions between different molecular entities [26, 27].

Besides pathways analysis, protein structure prediction also plays a key role in target identification. Protein structures prediction task provides insights into 3D conformation of proteins which reveals various functional regions [28] and cleavage sites of proteins [29]. Numerous AI applications including AlphaFold [30], RoseTTAFold [31], OpenFold [32], ESMFold [33], and OmegaFold [34] have a greatly streamlined process of protein structure prediction. For example, AlphaFold predicts 3D protein structures by utilizing evolutionary data from homologous proteins [30], while RoseTTAFold employs multiple sequence alignments to determine the spatial arrangement of proteins [31]. OpenFold builds upon the foundations of both AlphaFold and RoseTTAFold [32], whereas ESMFold uses a protein language model approach for structure prediction [33]. Additionally, OmegaFold incorporates nonprotein molecules to offer a comprehensive view of protein interactions within their functional environments [34].

11.2.2 Hit-to-Lead Compounds Identification

Hit-to-Lead (HTL) compounds identification process performs virtual screening that predicts potential drugs (lead compounds) from a large-scale compounds library [35]. Specifically, given a drug candidate compound and target protein pair, the main goal of this phase is to estimate whether drug candidate compound interacts/binds with target protein or not (drug-target interaction prediction). Then, it estimates drug candidate compound position and orientation on protein's binding site (compound binding pose prediction). Finally, it estimates the strength of binding between drug candidate compound and target protein (binding affinity (BA) prediction). To accomplish this, a traditional virtual screening process makes use of a computational technique called molecular docking that performs two steps: sampling and scoring [35]. The sampling step looks at whether drug candidate compounds can fit into the target protein's binding site and explores different ways (poses/orientations) they can bind to target protein. Then, the scoring step evaluates how strong these bindings are by measuring their binding strength or energy levels. Currently, most molecular docking tools employ classical scoring functions (SFs) including physical, empirical, and knowledge-based approaches. These SFs typically assume a predetermined functional form, usually a linear regression model to explain the relationships among descriptors utilized to characterize compound-target complex and binding affinity. However, this linear relationship assumption often leads to inaccurate predictions since it may not be valid for every complex [35].

Given AI-driven approaches potential to learn functional forms directly from data and capture nonlinear relationships, researchers have applied AI at both levels (sampling and scoring). Within the landscape of AI-empowered compounds virtual screening, predictive pipelines can be categorized into two primary categories: (1) structure-based virtual screening and (2) ligand-based virtual screening. Figure 11.2 illustrates a high-level graphical overview of both types of predictive pipelines. The most left part in Fig. 11.2 illustrates type of data utilized in both types of predictive pipelines. It can be seen that structure-based virtual screening operates on molecular descriptors extracted from compounds chemical structures and 3D structure of target protein, while ligand-based predictive pipelines are fed with compounds bioactivity data. At the next stage, predictive pipelines make use of distinct machine and

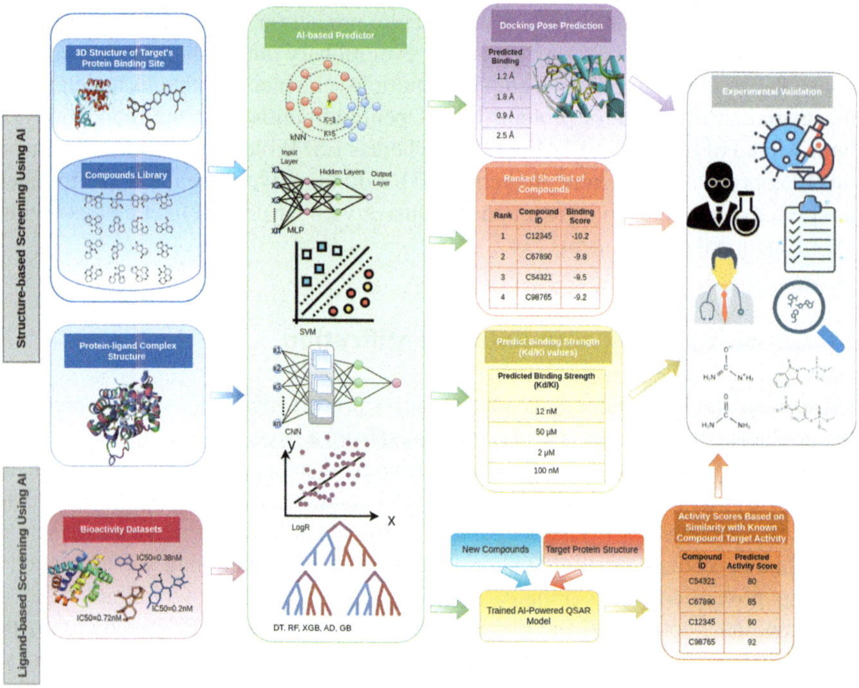

Fig. 11.2 Workflow of four different AI-driven tasks augmenting hit-to-lead identification

deep learning models to extract informative features from given data, and based on the extracted features, they predict the binding possibility, compound binding pose (only structure-based approach can), and binding affinity. Based on predicted binding affinities, compounds are ranked, and only the top-ranked compounds are considered highly promising leads for further experimental evaluation. To better describe the dynamics and appropriate applications of both structure-based and ligand-based virtual screening approaches, we provide key details of both methods in the following passages.

In structure-based virtual screening (SBVS) (also known as target-based or receptor-based virtual screening), compounds' chemical structures and already known target protein structure are used to determine the binding possibility (drug-target interaction), compound binding pose, and binding affinity (BA) on the basis of compounds complementarity to target protein structure [36]. Given true binders and false binders compound-target pairs, firstly, in AI-driven binding possibility prediction pipeline, compound statistical vectors based on their chemical structures are being learned using SMILES encoding, fingerprints, graph-based representation learning methods (atom- and bond-level molecular graph, drug-drug interaction network, and drug-disease association network, if available), or hybrid representation learning approaches, whereas target protein statistical vector is being learned using 3D voxelization, distance matrices, and graph-based methods (protein-protein interaction network [37] and protein-disease association network, if available). These statistical vectors are passed to deep learning models that learn useful patterns for accurate prediction of binding possibility [38].

Similar to binding possibility prediction, compound binding pose prediction is also treated as a binary classification task [35]. Positive and negative compound poses are generated using a specific docking tool and threshold criteria. For example, few researchers have performed compounds docking against target proteins using AutoDock or MedusaDock to get their poses [35]. Then, they computed Root Mean Square Deviation (RMSD) between generated poses and native poses. Afterward, only those poses are classified as positive poses whose RMSD values are less than 2 Å, whereas poses having RMSD values more than 4 Å are treated as negative poses, and all poses having RMSD in between are discarded from the dataset [35]. Unlike these two tasks, binding affinity prediction is a regression task as it predicts the strength of compound-target binding using a deep learning model. After performing inference of deep learning models on large-scale compounds against target protein, compounds are ranked on the basis of binding affinities, and only most promising compounds are analyzed and optimized through experimental testing.

Unlike structure-based virtual screening approaches that essentially require a detailed 3D structure of target protein, ligand-based virtual screening (LBVS) approaches rely on known active compounds that have been experimentally validated to show activity (binding) against target of interest [39]. LBVS approaches require known active compounds structural data like SMILES strings, molecular weight, and key group (e.g., hydroxyl group) and activity/functional data against target like inhibitory concentration IC% (concentration needed to reduce target activity), binding affinity constant (Ki), binding energy (energy released/required during binding), and inhibition at specific concentration. Researchers are also enriching this activity data by the addition of compound-drug interactions and compound-disease associations [39].

In AI-driven LBVS pipeline, firstly, active compound statistical vectors based on their structural and activity data are being learned using structure-based representation learning methods (SMILES encoding), graph-based representation learning methods (atom- and bond-level molecular graph, drug-drug interaction network, and drug-disease association network, if available), pattern-based encoding (e.g., fingerprint), transfer learning methods (pretrained QSAR models), or hybrid representation learning approaches, whereas target protein statistical vector is being learned using amino acid sequence-based representation learning methods, contact maps-based representation learning methods, or graph-based methods (protein-protein interaction network [40] and protein-disease association network [40], if available). Secondly, learned representations are passed to deep learning models that learn which features correlate with higher activity by identifying common structural and chemical patterns. After training, deep learning models are applied to screen large collection of compounds to predict their likelihood of being active (drug molecules) based on certain criteria like similarity to known actives. Model inference is based on the principle that compounds having similar structural and chemical properties may act in a similar manner. After model inference, only shortlisted compounds undergo ADMET (absorption, distribution, metabolism, excretion, and toxicity) analysis and optimization preceded by different bioassays before entering clinical trials.

11.2.3 Lead Optimization

Once the promising lead compounds are identified, the next critical step is to optimize them into potent and safe drug candidates. Effective lead optimization typically involves iterative cycles of synthesis, testing, and modification, each aimed at reducing toxicity, enhancing efficacy, and pharmacokinetic properties to maximize the therapeutic potential of a compound. AI has emerged as a powerful tool to streamline and accelerate this phase by accurately performing five key tasks including (1) property prediction, (2) drug-likeness prediction, (3) structure modifications prediction, (4) de novo molecule generation, and (5) retrosynthetic analysis. Figure 11.3 illustrates workflows of different tasks involved in the lead optimization phase.

Lead compounds must meet specific physicochemical, pharmacokinetic, and biological criteria to be considered as potential drug candidates. Lead optimization involves the systematic refinement of drug candidate lead compounds to enhance their drug-like properties including primary activity (target binding affinity and functional potency), physicochemical properties (aqueous solubility, chemical/metabolic stability, and permeability), and ADME/Tox properties (absorption, distribution, metabolism, excretion, and toxicity). Table 11.1 provides a comprehen-

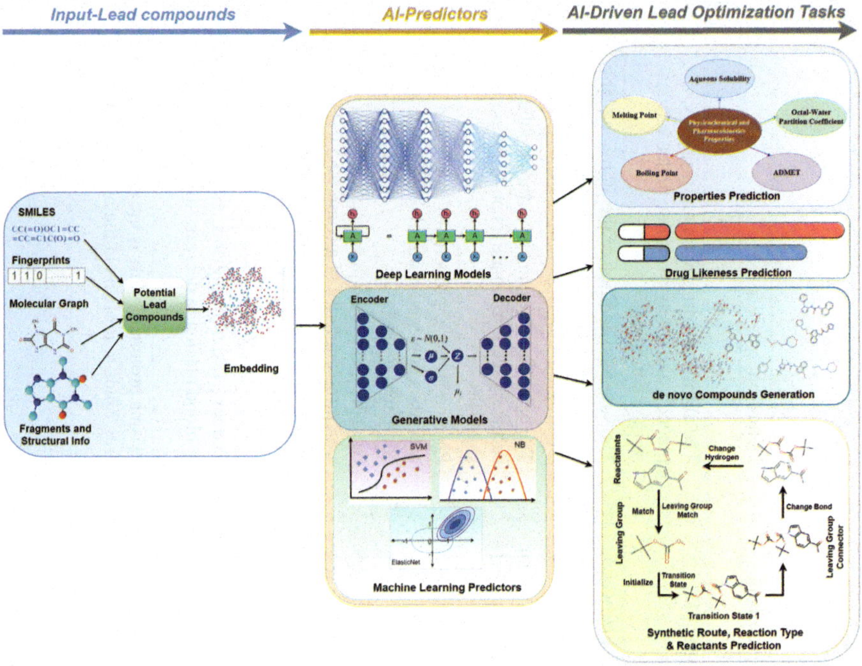

Fig. 11.3 AI-driven tasks augmenting lead optimization

sive overview of the key properties under their respective categories that are being predicted to identify promising lead compounds. By leveraging property prediction tasks, researchers can identify potentially less effective drug candidate compounds early in the drug discovery process. These predictions guide synthetic efforts toward drug candidate compounds with optimal pharmacokinetic profiles.

Another lead optimization paradigm makes use of deep learning for the prediction of structural modifications which refine the structures of existing compounds to enhance their drug-likeness, efficacy, selectivity, and pharmacokinetic properties. To achieve maximum therapeutic benefits while minimizing adverse effects, compounds efficacy and safety enhancement is a multiparameter optimization problem. It necessitates a thorough understanding of the compounds' mechanism of action, metabolic pathways, and its interactions with other biological molecules. Despite advancements in High-Throughput Screening (HTS) and other technologies, multiparameter optimization of drug candidate compounds at large scale remains inefficient, resource-intensive, and prone to false positives and negatives, which may result in high attrition rates in clinical trials. To address these concerns, researchers are using deep learning models like Recurrent Neural Networks (RNNs), Variational Autoencoders (VAEs), and Generative Adversarial Networks (GANs) to explore vast "chemical spaces" to create completely new chemical compounds. These models process simple text-based SMILES to complex 3D structural representations

Table 11.1 Summary of pharmacological and physicochemical properties of drugs

Property name	Definition	Property name	Definition
Physicochemical properties			
Aqueous solubility [41]	Maximum concentration of drug which can dissolve in a given volume of water under specific conditions (temperature, pressure)	Stability [42]	The ability of a drug to retain its quality, potency, and safety over time under various storage conditions
Octanol-water partition coefficient [43]	A critical measure that represents a drug's ability to dissolve preferentially in an organic phase (like octanol) versus an aqueous phase (water)	Molecular weight [44]	The sum total addition of the atomic weights of all the particular elemental atoms that make up the drug. Its unit is grams per mole (g/mol)
Melting point [43]	The temperature at which the solid form of the drug transitions to its liquid state	Topological polar surface area [44]	The calculated measure of a drug's polar surface, used to estimate its polarity and lipid solubility
Boiling point [43]	The temperature at which the vapor pressure of the liquid equals the pressure surrounding the liquid and the liquid changes into a vapor	Lipophilicity [45]	Physicochemical property of a drug that describes how well it dissolves in fat or oil compared to water
Vapor pressure [43]	It is the measure of the volatility of the drug and is defined as the maximum pressure that can be produced by the drugs at a given temperature and pressure	Drug-likeness [46]	Drug-likeness is the physicochemical or structural similarity of a compound to known drugs, used to evaluate its potential to successfully pass clinical trials
Pharmacokinetic properties			
Bioconcentration factor [43]	A ratio that measures how much of a chemical accumulates in an organism compared to the concentration of the chemical in the environment	Half-life [47]	The time taken for the concentration of the drug in blood plasma to decrease by half
Volume of distribution at steady state (VDSS) [48]	The volume of blood needed to maintain the drug at the same concentration observed in the vascular compartment	Plasma protein binding	Plasma protein binding is the degree at which a drug binds to proteins in the blood plasma

Cell effective permeability [49]	The ability of a drug to pass through biological membranes, which is crucial for its absorption and bioavailability	Drug score [44]	A value that combines various properties (i.e., drug-likeness, cLogP, logS, molecular weight, and toxicity risks) of a compound to assess its potential to become a drug
Drug metabolism and clearance			
Microsome drug clearance [50]	The rate at which metabolizing enzymes within liver microsomes eliminate a drug from the body	Hepatic drug clearance [51]	The rate at which the liver removes the drug from the blood, expressed as the volume of blood cleared of the drug per unit of time
Biological activity and safety			
Drug sensitivity [52]	The degree at which a specific cell type (i.e., cancer or tumor cells) responds to therapeutic agents	Acute toxicity (LD50) [53]	The amount of a substance that kills half of a group of test animals within a specific time frame
Bioactivity [54]	Ability of a drug to produce specific biological effects in the body by interacting with biological targets like receptors, enzymes, or DNA	Toxicity [55]	The harmful effects a drug can cause, depending on factors like dosage, duration of use, and individual sensitivity
Pharmacodynamics [56]	Biochemical, physiologic, and molecular effects of drugs on the body (i.e., receptor sensitivity, postreceptor effects, and chemical interactions)	Blood-brain barrier penetration [57]	The estimation of a drug's ability to penetrate the protective blood-brain barrier
Blood-brain barrier permeability [57]	The estimation of a drug's ability to cross the protective blood-brain barrier	Fungal activity [58]	The effectiveness of drug in inhibiting or killing fungal pathogens, which are organisms that can cause infections in organisms
Antibacterial activity [59]	A type of antimicrobial substance active against bacteria. It is an important type of antibacterial agent for fighting bacterial infections		

and incorporate multiple objectives like drug-likeness, synthetic accessibility, and target activity to generate desired functional compounds. AI-generated molecules are then validated through virtual screening and molecular dynamics simulations which reveal synthetic molecules binding affinity against target of interest [60].

Finally, retrosynthesis analysis evaluates potential drug candidates (leads) and refined them based on their synthetic accessibility.[1] This process involves working backward from a target molecule to determine the most efficient way to synthesize it by using commercially available chemical compounds.[2] Two primary tasks dominate this process: synthetic route prediction and reaction mechanism analysis [61].

The reaction mechanism analysis involves two distinct subtasks. The first task, reaction type identification, involves classification of specific chemical transformations required at each step of a synthetic pathway using AI models [62]. The second task, reactant identification, predicts specific commercially available chemical compounds required for each reaction step in a synthetic pathway. In this context, AI models including Graph Neural Networks (GNNs) and Transformers process molecular structures to identify potential reaction centers and suggest appropriate reactants. These models learn from large databases of known reactions, such as USPTO, to make accurate predictions. The benefits of AI-driven retrosynthesis are substantial: (1) It explores a larger chemical space than manual methods, (2) it identifies nonobvious synthesis routes, (3) it accelerates the optimization process by eliminating compounds that are impractical to synthesize, and (4) this systematic evaluation helps pharmaceutical companies to direct resources on the most promising and synthetically accessible drug candidates [61, 62].

11.2.4 Preclinical Testing

The preclinical development phase is crucial as it establishes the safety and potential efficacy of a drug candidate before human trials. It involves comprehensive laboratory testing, animal studies, and data analysis to understand drug properties, toxicity, and biological effects. Artificial intelligence (AI) is transforming this phase through several key applications such as in histopathological analysis, AI models detect quantification of subtle changes in tissue samples [63], which is crucial for assessing drug efficacy and safety [64]. AI-driven analysis aids in examining cell morphology and detecting cytotoxic effects in cultured cells. Preclinical trail also aims to support the design of clinical trials.

Primarily, preclinical investigations often include rodent and non-rodent mammalian models to evaluate the pharmacokinetic profile, general safety, and toxicity patterns [65]. These models determine the drug mean residence time in the body,

[1] https://grace.com/blog/could-ai-enabled-retrosynthesis-lead-to-the-next-big-breakthrough/

[2] https://www.merckgroup.com/en/research/science-space/envisioning-tomorrow/future-of-scientific-work/synthia.html

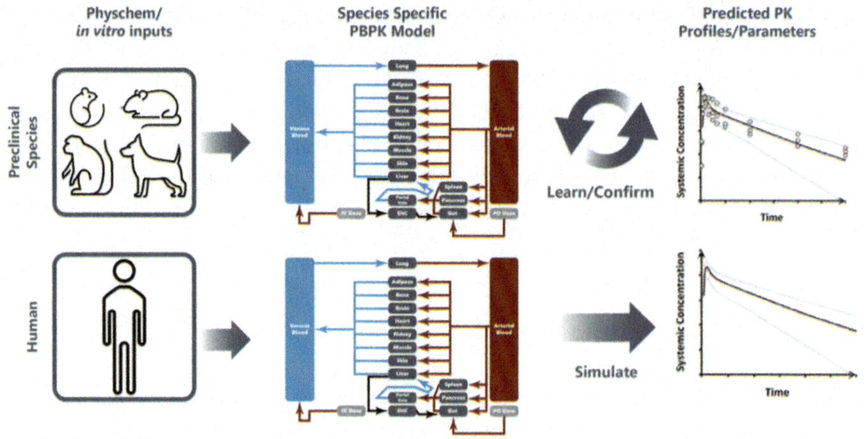

Fig. 11.4 Workflow of FIH dose estimation using AI-driven physiologically based pharmacokinetic modeling (https://www.certara.com/first-in-human-fih-dose-prediction/)

which depends on absorption, distribution, metabolism, and excretion properties. For instance, drug ability to cross the blood-brain barrier is assessed for brain-related conditions (Alzheimer's disease) [65]. Similarly, toxicology, and safety studies focus on potential adverse effects identification, identification of organ affected by toxicity, and establishment of therapeutic index (TI), which facilitates in initial clinical trial dosing, i.e., first-in-human (FIH) dose estimation [65]. AI-based prediction of first-in-human dose facilitates with more informed decision-making for clinical trial progression [66].

FIH dose prediction determines the initial safe dose of a new drug compound to be administered to humans during Phase 1 clinical trials.[3] As shown in Fig. 11.4, this critical task is performed during preclinical development to establish a dose that balances safety with potential therapeutic effect. The importance of FIH dose prediction lies in its fundamental role in ensuring safe and efficient clinical translation of new drugs. Starting with too high a dose could risk volunteer safety, while too low a dose may fail to provide meaningful data about drug effects. Accurate FIH dose prediction integrates various preclinical data sources including animal toxicology studies, pharmacokinetic data, safety margins, and biological responses to establish a scientifically justified starting dose. This prediction helps minimize risks to human volunteers while maximizing the likelihood of detecting drug effects. Advanced AI/ML approaches are increasingly being used to improve FIH predictions by better accounting for species differences and integrating complex datasets. These methods can potentially reduce reliance on animal testing while improving prediction accuracy. Accurate FIH dose prediction is also crucial for

[3] https://www.verisimlife.com/publications-blog/solving-first-in-human-dosing-challenges-with-ai

regulatory submissions and can significantly impact the efficiency of early clinical trials. Poor dose selection can lead to failed trials, safety issues, or the need for additional dose-finding studies, all of which increase development time and costs. Furthermore, well-predicted FIH doses help optimize clinical trial design by providing better estimates of the therapeutic window, potentially accelerating the drug development process. Ultimately, this critical step serves as the bridge between preclinical research and human trials, fundamentally influencing the success of clinical development.

11.2.5 Clinical Testing

Clinical testing is a four-phase process in drug discovery pipeline which serves as gold standard for ensuring new medications safety and effectiveness [67]. In Phase I, various parameters (including maximum tolerated dose, safety, and pharmacokinetics) are evaluated on a small group of 20–80 participants to understand the drug safety and tolerability [64]. Approximately 70% of drugs advance to the next phase. Phase II expands testing to several hundred patients to assess therapeutic efficacy and identify the optimal dose (approximately 33% progression rate) [64]. Phase III typically involves 1000–3000 patients and confirms drug therapeutic benefits and safety (Clinical Utility Validation) [64]. Approximately 25–33% of drugs successfully progress to the next phase. Phase IV occurs post-FDA approval, using a broad cohort of around 10,000 patients to monitor long-term safety, efficacy, and potential uses in other indications [68]. Throughout these phases, pharmacokinetic and biodistribution studies assess how human body absorbs, distributes, metabolizes, and excretes the drug. Toxicology studies play a key role in ensuring drug safety and identifying biomarkers for ongoing monitoring.

From economic perspective, Eroom's law highlights an alarming trend that drug development process costs double roughly every 9 years. By 2019, cost per approved drug surpassed 2.6 billion dollars [67], with clinical testing alone accounting for 50–70% of these expenses. This makes clinical trials a major economic bottleneck in the development process. Over the past two decades, complexity and scale of clinical trials, especially Phase III, have grown considerably. The number of procedures has increased from 86 to 167 [67], and participant requirements have risen by 50% since 2010. Phase III trials alone cost an average of 100–350 million dollars [67], and this expansion has extended timelines. It takes 6.8 years, on average, from the start of Phase I to receive regulatory approval. However, success rates remain low, with only 13.8% of drug candidates entering clinical trials receiving FDA approval.

Artificial intelligence has greatly advanced healthcare, particularly in drug development and clinical design [69]. The ability of AI applications to analyze vast datasets, recognize patterns, and generate data-driven predictions has transformed the pharmaceutical industry [69]. These abilities are essential in the complex realm of clinical research. AI is now increasingly applied in clinical trial procedures to effectively manage electronic health records [69]. Additionally, AI-generated study protocols can significantly improve the success rate of clinical trials. In the

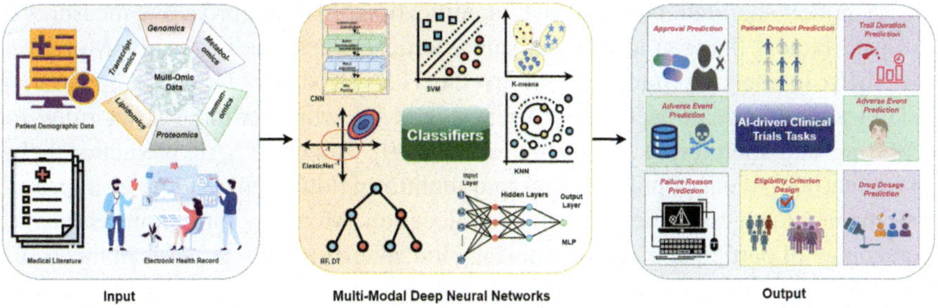

Fig. 11.5 Application of AI-driven acceleration of clinical trials through eight well-defined predictive tasks

clinical testing phase, AI applications are used to perform various tasks including (1) trial duration prediction [70], (2) patient dropout prediction [70], (3) serious adverse event prediction [70], (4) mortality rate prediction [70], (5) trial approval rate prediction [70], (6) trial failure reason identification [70], (7) eligibility criteria design [70], and (8) optimal drug dosage identification [70]. Figure 11.5 depicts an overview of all tasks involved in the AI-driven clinical testing phase.

Clinical trial duration is a critical challenge in drug development which typically spans over 7–11 years and costs billions. Primarily, trial duration [70] is the time needed for each clinical trial phase, from initial patient enrollment to data analysis and reporting. AI-driven trial duration prediction estimates this time based on a variety of features, including therapeutic area, disease complexity, eligibility criteria, patient population characteristics, and electronic health records. Clinical trial duration is closely linked to its cost, i.e., longer trial durations increase expenses due to prolonged use of personnel, facilities, and materials. AI-driven predictors' performance is further enhanced by optimizing resource allocation, ensuring precise staffing, budget planning, and clinical site operations. Accurate trial duration prediction facilitates earlier risk identification and proactive measures implementation to minimize delays. Another significant challenge is high patient dropout rates, which represents the proportion of enrolled participants withdrawing from clinical trials before completion [70]. It can undermine validity of outcomes, leading to higher costs and delays. Thus, AI technologies are incorporated to predict patient dropout events based on various features including trial duration, visit frequency, treatment complexity, and patient demographics. AI-enabled predictors for patient dropout rates demonstrate significant potential for enhancing clinical trial efficiency. Additionally, they can reduce the need for additional participant recruitment and ultimately save both time and resources.

Besides this, ensuring participant safety is the foremost priority during clinical trials, which highlights the importance of serious adverse event (SAE) and mortality rate prediction [70]. SAE prediction focuses on forecasting the probability of serious adverse effects using features including eligibility criteria, drug molecular properties, therapeutic indication, patient characteristics, and comorbidities. AI-based models facilitate potential risk identification, enabling proactive measures to mitigate factors that can affect trial safety, efficacy, and study success. Moreover, they help in meeting standards set by the U.S. Food and Drug Administration (FDA) and European Medicines Agency (EMA), which enforce strict rules for reporting and monitoring of adverse events [70]. Accurate SAE prediction helps in estimating patient population size and monitoring duration, leading to improved risk management and efficient resource allocation. Mortality rate, representing the percentage of participants who die during clinical testing [70], is a crucial metric for evaluating the safety and potential risks associated with treatments or interventions under investigation. Mortality events are particularly critical as they can lead to premature trial termination. These events significantly impact drug development timelines and financial planning while necessitating immediate measures to ensure patient safety. AI-driven predictors utilize multiple features (including drug molecular characteristics, therapeutic indication, eligibility criteria, and patient risk factors) to predict mortality likelihood [70]. Leveraging artificial intelligence to identify trials at risk of such events can greatly improve risk management and ensure efficient resource utilization. Furthermore, it ensures compliance with regulatory standards and fosters public trust and ethical practices.

Clinical trials often suffer from low approval rates that emphasize the need for AI-based predictive models to assess trial success before commencement [70]. This approach enables decision-makers to prioritize trials with higher success potential and avoid those with lower approval prospects. This strategic approach improves overall efficiency and profitability by directing resources toward the most promising projects and minimizing investments in less viable studies. Ultimately, it facilitates more informed decision-making in the clinical trial process [70]. Trial approval determines whether a drug progresses through a specific phase by analyzing various attributes including drug molecular composition, therapeutic indication, and eligibility criteria. Additionally, this approach streamlines drug development, improves patient outcomes, and enhances both regulatory compliance and investor confidence, contributing to a more efficient and impactful healthcare landscape. Trial approval outcome prediction also helps identify root causes of trial failures that often lead to significant resource wastage and delays in new treatment development. Clinical trial failures are commonly attributed to four primary factors: strategic business decisions, inadequate patient enrollment, safety concerns, and lack of efficacy [70]. While strategic business decisions involve complex market factors, other failure modes can be analyzed and potentially mitigated through predictive analytics. To address these challenges, AI-driven predictive models are employed to identify potential risks at early stages and categorize them based on specific factors including enrollment challenges, safety issues, and efficacy concerns. Accurate failure reason prediction accelerates effective treatment delivery to patients. Furthermore, well-designed trials with higher predicted success rates can foster increased stakeholder confidence and improve patient outcomes and participation in clinical research.

A cornerstone of clinical trials is the design of effective eligibility criteria, which directly influences trial outcomes and success rates [70]. Ineffective criteria, whether overly strict or too lenient, can lead to poor enrollment rates, insufficient sample sizes, and inconclusive results. Historically, these criteria have been designed based on the subjective expertise of physicians, potentially introducing bias and inconsistency. To address these challenges and improve efficiency, AI-driven applications can be employed to generate objective and evidence-based eligibility criteria [70]. Large language models and machine learning approaches can streamline patient screening, improve eligibility prediction, and accelerate recruitment by leveraging historical clinical trial data, medical literature, and specific trial parameters including therapeutic indication, trial phase, and molecular characteristics [70]. This innovative approach has the potential to reduce costs, improve clinical trial quality, and expedite the development of new therapies. Additionally, a key goal of clinical trials is to determine the appropriate drug dosage to maximize therapeutic benefits while minimizing adverse effects [70]. AI-driven predictors leverage insights from molecular structures, pharmacokinetic properties, and disease mechanisms to accurately estimate optimal drug dosages. Integration of in-depth analysis of dose-response relationships with safe and effective dosage range identification helps optimize clinical trial design. This approach streamlines the drug discovery process by maximizing therapeutic outcomes while minimizing risks. Ultimately, AI-driven approaches not only accelerate drug development but also increase the likelihood of successful, safe, and effective treatments reaching the market.

11.3 Drug Repurposing: A Strategic Approach to Drug Discovery

Drug repurposing identifies new therapeutic applications for approved drugs by leveraging their ability to interact with multiple biological targets. This approach offers significant advantages over traditional drug development. Traditional development requires approximately $2.5 billion and 10–15 years with success rates below 10%, while repurposing needs only $300 million and 6–7 years. This efficiency stems from available safety profiles and established toxicological data [71]. The approach is particularly valuable for rare disease treatment. Currently, 95% of 7,000 known rare diseases lack FDA-approved treatments. Complex mechanisms and small patient populations make conventional drug discovery ineffective for rare diseases [71]. Drug repurposing addresses this through two strategic approaches: on-target repurposing, and off-target repurposing [72]. On-target drug repurposing finds new therapeutic uses for existing drugs by utilizing their known primary mechanism of action but applying it to treat a different disease [73], for example, using a drug that inhibits a specific protein to treat another condition where the same protein is involved [40, 74]. On-target drug repurposing greatly benefits from established safety profiles of drugs. Conversely, off-target drug repurposing identifies and exploits secondary or previously unknown mechanisms of action of existing drugs, essentially using the drug's "side effects" as the primary therapeutic effect for treating a different disease [72].

11.4 Current Trends in AI-Driven Drug Discovery

The evolution of drug discovery paradigms demonstrates a systematic transition from traditional methodologies to AI-driven approaches through three fundamental technological advances: knowledge graphs, AI algorithms, and validated therapeutic applications. Modern knowledge graphs constitute the cornerstone of computational drug discovery through integration of diverse biomedical resources. PrimeKG integrates 20 biomedical databases encompassing 17,080 diseases across 4 million relationships [75]. Similarly, Scalable Precision Medicine Oriented Knowledge Engine (SPOKE) repository [76] maintains 27 million nodes through 53 million connections across 41 databases. Genetic and Rare Diseases (GARD) Knowledge Graph [77] further augments this landscape by mapping 2,000 FDA orphan designations and 180,000 drug-indication associations specifically for rare disease research.

Contemporary AI algorithms harness these knowledge graphs through Knowledge Graph Embedding Models (KGEMs) and GNNs. KGEMs transform complex biological relationships into mathematical representations while preserving hierarchical structures and interaction patterns. GNNs enhance this framework through iterative learning processes that capture molecular properties and pathway interactions across multiple biological scales. These computational approaches enable identification of nonobvious therapeutic applications through automated analysis of higher order network structures.

Recent implementations demonstrate significant therapeutic advances across multiple domains. The Every Cure LinkMap's evaluation of 3,000 drugs against 12,000 diseases has yielded novel therapeutic applications. For example, it has identified adalimumab for treating idiopathic multicentric Castleman disease [78]. AI applications in rare diseases led to the discovery of vorinostat for Rett Syndrome through transcriptomic signature analysis [79]. Network-based screening during the COVID-19 pandemic revealed plitidepsin's therapeutic potential and facilitated Paxlovid development through the mapping of 332 protein-protein interactions [80]. Deep learning approaches identified halicin as a novel broad-spectrum antibiotic and enabled evaluation of 39,000 compounds for antibiotic activity against resistant pathogens [81]. While these developments show promise, ongoing clinical validation and methodological refinement remain crucial for establishing robust therapeutic applications [82].

11.5 Artificial Intelligence Transforming Pharmaceuticals: From Lab to Life

This section explores how AI is transforming pharmaceutical industry to streamline drug development process. This study examines 32 AI pharmaceutical companies identified from three distinct review articles [83–85]. Moreover, to provide a comprehensive and up-to-date overview of industrial landscape, an additional 23 pharmaceutical companies are manually identified and included in the analysis. Following subsection provides an analytical overview of 55 AI pharmaceutical companies and 196 AI-designed drugs developed by these companies.

11.5.1 AI Pharmaceutical Companies

This section offers valuable insights into the global landscape of AI-driven innovation in drug development by providing a comprehensive overview of existing 55 AI-driven pharmaceutical companies. It explores various dimensions of companies such as the country of origin, the specific services they offer within the drug development life cycle, organizations and other pharmaceutical companies they have collaborated with, and the unique tools they have developed as illustrated in Table 11.2.

It is evident from Table 11.2 that United States clearly excels with 55 AI-driven pharmaceutical companies and also appears to be the leading country in AI pharmaceutical collaborations. These pharmaceutical companies have collaborated with over 30 different global pharmaceutical firms, academic institutions, and research organizations. The dominance of the United States in AI-driven pharmaceutical companies can be attributed to several factors such as robust environment for technology, significant capital funding, large pharmaceuticals market, leading research institutions, and pharmaceutical companies. Canada and the UK's strong showing can be linked to their excellent research institutions, supportive government policies for AI and biotech development, and strong ties to the US market. Germany's presence reflects its strong pharmaceutical industry and increasing focus on AI technologies. The relatively fewer pharmaceutical companies in other countries might be due to factors such as smaller domestic markets, less developed venture capital ecosystems, or stricter regulations around AI and drug development.

A comprehensive analysis of Table 11.2 reveals a diverse landscape of services and tools in drug discovery and development provided by AI-driven pharmaceutical. The majority of companies (55) focus on core services like target identification, candidate drug generation, and drug repurposing, with firms such as Deep Genomics, Insilico Medicine, and Cyclica leading in these areas. Many companies also offer ADMET prediction and precision medicine services. Some firms provide specialized services, like April19 Discovery's novel psychedelics development and TK Analytics' microscopy image analysis. The variety of tools and platforms developed by these companies are notable, with Insilico Medicine, PathAI, and Cyclica standing out for their multiple platforms covering various aspects of drug discovery. However, some companies like Seoul National University Hospital and Gilead Sciences focus on niche services without proprietary tools. This diversity in services and tools reflects different strategic approaches within the industry, with some companies adopting a platform-centric model, while others specialize in specific aspects of pharmaceutical R&D or operate as service providers. Overall, the AI-driven pharmaceutical landscape demonstrates a rich ecosystem of innovative

Table 11.2 List of 55 AI pharmaceutical companies

Country	Company name	Services	Collaboration	Tool/platform
Canada	BenchSci	Preclinical AI Assistance	–	ASCEND
	Cyclica12.	Drug-target interaction prediction, multi-parametric drug design, ADME/T properties prediction	–	1. Ligand express® . 2. Ligand design™ . 3. MatchMaker™ . 4. Poem™
	Deep Genomics	Novel Targets Identification, Therapeutic Drugs Identification, Drug Discovery	1. BioMarin Pharmaceutical Inc. 2. Mila—Quebec Artificial Intelligence Institute	BigRNA Platform
	Entos Pharmaceuticals	Genetic Medicines Delivery, Genetic Therapies Development	Aegis, Bill Melinda Gates Foundation, Lilly, Oisin Biotechnologies, OncoSenx	1. Fusogenix PLV delivery platform. 2. Fusogenix AIM
	Gandeeva Therapeutics	Molecular Generation, Antibodies Development, Precision Medicine, Target Identification	–	1. SPOTLIGHT. 2. HYPERFOCUS. 3. CRYO-CADD
China, United States	Insilico Medicine	Novel Targets Identification, Disease Identification, Repurpose Drugs Identification, Target Attention Prediction, Omic Data Analysis, Pathway Identification, Lead Identification and Optimization, ADME/T Properties Prediction, Relative Binding Free Energy Estimation, Kinome Activity Prediction	Taisho, Astellas, Boehringer Ingelheim, Beijing Tide Pharmaceutical, Pfizer	1. PandaOmics®2. Generative Biologics. 3. Chemistry42® 4. InClinico® 5. Science42: DORA
France	Iktos	Retrosynthesis Analysis	–	1. Makya Generative AI SaaS Platform. 2. Spaya Retrosynthesis AI

Germany	Cortex Discovery	Therapeutic effects prediction, off-target effects and pharmacological properties prediction, protein affinity prediction, pathways prediction, activation/inhibition prediction, complex biochemical interactions prediction, high-quality compounds identification	–	–
	Evotec	Early formulation, Safety Assessment, Abuse Liability, Bioanalysis, Pharmacometrics, Regulatory Support, Clinical Assessment, Cell Therapies development and clinical manufacturing, ADME PK/PD properties Assessment	Bristol Myers Squibb, Sandoz, Just—Evotec Biologics, Boehringer Ingelheim, Lilly, Pfizer, Merck, Sernova, Alpine Immune Sciences, Kazia Therapeutics, US Department of Defense	1. PanOmics. 2. Bioprocess Platform
Hungary	Turbine AI	Antibody-drug conjugate selection, patient response identification, DNA damage response inhibitor	Bayer, Accenture, AstraZeneca, Harmonic Discovery, Cancer Research Horizon, Debiopham, Ono Pharma	Simulated cell™
India	Prescience Insilico	Drug Discovery, Advanced Screening Methodologies, In Silico Drug Candidate Assessment, Molecular Docking, Molecular Dynamics, Meta-Dynamics, Synergistic Molecule Identification, Advanced Scoring Functions for Drug Discovery, Drug Repurposing	a. Swami Vivekananda University. b. Central Food Technological Research Institute (CFTRI). c. Indian Institute of Science Education and Research (IISER) Kolkata. d. Indian Institute of Technology (IIT) Madras. e. National Institute of Technology (NIT) Warangal	PRinS3 Platform

(continued)

Table 11.2 (continued)

Country	Company name	Services	Collaboration	Tool/platform
London	BenevolentAI	Target Identification, Chemical Designing	AstraZeneca, Lilly, Merck, DNDi, Cancer Research UK, University of San Diego, University of Glasgow, Sitran, Queen Mary University of London, Stanford, National Neuroscience Institute, Sano, University of Southampton	Benevolent platform
Netherlands, Switzerland	Cradle	Protein Target Identification, Advanced Antibody Discovery	–	Private
Republic of Korea	Pharos iBio	New Drugs Designing, Clinical Trials Strategies	–	Chemiverse® Platform
Slovenia	TK Analytics	Microscopy Image Analysis, High-Content Screening Data Analysis	–	CellusAI Platform
South Korea	Seoul National University Hospital	Consultation Service	–	–
Spain	Devs-Health	Anti-infective Drug Development	1. Fight Infections Foundation. 2. Foundation for Neglected Disease Research (FNDR). 3. Germans Trias i Pujol Research Institute (IGTP). 4. Aragen	Devs-Health OS
Switzerland	Hoffmann-La Roche	New Medicines Development, Diagnostic Tests Development	BioNTech, Weill Neurohub, C4 Therapeutics, CEVEC Pharmaceuticals, Dyno Therapeutics, Sarepta Therapeutics, Spark Therapeutics, PTC Therapeutics, SMA Foundation, Skyhawk Therapeutics, Atalanta Therapeutics, RC2NB, HARMAN, Genentech, Hitachi, Chugai, Inven2, Recursion	–

United Kingdom	April19 Discovery	Novel Psychedelics Development, Virtual Screening and Chemical Space Exploration, De Novo Molecular Design, Synthetic Accessibility Prediction, Molecular Target Identification, Hit and Lead Identification and Optimization	CAS (a division of the American Chemical Society)	–
	Healx	Drug-Disease Relationship Prediction, De-risked Drug Discovery	–	Healnet
	Pangea Botanica	Neurological and Neuropsychiatric Disorders Drugs Development	–	PangeAI™ Discovery Platform
United Kingdom, United States	Exscientia	Precision Target, Precision Design, Precision Experiment, Precision Medicine, Precision Data Generation	Sumitomo Dainippon Pharma Co. Ltd, Evotec, Sanofi, Bristol Myers Squibb, Merck KGaA, The University of Texas MD Anderson Cancer Center	Patient-First AI. i. Precision Target. ii. Precision Design. iii. Precision Experiment. iv. Precision Medicine
United States	A2A Pharmaceuticals	Novel Therapeutics Drugs Designing, Protein-Protein Interactions Prediction, Antibiotics Development	–	Sculpt™
	Ailynix	Desired Target Identification, Biological Activity Prediction	1. YVLabs 2. University of L'Aquila	1. GPCRMIDAS. 2. ZivCADD. 3. AIDRRP. 4. CAViDD
	Araceli Biosciences	Therapeutic Discovery	Biosero, Inc	1. Endeavor®2. EndeavorGT™. 3. Clairvoyance™
	Aria Pharmaceuticals	–	–	–

(continued)

Table 11.2 (continued)

Country	Company name	Services	Collaboration	Tool/platform
	Atomwise	Potential Drug Candidates Identification, Small Molecule Drug Repurpose Identification	–	AtomNet®
	Auransa	Drug Candidates Identification, Development of Novel Personalized Therapies for Cancer Patients	–	SMarTR ENGINE®
	Berg/BPGBio	Biomarkers Identification, Prostate Cancer Diagnostic Test, Breast Cancer Target Discovery and Diagnostic Development, Breast Cancer Identification, 34-Gene Panel Test for Breast Cancer, Parkinson's Disease Novel Biomarkers Identification	–	The NAi Interrogative Biology® Platform
	BioXcel Therapeutics	Novel Drug Candidates Identification, Repurpose Existing Drugs	1. University of North Carolina at Chapel Hill (UNC). 2. Nektar Therapeutics	NovareAI™ Platform
	Collaborations Pharmaceuticals	Toxicology Assessment, Properties Prediction	1. MegaPredict. 2. MegaSyn. 3. AssayCentral. 4. MegaTox. 5. MegaTrans	–
	DeepChem	Drug Discovery, Materials Science, Biology Applications	–	–
	Denovicon Therapeutics	Biological Pathways and Target Selection, Hits and Lead Identification and Optimization, Clinical Trials	–	–
	Exelixis	Cancer Treatment, Biotherapeutics, In-House Small Molecule Discovery, Targeted Bispecific Antibodies, Antibody-Drug Conjugates, Clinical Trials	ADAGENE, Ajinomoto, Bristol Myers Squibb, Catalent, Daiichi-Sankyo, ICONIC Therapeutics, Insilico Medicine, Invenra, IPSEN, Roche Genentech, RYVU Therapeutics, Sairopa, Takeda	–

	Frontier Medicines	Drug Discovery, Targets Identification, Patient Subtypes Identification, Biomarker Strategies Implementation, Covalent Drug-Target Binding Identification	AbbVie	1. Frontier™ Platform. 2. AutoTAC™ Platform
	Genesis Therapeutics	Drug Candidates Identification	–	GEMS
	Gilead Sciences	Viral Diseases Therapies	–	–
	Insitro	Decodes Human Biology, Multimodal Phenotypic Cellular Data Generation, Causal Intervention Points Identification	Lilly, Bristol Myers Squibb, Biobank Genomics England	–
	Nimbus Therapeutics	Drug Discovery, Target Selection, Molecular Docking	Gilead, Celgene, Lilly, Takeda, Access Biotechnology, Atlas Venture, Commodore, GV, Lightstone Venture, Bain Capital Life Science, Logos Capital, Pfizer, RA Capital, SR One, Surveyor, SV Health Investor	–
	PathAI	Clinical Trials	1. Roche Tissue Diagnostics (RTD). 2. Bristol Myers Squibb. 3. Cleveland Clinic 4. Crohn's Colitis Foundation.	1. AISight™. 2. PathExplore™. 3. PathExplore™ IOP. 4. IHC Explore™. 5. PathExplore™ Fibrosis. 6. IBD Explore™. 7. Liver Explore™
	Pathos	Disease Identification, Biomarker-Driven Development Paths Identification	1. Prelude Therapeutics. 2. Novo Nordisk	PathOS™ Platform
	Recursion Pharmaceuticals	Target Identification, Phenotypic Screening, Biomarker Discovery, Drug Candidate Identification	Roche and Genentech, Bayer Pharmaceuticals, NVIDIA, Tempus, Helix, Enamine	Recursion Operating System

(continued)

Table 11.2 (continued)

Country	Company name	Services	Collaboration	Tool/platform
	Relay Therapeutics	Undruggable Protein Target Identification	Pfizer to Evaluate Atirmociclib in Combination with RLY-	Dynamo Platform
	Schrödinger Inc	Small Molecule Discovery, Structure Prediction, Hit and Lead Identification and Optimization, Antibody Designing	Ajax Therapeutics, Bright Angel Therapeutics, Bristol Myers Squibb, Gilead, Lilly, Nimbus Therapeutics, Novartis, Otsuka, Structure Therapeutics, Takeda	–
	Seismic Therapeutic	Biologics Developability, T Cell Invisibility, B Cell Invisibility, Functional Activity	1. Amgen Ventures. 2. Bessemer Venture Partners	IMPACT Platform
	Strateos	Drug Discovery, Antibody Discovery, Small Molecule Drug Discovery	–	Strateos™ Platform
	Structure Therapeutics	Small Molecule Drugs Design	–	–
	Valo Health	Target Discovery, Novel Human Disease-Associated Targets and Biomarkers, Molecule Designing, Clinical Development	–	Opal Platform
	Vevo Therapeutics	Novel Targets and Drug Candidates Identification	–	Mosaic
United States, China	XtalPi	Drug Discovery	1. Collaboration with Lilly, Using AI + Robotics to Uncover First-in-Class Therapeutics. 2. Research Collaboration with Janssen. 3. XtalPi and Signet Expand AI Drug Discovery Collaboration to Novel Cancer Target	1. ID4Inno™. 2. XupremAb™

approaches to drug discovery and development, leveraging artificial intelligence to address various challenges in the field.

In a nutshell, AI pharmaceutical companies are transforming drug discovery and development through a variety of tools and collaborative efforts. Leading pharmaceutical companies are particularly skilled at creating multi-platform solutions, whereas others focus on niche services.

11.5.2 AI-Designed Drugs

This section provides a holistic view of AI-designed drugs focusing on two core aspects: (1) Which therapeutic areas are being addressed by AI-driven drugs? and (2) where do AI-designed drugs stand in their development journey?

(1) Which therapeutic areas are being addressed by AI-driven drugs?

The selection of therapeutic domains for AI-driven drug discovery focuses on operational efficiency and risk mitigation. Research initiatives predominantly concentrate on well-established therapeutic areas with robust, high-quality datasets. Pharmaceutical start-ups initially prioritize therapeutic areas with substantial market potential due to resource constraints. These enterprises systematically diversify their therapeutic portfolio as they mature and secure larger funding. This strategic evolution enables them to address critical unmet clinical needs while maintaining balance between commercial feasibility and scientific advancement. This study has identified approximately 196 AI-designed drugs targeting almost 14 therapeutic areas including virology, cardiovascular, dermatology, gastroenterology, hematology, immunology, musculoskeletal, nephrology, neurology, oncology, ophthalmology, pulmonology, systemic, and others that are briefly described in Table 11.3.

Figure 11.6 depicts the distribution of 196 AI-designed drugs across 14 therapeutic areas. Oncology emerges as the predominant focus with 92 drugs in development or clinical trials, followed by neurology (21 drugs) and virology (16 drugs). Pulmonary and immunology fields also demonstrate substantial research activity.

A leading therapeutic area Oncology's prominence in AI-driven drug discovery stems from cancer's inherent complexity and molecular heterogeneity across patients, tumor types, and intratumoral regions. AI approaches enable identification of specific genetic markers and facilitate personalized treatment strategies. Neurology represents another key therapeutic domain where AI analyzes brain imaging, genetic data, and patients' histories to advance treatments for disorders like Alzheimer's, Parkinson's, and epilepsy. Furthermore, the infectious disease sector shows significant AI integration, targeting conditions such as COVID-19, malaria, tuberculosis, and HIV. This widespread adoption of AI across therapeutic areas signals its transformative role in modern drug discovery.

(2) Where do AI-designed drugs stand in their development journey?

This section highlights progress of AI-designed drugs across drug development pipeline. Primarily, this pipeline comprises five different phases including target identification and validation, hit-to-lead identification, lead optimization, preclinical

Table 11.3 Therapeutic areas and their specializations

Therapeutic area	Specialization	Therapeutic area	Specialization
Virology	Treatment and prevention of infections caused by bacteria, viruses, fungi, and parasites	Cardiovascular	Heart and blood vessel diseases, including hypertension and coronary artery disease
Dermatology	Diagnosis and treatment of conditions related to skin, hair, and nails	Gastroenterology	Disorders of the digestive system
Hematology	Blood-related health issues such as anemia, clotting disorders, and leukemia	Immunology	Immune system-related conditions like autoimmunity, allergies, and immunodeficiencies
Musculoskeletal	Conditions affecting the musculoskeletal system, including arthritis and osteoporosis	Nephrology	Study and treatment of kidney function and diseases
Neurology	Disorders of the nervous system, including conditions affecting the brain, spinal cord, and nerves	Oncology	Prevention, diagnosis, and treatment of cancers and tumors
Ophthalmology	Eye health issues, including vision disorders, cataracts, and glaucoma	Pulmonology	Lung- and respiration-related conditions such as asthma, COPD, and pneumonia
Systemic	Diseases affecting multiple organs or the entire body, such as lupus or systemic infections	Others	Therapeutic areas not classified under specific categories, including rare or niche fields

testing, and clinical testing. Figure 11.7 highlights the progress of 196 AI-designed drugs across five different phases of drug development pipeline.

It is evident in Fig. 11.7 that early discovery phases show substantial AI implementation, with 28 drugs in hit-to-lead identification followed by 14 in lead optimization and 2 in target identification and validation. The preclinical phase demonstrates the highest activity with 43 drugs, reflecting intensive AI application in safety and efficacy evaluations. Clinical trials comprise 58 drugs distributed across various phases: 32 in Phase 1, 5 in Phase 1/2, 17 in Phase 2, 1 in Phase 2/3, and 2 in Phase 3, with 1 drug spanning Phases 1/2/3. This significant presence in clinical trials underscores AI's potential to optimize trial design, patient stratification, and outcome prediction. Recent analyses [86–88] indicate that AI integration can reduce R&D time and costs by 25–50%. The economic impact is substantial, with AI applications in clinical trials projected to generate $13–25 billion in value, while overall generative AI implementation across drug development could yield $15–28 billion annually [89].

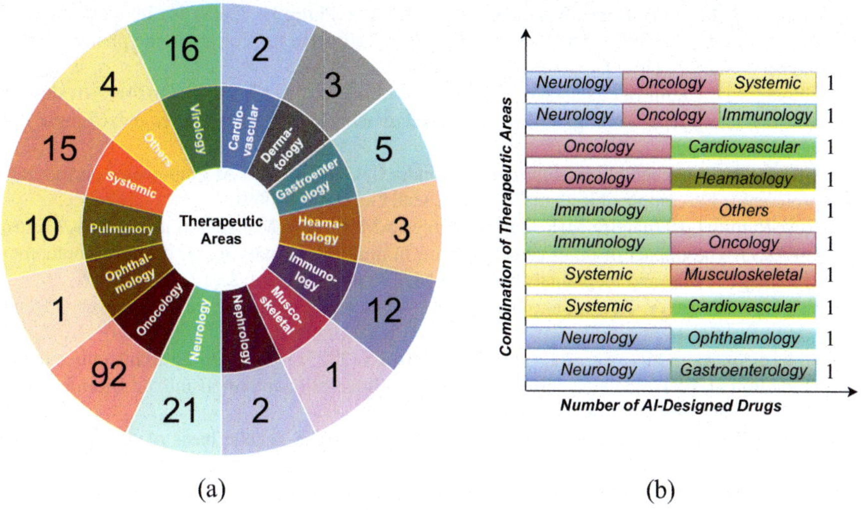

Fig. 11.6 Number of AI-designed drugs for different therapeutic areas. (**a**) Distribution of AI-designed drugs for individual therapeutic area. (**b**) Distribution of AI-designed drugs for two or more therapeutic area

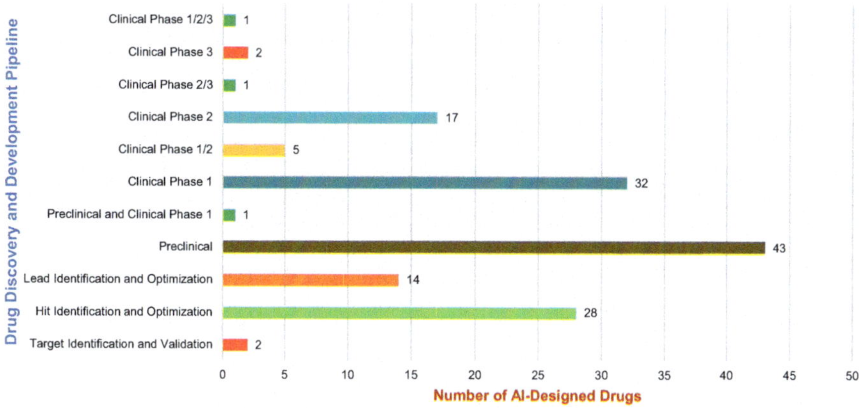

Fig. 11.7 Number of AI-designed drugs at different phases of drug discovery and development life cycle

References

1. Buckley LA , Bebenek I, Cornwell PD , Hodowanec A, Jensen EC, Murphy C, Ghantous HN (2020) Drug development 101: a primer. Int J Toxicol 39(5):379–396
2. Vivek S (2023) The next generation of evidence-based medicine. Nat Med 29(1):49–58
3. Duxin S, Wei G, Hu H, Simon Z (2022) Why 90% of clinical drug development fails and how to improve it? Acta Pharmaceut Sinica B 12(7):3049–3062

4. Francois P, Mario B, Harrie B, Andreas C, Gordon D, Philip H, Tomas M, Teija O, Adrian R, Thomas S-H et al (2023) The evolving role of investigative toxicology in the pharmaceutical industry. Nat Rev Drug Discovery 22(4):317–335
5. Gholap AD, Uddin MJ , Faiyazuddin M, Omri A, Gowri S, Khalid M (2024) Advances in artificial intelligence in drug delivery and development: a comprehensive review. Comput Biol Med178:108702
6. Qian L, Sun R, Aebersold R, Bühlmann P, Sander C, Guo T (2024) AI-empowered perturbation proteomics for complex biological systems. Cell Genom 4:100691
7. Ruiz-Ramos J, Hernández MH, Juanes-Borrego AM, Milà R, Mangues-Bafalluy MA, Mestres C (2021) The impact of pharmaceutical care in multidisciplinary teams on health outcomes: systematic review and meta-analysis. J Am Med Directors Assoc 22(12):2518–2526
8. Pun FW, Ozerov IV, Zhavoronkov A (2023) Ai-powered therapeutic target discovery. Trends Pharmacol Sci 44:561–572
9. Abbasi AF, Asim MN , Trygg J, Dengel A, Ahmed S (2023) Deep learning architectures for the prediction of YY1-mediated chromatin loops. In: International Symposium on Bioinformatics Research and Applications. Springer, pp 72–84
10. Jackson M, Marks L, May GHW, Wilson JB (2018) The genetic basis of disease. Essays Biochem 62(5):643–723
11. Joanna O (2007) Determining druggability. Nat Rev Drug Discovery 6(3):187–187
12. Lin Q, Tam PK-H, Tang CS-M (2023) Artificial intelligence-based approaches for the detection and prioritization of genomic mutations in congenital surgical diseases. Front Pediatr 11:1203289
13. Lei C, Wu Y, Jingyang G (2019) DeepSV: accurate calling of genomic deletions from high-throughput sequencing data using deep convolutional neural network. BMC Bioinf 20:1–17
14. Wu Y, Xie L (2024) Ai-driven multi-omics integration for multi-scale predictive modeling of causal genotype-environment-phenotype relationships. arXiv preprint arXiv:2407.06405
15. Mamoshina P, Volosnikova M, Ozerov IV , Putin E, Skibina E, Cortese F, Zhavoronkov A (2018) Machine learning on human muscle transcriptomic data for biomarker discovery and tissue-specific drug target identification. Front Genet 9:242
16. DePristo M, Poplin R (2017) DeepVariant: highly accurate genomes with deep neural networks. Google AI Blog
17. Metallo CM, Vander Heiden MG (2013) Understanding metabolic regulation and its influence on cell physiology. Mol Cell 49(3):388–398
18. Asim MN, Ibrahim MA, Malik MI, Dengel A, Ahmed S (2020) Enhancer-DSNet: a supervisedly prepared enriched sequence representation for the identification of enhancers and their strength. In: Neural Information Processing: 27th International Conference, ICONIP 2020, Bangkok, Thailand, November 23–27, 2020, Proceedings, Part III 27. Springer, pp 38–48
19. Asim MN, Ibrahim MA, Zehe C, Cloarec O, Sjogren R, Trygg J, Dengel A, Ahmed S (2021) L2S-MirLoc: a lightweight two stage MiRNA sub-cellular localization prediction framework. In: 2021 International Joint Conference on Neural Networks (IJCNN). IEEE, pp 1–8
20. Asim MN, Ibrahim MA, Imran Malik M, Dengel A, Ahmed S (2022) Circ-LocNet: a computational framework for circular RNA sub-cellular localization prediction. Int J Mol Sci 23(15):8221
21. Asim MN, Ibrahim MA, Imran Malik M, Dengel A, Ahmed S (2021) Advances in computational methodologies for classification and sub-cellular locality prediction of non-coding RNAs. Int J Mol Sci 22(16):8719
22. Asim MN, Ibrahim MA, Malik MI, Dengel A, Ahmed S (2022) ADH-PPI: an attention-based deep hybrid model for protein-protein interaction prediction. iScience 25(10):105169
23. Asim MN, Ibrahim MA, Malik MI, Razzak I, Dengel A, Ahmed S (2023) Histone-net: a multi-paradigm computational framework for histone occupancy and modification prediction. Compl Intell Syst 9(1):399–419
24. Stricker M, Asim MN, Dengel A, Ahmed S (2022) CircNet: an encoder–decoder-based convolution neural network (CNN) for circular RNA identification. Neural Comput Appl34:1–12

25. Asim MN, Ibrahim MA, Malik MI, Zehe C, Cloarec O, Trygg J, Dengel A, Ahmed S (2022) EL-RMLocNet: an explainable LSTM network for RNA-associated multi-compartment localization prediction. Comput Struct Biotechnol J 20:3986–4002
26. Sudharshan P (2020) Ai-driven metabolomics: Uncovering metabolic pathways and biomarkers for disease diagnosis and treatment. Distribut Learn Broad Appl Sci Res 6:354–391
27. Asim MN , Ibrahim MA , Zehe C, Trygg J, Dengel A, Ahmed S (2022) BoT-Net: a lightweight bag of tricks-based neural network for efficient lncRNA–miRNA interaction prediction. Interdiscipl Sci Comput Life Sci 14(4):841–862
28. Asim MN, Ibrahim MA, Malik MI, Dengel A, Ahmed S (2021) ChrSLoc-Net: machine learning-based prediction of channelrhodopsins proteins within plasma membrane. In: 2021 IEEE EMBS International Conference on Biomedical and Health Informatics (BHI). IEEE, pp 1–4
29. Jinghan Y, Zhiqiang G, Xiuhan R, Jie S, Xu P, Cheng C, Fu Y (2021) DeepDigest: prediction of protein proteolytic digestion with deep learning. Analy Chem 93(15):6094–6103
30. Jumper J, Evans R, Pritzel A, Green T, Figurnov M, Ronneberger O, Tunyasuvunakool K, Bates R, Žídek A, Potapenko A et al (2021) Highly accurate protein structure prediction with AlphaFold. Nature 596(7873):583–589
31. Baek M, DiMaio F, Anishchenko I, Dauparas J, Ovchinnikov S, Lee GR , Wang J, Cong Q, Kinch LN, Schaeffer RD et al (2021) Accurate prediction of protein structures and interactions using a three-track neural network. Science 373(6557):871–876
32. Gustaf A, Nazim B, Christina F, Sachin K, Qinghui X, William G, Timothy JO et al (2024) OpenFold: Retraining AlphaFold2 yields new insights into its learning mechanisms and capacity for generalization. Nature Methods 21(8):1514–1524.
33. Zeming L, Halil A, Roshan R, Brian H, Zhongkai Z, Lu W, Nikita S, Robert V, Ori K, Yaniv S et al (2023) Evolutionary-scale prediction of atomic-level protein structure with a language model. Science 379(6637):1123–1130
34. Wu R, Ding F, Wang R, Shen R, Zhang X, Luo S, Su C, Wu Z, Xie Q, Berger B et al (2022) High-resolution de novo structure prediction from primary sequence. BioRxiv, 2022–07
35. Vittorio S, Lunghini F, Morerio P, Gadioli D, Orlandini S, Silva P, Martinovic J, Pedretti A, Bonanni D, Bue AD et al (2024) Addressing docking pose selection with structure-based deep learning: recent advances, challenges and opportunities. Comput Struct Biotechnol J 23:2141–2151
36. Lima AG, Penteado AB, de Jesus JG, de Paula VJR, Ferraz WR, Trossini GHG (2024) Structure-based virtual screening: successes and pitfalls. J Brazil Chem Soc 35(10):e–20240112
37. Asim MN, Fazeel A, Ibrahim MA , Dengel A, Ahmed S (2022) MP-VHPPI: meta predictor for viral host protein-protein interaction prediction in multiple hosts and viruses. Front Med 9:1025887
38. Gentile F, Yaacoub JC, Gleave J, Fernandez M, Ton A-T, Ban F, Stern A, Cherkasov A (2022) Artificial intelligence–enabled virtual screening of ultra-large chemical libraries with deep docking. Nat Protocols 17(3):672–697
39. Wu H, Liu J, Zhang R, Lu Y, Cui G, Cui Z, Ding Y (2024) A review of deep learning methods for ligand based drug virtual screening. Fundamental Res 4:715–737
40. Asim MN, Ibrahim MA, Malik MI, Dengel A, Ahmed S (2022) LGCA-VHPPI: a local-global residue context aware viral-host protein-protein interaction predictor. Plos one 17(7):e0270275
41. Francoeur PG, Koes DR (2021) SolTranNet–a machine learning tool for fast aqueous solubility prediction. J Chem Inf Model 61(6):2530–2536
42. Ahuja S, Dong M (2005) Handbook of Pharmaceutical Analysis by HPLC. Elsevier, Amsterdam
43. Zang Q, Mansouri K, Williams AJ, Judson RS, Allen DG, Casey WM, Kleinstreuer NC (2017) In silico prediction of physicochemical properties of environmental chemicals using molecular fingerprints and machine learning. J Chem Inf Model 57(1):36–49
44. Shefin B, Sreekumar S, Biju CK (2022) In silico evaluation of anti-SARS-CoV-2 activity of *Punica granatum* L. phytochemicals. In: Coronavirus Drug Discovery. Elsevier, Amsterdam, pp 335–354

45. Chandrasekaran B, Abed SN, Al-Attraqchi O, Kuche K, Tekade RK (2018) Computer-aided prediction of pharmacokinetic (ADMET) properties. In: Dosage form Design Parameters. Elsevier, Amsterdam, pp 731–755
46. Wenyu Z, Yanxing W, Yan N, Liangren Z, Zhenming L (2023) Current trends and challenges in drug-likeness prediction: are they generalizable and interpretable? Health Data Sci 3:0098
47. Smith DA , Beaumont K, Maurer TS, Di L (2017) Relevance of half-life in drug design: miniperspective. J Med Chem 61(10):4273–4282
48. Yates JWT, Arundel PA (2008) On the volume of distribution at steady state and its relationship with two-compartmental models. J Pharmaceut Sci 97(1):111–122
49. Lennernäs H, Crison JR, Amidon GL (1995) Permeability and clearance views of drug absorption: a commentary. J Pharmacokinet Biopharmaceut 23:333–337
50. Hosey CM, Benet LZ (2017) Experimental ADME and toxicology. In: Chackalamannil S, Rotella D, Ward S (edn) Comprehensive Medicinal Chemistry III, 3rd edn. Elsevier, Amsterdam
51. Taft DR (2009) Drug excretion. In: Pharmacology. Elsevier, Amsterdam, pp 175–199
52. Yi-Ching T, Assaf G (2021) Explainable drug sensitivity prediction through cancer pathway enrichment. Sci Rep 11(1):3128
53. Morris-Schaffer K, McCoy MJ (2020) A review of the ld50 and its current role in hazard communication. ACS Chem Health Safety 28(1):25–33
54. Wallach I, Dzamba M, Heifets A (2015) AtomNet: a deep convolutional neural network for bioactivity prediction in structure-based drug discovery. arXiv preprint arXiv:1510.02855
55. Andreas M, Günter K, Thomas U, Sepp H (2016) DeepTox: toxicity prediction using deep learning. Front Environ Sci 3:80
56. Mark M, Zohaib J, Patrick MZ, (2023) Pharmacodynamics. In: StatPearls [Internet]. Treasure Island (FL): StatPearls Publishing; 2025 Jan. PMID: 29939568
57. Shaker B, Yu M-S, Song JS, Ahn S, Ryu JY, Oh K-S, Na D (2021) LightBBB: computational prediction model of blood–brain-barrier penetration based on lightGBM. Bioinformatics 37(8):1135–1139
58. Zonglin Y, Yaqi S, Pinglin L, Kanghong P, Guoqiang L, Xianguo L, Shuo Y, Dahai Z (2022) Application of principal component analysis (PCA) to the evaluation and screening of multiactivity fungi. J Ocean Univer China 21(3):763–772
59. Stokes JM , Yang K, Swanson K, Jin W, Cubillos-Ruiz A, Donghia NM , MacNair CR, French S, Carfrae LA, Bloom-Ackermann Z et al (2020) A deep learning approach to antibiotic discovery. Cell 180(4):688–702
60. Rajkumar C, Yasha H (2023) Utilizing deep learning to explore chemical space for drug lead optimization. Expert Syst Appl 229:120592
61. Yu W, Chao P, Yuzhe W, Junru J, Jingjie Z, Xiangxiang Z, Su R, Quan Z, Leyi W (2023) Retrosynthesis prediction with an interpretable deep-learning framework based on molecular assembly tasks. Nat Commun 14(1):6155
62. Yinjie J, Yu Y, Ming Yu K, Mei Yuan L, Zhengxing H, Kun K, Zhihua W, Huaxiu Y, James Z et al (2023) Artificial intelligence for retrosynthesis prediction. Engineering 25:32–50
63. Abbasi AF, Asim MN, Ahmed S, Dengel A (2024) Long extrachromosomal circular DNA identification by fusing sequence-derived features of physicochemical properties and nucleotide distribution patterns. Sci Rep 14(1):9466
64. Shegokar R (2020) Preclinical testing—understanding the basics first. In: Drug Delivery Aspects. Elsevier, Amsterdam, pp 19–32
65. Steinmetz KL, Spack EG (2009) The basics of preclinical drug development for neurodegenerative disease indications. BMC Neurol 9(Suppl 1):S2
66. Vijayan R, Kihlberg J, Cross JB, Poongavanam V (2022) Enhancing preclinical drug discovery with artificial intelligence. Drug Discovery Today 27(4):967–984
67. Matthew H (2024) How AI is being used to accelerate clinical trials. Nature 627(8003):S2–S5
68. Viraj S (2010) Phase iv of drug development. Perspect Clin Res 1(2):57–60
69. Mourya A, Jobanputra B, Pai R (2024) Ai-powered clinical trials and the imperative for regulatory transparency and accountability. Health Technol 14:1–11

70. Chen J, Hu Y, Wang Y, Lu Y, Cao X, Lin M, Xu H, Wu J, Xiao C, Sun J et al (2024) TrialBench: multi-modal artificial intelligence-ready clinical trial datasets. arXiv preprint arXiv:2407.00631
71. Natalia P (2024) Ai-driven drug repurposing for novel therapeutic applications: utilizes AI algorithms to identify existing drugs with potential therapeutic applications in new disease areas. J Artif Intell Res Appl 4(1):182–190
72. Porkodi A, Umamaheswari S (2022) Repurposing-second life for drugs. Pharmacia 69(1):51–59
73. Asim MN , Malik MI , Zehe C, Trygg J, Dengel A, Ahmed S (2020) A robust and precise ConvNet for small non-coding RNA classification (RPC-snRC). IEEE Access 9:19379–19390
74. Asim MN, Ibrahim MA, Fazeel A, Dengel A, Ahmed S (2023) DNA-MP: a generalized DNA modifications predictor for multiple species based on powerful sequence encoding method. Brief Bioinf 24(1):bbac546
75. Chandak P, Huang K, Zitnik M (2023) Building a knowledge graph to enable precision medicine. Sci Data 10:67
76. Morris JH, Soman K, Akbas RE, Zhou X, Smith B, Meng EC, Huang CC, Cerono G, Schenk G, Rizk-Jackson A, Harroud A, Sanders L, Costes SV, Bharat K, Chakraborty A, Pico AR, Mardirossian T, Keiser M, Tang A, Hardi J, Shi Y, Musen M, Israni S, Huang S, Rose PW, Nelson CA, Baranzini SE (2023) The scalable precision medicine open knowledge engine (spoke): a massive knowledge graph of biomedical information. Bioinformatics 39:btad080
77. Qian Z et al (2020) An integrative knowledge graph for rare diseases, derived from the genetic and rare diseases information center (gard). J Biomed Semant 11(1):13
78. Sperry M, Ingber DE (2024) Drug repurposing strategies, challenges and successes. Technology Networks
79. Novak R, Lin T, Kaushal S, Sperry M, Vigneault F, Gardner E, Loomba S, Shcherbina K, KeshariV, Dinis A, Vasan A, Chandrasekhar V, Takeda T, Turner JR, Levin M, Ingber DE (2022) Target-agnostic discovery of Rett syndrome therapeutics by coupling computational network analysis and CRISPR-enabled in vivo disease modeling. bioRxiv
80. GordonDE, Jang GM , Bouhaddou M, et al (2020) A sars-cov-2 protein interaction map reveals targets for drug repurposing. Nature 583:459–468
81. Stokes JM, Kevin Y, Kyle S, Wengong J, Andres C-R, Donghia NM, MacNair CR, French S, Carfrae LA, Bloom-Ackermann Z, Tran VM, Chiappino-Pepe A, Badran AH, Andrews IW, Chory EJ, Church GM, Brown ED, Jaakkola TS, Regina B, Collins JJ (2020) A deep learning approach to antibiotic discovery. Cell 180(4):688–702.e13
82. Wong F, Zheng EJ, Valeri JA et al (2024) Discovery of a structural class of antibiotics with explainable deep learning. Nature 626:177–185
83. Hasselgren C, Oprea TI (2024) Artificial intelligence for drug discovery: are we there yet? Ann Rev Pharmacol Toxicol 64(1):527–550
84. Chakraborty C, Bhattacharya M, Lee S-S, Wen Z-H, Lo Y-H (2024) The changing scenario of drug discovery using AI to deep learning: recent advancement, success stories, collaborations, and challenges. Mol Therapy Nucleic Acids 35(3):102295
85. Mak K-K, Wong Y-H, Pichika MR (2024) Artificial intelligence in drug discovery and development. Drug Discov Evaluat Safety Pharmacokinet Assays 26:1461–1498
86. Mariam Z, Niazi SK, Magoola M (2024) Unlocking the future of drug development: generative AI, digital twins, and beyond. BioMedInformatics 4(2):1441–1456
87. Adriano FC (2024) Impact of artificial intelligence in the pharmaceutical business: a bibliometric analysis. Master's Thesis

88. Bharadwaj S, Deepika K, Kumar A, Jaiswal S, Miglani S, Singh D, Fartyal P, Kumar R, Singh S, Singh MP et al (2024) Exploring the artificial intelligence and its impact in pharmaceutical sciences: insights toward the horizons where technology meets tradition. Chem Biol Drug Design 104(4):e14639
89. Proudman D, Desai U, Gomes DA , Khomenko M, Martin S, Nellesen D, Neumann U, Grabowski H (2024) Public sector replacement of privately funded pharmaceutical R&D: cost and efficiency considerations. J Med Econ 27(1):1253–1266

Why AI Progress Will Necessitate Harnessing Synthetic Biology to Leverage the Ground Truth of Intelligence 12

Brett J. Kagan and Andy C. Kitchen

Contents

Abstract

Interest and investment in AI research is at a historical peak. Yet, despite the performance obtained, current artificial intelligence systems have adopted a narrow approach with distinct limitations. One source of these limitations is the reliance on extrinsic learning mechanisms, with gradient descent and its variants serving as the primary means of model optimization. However, intelligent systems arising from biological brains display intrinsic learning through integrated processes. Since biological brains operate with power and data efficiencies

B. J. Kagan (✉)
Cortical Labs, Melbourne, VIC, Australia

Department of Biochemistry and Pharmacology, University of Melbourne, Parkville, VIC, Australia
e-mail: brett@corticallabs.com

A. C. Kitchen
Cortical Labs, Melbourne, VIC, Australia

U. A. K. Betz (ed.), *Science for a Better Tomorrow*,
https://doi.org/10.1007/978-3-031-93623-4_12

otherwise unparalleled, along with the capability for rapid real-time adaption, it may be beneficial to reconsider neural systems in the current search for AI. There are two mutually inclusive possibilities this may take: (1) using biological cells as the substrate for information processing and intelligence; (2) developing new AI algorithms inspired by these biological systems. In either of these cases, exploiting synthetic biological methods to explore the underlying mechanisms of how neural systems process information and exhibit intelligence will be necessary. By reviewing a recent proof-of-concept for Synthetic Biological Intelligence and the corresponding implications in this context, potential methods for both these possibilities are discussed. Ultimately, this work aims to encourage drawing inspiration from natural intelligence to foster more robust, efficient, and adaptive AI.

Keywords

Artificial intelligence · Learning · Synthetic biology · Intelligence · Neuroscience · Criticality · Functional connectivity · Neural cells

12.1 A Search for Intelligence

The idea of harnessing non-human generalized intelligence for a variety of purposes has fascinated humankind throughout history. For thousands of years before the idea of silicon computers were even theorized, stories of golems, jinn, spirits, and a plethora of other divine or magical beings pervaded almost every human culture. While each had distinct properties, in one way or another, all displayed a form of intelligence and most did work for or with humans—although not always with the intended consequences. In modern times, however, this focus for an external arbiter of non-human work has shifted from ideas now considered mythology toward one once held to be science fiction: that of artificial intelligence (AI). In most ideations, AI is defined as the maximal capacity of an artificial system to successfully achieve a novel goal through computational algorithms [1]. Yet, AI is often postulated as merely an intermediary step to Artificial Generalized Intelligence (AGI). The distinction is that AGI includes an AI with the ability to perceive, possess knowledge, understand, learn, and function either partially or completely autonomously in a variety of environments and tasks [1]. Work exploring AI has rapidly accelerated, with huge commercial, scientific, and public interest in the tools being generated [2]. Yet, contrary to the aforementioned beings spoken of by our ancestors, AI cannot intrinsically learn in dynamic environments or accurately model the world outside an established dataset. Instead, the field of AI has been dominated by extrinsic learning approaches, where models are trained using optimization algorithms like gradient descent. This method, while effective for specific tasks, fundamentally separates the process of learning from the behavior of the model itself. Learning is treated as a discrete, externally imposed phase, distinct from the model's operation once training is complete. This artificial separation creates significant limitations, particularly in terms of adaptability, efficiency (both power and data), and the alignment of AI systems with natural intelligence—potentially preventing the achievement of AGI with these methods. This chapter aims to firstly discuss the limitations of existing AI research approaches and key outcomes that may arise. These conclusions will support an argument for why exploring biological systems as a ground truth for generalized intelligence is a sensible direction. Secondly, why synthetic biological methods will be important to understand how neural systems are able to display their currently unique information and intelligence processing ability will be discussed. The importance of moving beyond relatively simple pairwise models of neural systems will be highlighted here. To emphasize the potential of this approach, a proof of concept for how biological brain cells in a dish may be leveraged as information processing and intelligence functions will be presented. Finally, a proposal for how concepts learned from these biological systems can be adapted to improve silicon-based AI research will be presented.

12.1.1 Limitations of Existing AI Approaches to Achieve AGI

Machine learning methods such as reinforcement learning (RL) can produce powerful results but have generally been recognized to have several limitations. RL algorithms require enormous data to converge on ideal solutions [3], remain highly fragile and prone to failures when minor changes are introduced [4], and require considerable power [5]. Recent breakthroughs in generative AI models, including generative pre-trained transformer (GPT) based large language models (LLMs) for text and diffusion models for images have addressed some of these concerns, while amplifying others [6]. Tools such as LLMs have demonstrated significant advancements in the realm of AI, particularly with their ability to perform "in-context learning" [7]. This form of learning allows models to adapt their responses based on the context provided in a conversation or task, simulating a level of real-time adjustment and flexibility. These improved abilities have already been identified as likely to have a significant impact in the world economy [8]. Yet, while in-context learning represents a shift away from the rigid, pre-trained behaviors of earlier models, it still falls short of true intrinsic learning. The underlying mechanics of LLMs remain fundamentally symbolic and static, limiting their adaptability and preventing genuine integration of learning into their operational dynamics. The models are symbolic as they do not internalize new knowledge or adjust its dynamics based on new inputs; rather, it generates responses by leveraging pre-trained patterns and statistical correlations learned during massive training phases. The system is statics as it does not truly adapt to novel information in real time but instead performs a type of pattern-matching within a fixed knowledge base. This token-based, non-integrated approach means that while LLMs can produce responses that seem contextually relevant, they are bound by the constraints of their pre-trained state and lack the ability to evolve with experience. As such, these AI tools most probably do not possess world models, even though in-context learning functions still occurring through next-token-prediction based approaches [6]. The token prediction approach is a primary limitation of this method due to a rigid reliance on pre-defined structures. In-context learning simulates flexibility by shifting focus based on input, but the system's behavior is still driven by static representations learned during the original training. The model operates within the bounds of its existing knowledge, drawing on token-level correlations rather than developing a deeper understanding or adapting its internal architecture to new information. This creates a symbolic façade of learning without the underlying dynamism that characterizes intrinsic learning processes in natural systems. Therefore, at its core, in-context learning in LLMs is not "learning" in the biological sense.

Moreover, in-context learning lacks the continuous, feedback-driven integration seen in natural intelligence. In humans, learning is a constant process intertwined with cognition, where feedback from actions and environmental interactions directly influences ensuing behavior. LLMs, on the other hand, operate in a distinct separation between training and inference. The model can respond contextually, but once the interaction ends, no information is retained or incorporated into its future decision-making processes. This disconnect between inference and learning fundamentally limits the capacity of LLMs to develop true intelligence. As such, current AI tools have not achieved AGI and despite numerous use cases, retain key limitations [9]. Likewise, state-of-the-art LLMs require trillions of tokens and massive computational resources to achieve their levels of performance on benchmark tasks [10], in comparison to the adaptive intelligence exhibited by humans, who learn from far fewer experiences. The over-reliance on brute force scaling has led to a fixation on larger datasets and more powerful hardware, rather than a reconsideration of the foundational principles of learning itself. In short, this means that extremely high dimensional models can predict what patterns of text should be generated following a prompt with sometimes remarkable results. Yet these models are currently not able to understand and adjust their own behavior according to any given fundamental principle that may underpin either the original provided context or a dynamic world model.

12.1.2 Two Outcomes of Current AI Research and Why Biology May Be the Next Frontier Regardless

This current predicament has resulted in two opposing views. The first view holds that scaling brute force approaches has unlocked significant utility in existing AI methods, therefore further scaling is all that is required to unlock additional capability. This view particularly holds that by scaling the amount of training data, model size, optimizing hardware, and additional fine tuning, the potential for current AI methods has no practical short-term limitations [11–13]. The smaller (or at least less vocal and certainly less well funded) perspective argues that the fundamental nature of current AI methods means the function of these systems cannot continue to indefinitely grow simply by increasing model size and other minor optimisations [14, 15]. One key limitation is that current models are already trained models on practically all available digitized information. Future approaches that seek to optimize with synthetic data may be ineffective along with having significant risks due to propagating bias and misinformation [16, 17]. Indeed, this perspective would accord with the notion that reliance on external optimization algorithms has driven AI research and development into a local minimum bounded by existing information resources and methods, stifling progress toward AGI. The inherent separation between model and optimizer is not only artificial but detrimental, preventing the development of systems that can learn and adapt intrinsically, as natural intelligent systems do [18]. In natural systems, learning is an intrinsic, continuous process, inseparable from cognition. The human brain, for example, does not undergo a distinct "training" phase with an external optimizer; instead, learning occurs as part of an ongoing interaction with the environment. This intrinsic learning is not only more efficient but also more flexible, allowing for rapid adaptation in the face of novel situations. In stark contrast, modern AI systems are rigid, requiring vast amounts of data and computation to achieve limited forms of adaptation. These systems often fail to generalize beyond their training data, highlighting the inefficiency of extrinsic learning approaches. One solution to move forward is that AI research should embrace intrinsic learning—where learning is embedded within the dynamics of the system as a foundational property—enabling more robust, adaptive, and efficient models that better reflect the processes of natural intelligence. Interestingly, regardless of which view is correct, it may be that the future steps of humankind could remain similar.

Should those who predict that current AI methods are not sufficient to achieve AGI be accurate, it may also be reasonable to suggest that this represents a maximal exploitation of silicon computing. After all, the overwhelming majority of current AI research (and all mainstream commercial implementations) occurs on Von Neumann architecture, which despite its undeniable enabling capacity has inherent limitations [19, 20]. Key limitations include the separation between memory storage and information processing, the result of which contribute to the power consumption and sample inefficiency of current AI methods [21]. In the event these limitations were to impede the advancement of new intelligence capabilities, it will require a fundamental rethinking of AI architectures, shifting the focus from extrinsic optimization to systems capable of intrinsic, motivated learning. In this case, a logical approach is to explore the only known substrate that can give rise to generalized intelligence: biological neural systems. Already many advancements in AI research have been inspired by biological neural systems [22, 23]. Potentially by exploring the mechanisms that allow biological systems to process information and produce intelligence in such diverse settings, more powerful AI algorithms can be produced. However, there is also reason to believe that a biological substrate may produce functions not achievable with silicon Von Neumann based computing alone. Biological neurons likely exhibit higher-level complexity than silicon transistors, being able to display at least third order complexity, compared to the first order (due to binary encoding) that transistors display and some second order capabilities in many neuromorphic systems [24]. Aided by complex dendritic trees [25, 26], multiple neural subtype specialization [27], and the ability to operate in a parallelized non-linear fashion [28], the biological components of neural systems would be well suited for exploration as substrates to achieve generalized intelligence. After all, unlike classic silicon approaches, there is a ground truth that neural systems lead to generalized intelligence; every biological creature driven by biological neural systems displays generalized intelligence to some degree.

However, if scaling proponents are correct, given the overwhelming resources at their disposal [29], continually advancing capabilities will be harnessed by current methods and AGI will be achieved in the near future. If (a critical conditional) this occurs, the idea of a singularity event arising, whereby these advanced intelligence systems are able to recursively improve their function as they surpass human capability [30], seems a plausible outcome (even if the premise occuring may be considered unlikely). Despite the significant risks such an entity would create for humankind as it now stands, assuming a cooperative (or at least peaceful) coexistence, human ingenuity would have transformed raw inorganic material into truly generally intelligent systems, also likely exploiting the fullest potential of silicon computing. Yet, there is no reason that human innovation (AGI assisted or not) or desire for discovery would cease due to the arrival of AGI. Instead, new horizons will be sought and having seemingly conquered the inorganic world, the natural evolution would be to focus on the organic. In particular, given the ground truth of biology as the original generalized intelligence systems, the neural mechanisms which give rise to this ability will be of particular interest. Without doubt, exploring, understanding, and perhaps even enabling the fundamental principles of what allows biological intelligence to operate so efficiently through neural systems will have profound implications. At the very least these efforts will unlock methods and tools to better explore drug discovery and diseases that impact humans, perhaps advancing novel treatments. Yet, more than that, perhaps integrating the biological and artificial intelligence may also unlock further capabilities—or improve data and energy efficiencies—for AGIs. Such an advancement may herald even more advanced abilities for our civilization than achievable with silicon alone.

12.2 Synthetic Biology to Generate Neural Systems for Understanding Intelligence

Based on the above reasoning, regardless of whether AGI is achieved with current AI methods, there are compelling reasons to consider the value of enabling better neuroscience research tools to understand how biological neural activity gives rise to intelligent functions and perhaps even leverage these capabilities. Yet practical and ethical reasons exist so that most advancements in this area will require synthetic biological methods to enable neuroscientific research and development. Firstly, neural cell culture approaches will be necessary to understand the fundamentals of what allows neural systems to process information and produce intelligent outcomes. Research in humans or animals as to how intelligence arises at a fundamental level is difficult. The emergence of any complex traits from an in vivo neural system is likely a property of a part-whole relationship within the wider dynamic (potentially chaotic) system as a whole [31]. In living humans, it is not ethically permissible to remove working parts from the neural system to determine the effect on the function of the whole. Even if it was possible, as indeed is done sometimes when animals are used for testing, the integrated nature of in vivo brains makes it difficult to isolate distinct key effects of individual system components. Much can still be learned from in vivo methods exploring how this complex neural systems process information and give rise to intelligence. Indeed, iterating across the different levels of complexity will be necessary for understanding key emergent properties [32]. However, these methods alone will be insufficient for determining how to harness biological neural systems for information processing and intelligence. Previous work dissociated neural tissue from primary sources to explore intelligence and information processing in cortical [33–36] and hippocampal [37, 38] neural cultures. However, primary cell culture is neither scalable enough nor specific enough to be utilized as a meaningful substrate for information processing and intelligence [39].

The more compelling alternative is to use pluripotent stem cells and differentiate these cells to the desired neural subtype. In particular, the production of induced pluripotent stem cells is of particular interest. Human induced pluripotent stem cells (hiPSCs) are produced from consenting adult donors, typically from either blood or skin, and can be expanded almost indefinitely [40]. As these hiPSCs can be harvested from nearly any consenting adult, this provides access to a wide range of genetic tissue, ideal for exploring genetic variations in neurological disease or health [41, 42]. A range of methods have already been developed to differentiate hiPSCs into a variety of neural cell types, although each have distinct advantages and disadvantages depending on the intended purpose. It is possible to leverage natural genetic mechanisms by applying small molecules to direct hiPSCs down key differentiation pathways [43, 44]. This ontogenetic approach can result in a variety of neural subtypes for the discrete brain region of interest [45–48]. While these methods are technically challenging, they do produce a range of neural subtypes for a given brain region and are useful in capturing a component of the natural complexity of neural systems. However, this variety in cell types also means a higher degree of variation in the proportion of these subtypes, especially as these protocols may take significant time (sometimes months) to perform. Alternatively, using either viral induction or genetic modifications to upregulate key genes can also give rise to directly differentiated neural cultures of highly pure and specific subtypes [49–51]. While these direct differentiation protocols are shorter and have less variability than those using signaling pathways, in vivo neural systems do not consist of only a few cell types but rather highly integrated and organized multicellular units that form intricate nested structures. While presumably a minimal viable brain (MVB) exists with a (relatively) small number of neural subtypes, the composition and proportions of this MVB are unknown and subject to future research. For instance, even considering one of the smallest complex nervous systems—*Caenorhabditis elegans (C. elegans)*—multiple neural subtypes and neural connections exist, leaving significant questions as to how function arises [52]. It may be reasonable to propose that these small creatures represent an MVB, but if so, significant work remains to develop genetic engineering tools for the wide variety of subtypes even these simple organisms possess. Despite these limitations, it is not necessary to fully understand the properties of neural systems to begin to explore how information processing and intelligence arises within them. Rather, it is only by exploring using the tools we have available, that the necessary insights may become available.

12.2.1 Beyond Hebbian Plasticity

Mechanisms of learning, information processing, and intelligence underpinning neural systems are notoriously difficult to unravel. Hebbian plasticity, often summarized by the phrase "cells that fire together, wire together," is one of the few foundational principles in neuroscience that has captured a mechanism associated with learning [53]. Hebbian plasticity can function through a variety of neural circuit types, including feedforward excitation or inhibition, along with feedback inhibition, lateral inhibition, or mutual inhibition [54]. The relative simplicity of this approach means that Hebbian learning has also inspired various approaches in artificial intelligence [55]. However, Hebbian plasticity seems insufficient to explain all capabilities exhibited by biological neural systems [56]. As such, while this form of learning captures an essential aspect of synaptic plasticity in biological brains [57], the current implementations in AI systems are too simplistic and limited in their scope. The application of Hebbian-like mechanisms has failed to deliver the flexibility, generality, and adaptability seen in natural intelligence [56]. A key limitation of most Hebbian-inspired learning algorithms is their reliance on pairwise correlations between neurons or units. This approach, while elegant in its simplicity, lacks the sophistication needed for complex task-solving and robust generalization. Hebbian rules, as implemented in many artificial systems, typically capture only low-order associations, making them prone to overfitting and incapable of handling more abstract, higher-order learning. Furthermore, these simple Hebbian mechanisms often fall short when learning from sparse feedback is required. As a result, they typically perform poorly compared to optimization-based learning methods. Furthermore, in some cases Hebbian learning rules are functionally equivalent to gradient descent [58]. This equivalence undermines the goal of moving beyond extrinsic learning frameworks, as it suggests that Hebbian approaches, when used in isolation, offer little advantage over existing optimization-based methods. When these systems largely approximate gradient-based learning, they inherit the same weaknesses, such as the need for large amounts of data, without delivering the full benefits of intrinsic learning.

In addition to these limitations in learning protocols, there is a practical challenge with applying Hebbian plasticity directly to biological systems as a method to elicit complex information processing and intelligence. Even though the synthetic biological methods used to neural cultures results in relatively simple networks, the complexity of these systems is still far in excess of what can be interacted with at the level of pairwise Hebbian plasticity. Along with other methods of communication, such as chemical, neural systems distinctly use electrophysiological activity to convey information. As electricity forms the basis of communication in silicon systems, when it comes to interacting with cultured neural systems, most work occurs through multi-electrode arrays (MEA) [33, 34, 59–61]. Neural cells can be recorded to ascertain when activity occurs, and electrical stimulation can be delivered to evoke further activity. This can be simply thought of as an input/output bridge, where the neural system is then able to alter activity in response to incoming (electrical) information. Yet even high-density MEA that can allow recording from every neuron in a given area cannot also stimulate every neuron. Indeed, most high-density MEA are fairly limited with how many stimulation sites can be accessed at any one moment. Such sparse control would likely be insufficient to drive complex behavior through Hebbian plasticity alone. For instance, in the simple *C. elegans* system mentioned above, despite containing only 188 neurons in the head, at least 23,433 pairs of neurons have been measured, even then only accounting for approximately 80% of pairwise connections [52]. As such, even in this most simple system, interacting at the level of paired neuron is not feasible. Moreover, even in biological systems, Hebbian plasticity does not act alone and must be balanced against compensatory mechanisms such as homeostatic plasticity across multiple timescales [62]. Other mechanisms, such as rule-based decision-making can also warp the association between the stimulus and key responses [63]. Evidence also suggests that the population-wide activity of neural systems can alter how the system as a whole will behave under a variety of circumstances [64]. Similarly, the geometric properties of a neural system at both the macro and micro levels likely also play a role [65, 66]. How all these features operate despite ongoing neural oscillations at multiple levels [67] also remains largely an open question despite some meaningful advancements.

What is needed is a more powerful and expansive approach to learning that goes beyond the simple low-order interactions captured by Hebbian learning. Biological systems demonstrate rich, multi-level learning processes that adapt dynamically, continuously integrating feedback from complex environments in ways that simple Hebbian rules cannot mimic. To move toward systems that exhibit intrinsic learning, investigators must accept the limitations of current Hebbian approaches and develop models that integrate broader, more dynamic forms of synaptic plasticity and adaptation. This will involve mechanisms that incorporate temporal, contextual, and global feedback processes, enabling a more powerful and flexible form of learning that can adapt to a wide variety of tasks and environments. Therefore, to explore higher-order learning algorithms that may underlie biological intelligence, a new approach is needed.

12.2.2 A Proof-of-Concept for In Vitro Learning and Intelligence

The ability to use synthetic biology method enables the generation of an ethically sourced, renewable, and personalized cell source by differentiation hiPSCs to neural subtypes offers both specificity and scalability. In line with past work, MEA can be used to explore the capacity of these synthetic biological neural cultures for information processing and intelligence—previously termed Synthetic Biological Intelligence (SBI) [61]. However, as neural systems require tightly coupled information flows both in and out of the system, as even mild disruptions can cause significant aberrations [68]. To mitigate this, a neural system will require a real-time closed-loop flow of information to be able to effectively process and respond in a structured information landscape [101]. In our recent work [61], we managed to implement such a real-time closed-loop system via an MEA to simulate a simplified game of "pong" that would be compatible for both primary mouse neurons and human neurons differentiated from hiPSCs. In this system (termed *DishBrain*) we delivered information into the neural cultures through electrical stimulation using both the place and rate that the stimulation was delivered to designate the x and y position of a ball relative to the paddle. Activity was then read from discrete (counterbalanced) regions and used to move the paddle. This set up allowed the outgoing activity to change incoming information dynamically by moving the virtual paddle. Yet it was necessary to test a method of interacting with the neural systems to achieve a goal-directed or "intelligent" response.

As described above, we do not have a feasible capacity to directly program the entire network of neurons via Hebbian plasticity, nor would such a method be a scalable or robust approach. Other methods such as RL used in AI research were not feasible. Neural cultures in a dish are unlikely to have well-structured naturally occurring reward (e.g., dopaminergic) pathways like that seen in animals or humans [102]. Even if they do, there is no special privileged input to these networks that would allow a real-time reward or punishment to be administered to these networks—as done in most machine learning algorithms—with any degree of reproducible control [69]. However, biological intelligence is assumed to follow inherent predispositions to optimize for one behavior or activity over another. These predispositions likely act as fundamental drivers that may explain how intelligence arises and will be necessary to uncover to progress SBI. Recently we opted to explore this phenomenon with implications from the Free Energy Principle [70]. Simply put, free energy minimisation is a drive to reduce uncertainty or minimize surprise [71]. In a neural system this could occur between an internal model of the world and the actual world. Making our internal model (beliefs) match the external real world is necessary for us to be able to exist—it would be impossible for a biological system to survive in a changing world without making viable predictions at some level [72]. This can be done in two ways, either by predicting the environment or by acting to make the environment more predictable. In our work we removed one of these options, providing a random feedback stimulation in response to activity that resulted in undesirable behavior and a predictable stimulation following desirable activity. We hypothesized that following the Free Energy Principle, if the system is going to self-organize to minimize the amount of incoming random information, it would alter its activity in otherwise arbitrary ways to avoid this random stimulation.

Overall, the results strongly supported the above hypothesis. Both mouse cortical cells and human cortical cells altered activity over time to have a higher hit to miss ratio in the last 15 min of gameplay compared to the first 5 min. This trend of improved performance over time was observed over multiple metrics. Multiple controls were run to determine this effect was not due to independent factors in either the software, hardware, or wetware. Additional controls were run to compare how the neural cells would perform in an open-loop setup, or if the feedback protocol was changed, each variation reducing performance. However, perhaps the most compelling evidence was the nuanced electrophysiological changes observed in the cultures. These activity changes went beyond simple changes in the amount of activity in key regions. The correlations between the cellular activity in regions where information about ball position was input and cellular activity that moved the paddle increased over time. However, the correlation between the two counterbalanced motor regions decreased over time. This was coherent, as nuanced control of the paddle would ideally see activity increase in motor regions controlling only one direction at the time, so the system activity was coherent. To go further, follow up work established that lower-dimensional representations of neural activity were dramatically different during the first and second halves of gameplay; something not observed when neurons were not engaged in gameplay [73]. These representations appeared to relate to nuanced changes in the functional connectivity maps that neural cultures had during gameplay compared to spontaneous activity when unstimulated. Interestingly, when looking at the first 2 min of gameplay, neural cultures activity was generally the same as during unstimulated spontaneous activity, yet by the end of the gameplay, significant differences arose across every metric. Finally, in depth analysis clustering functional connectivity patterns in the highest vs lowest performing neural cultures showed stark results. Activity metrics linked to functional connectivity clustered counterbalanced motor regions that were spatially distant but functionally related significantly closer than the functionally opposing, but spatially proximal regions [74]. This shows that the system managed to rapidly reorganize the activity patterns in a network in an otherwise arbitrary manner but with meaningful control and nuance.

12.3 Insights for AI Research Based on Neural Criticality and Intrinsic Learning

The *DishBrain* proof of concept represented a significant example of controlled intrinsic learning in biological neural networks, offering insights directly relevant to the development of intrinsic learning systems [75]. This work demonstrated that neurons, when integrated into a structured environment, could exhibit adaptive, goal-directed behavior with minimal external intervention. This real-time, closed-loop interaction between neurons and a simulated game-world highlights the potential for systems to self-organize and learn intrinsically through direct feedback. Indeed the rapid learning observed in *DishBrain* even showed significantly more sample efficiency than multiple machine learning paradigms [76]. This shows how biological networks can adapt efficiently when provided structured information environments that target latent predispositions of the system, even with sparse sensory inputs. This contrasts starkly with the rigid, task-specific optimization seen in modern AI, where learning is often slow and dependent on vast datasets. By embedding learning within the dynamics of the system itself, *DishBrain* exemplifies a model of intrinsic learning where the boundaries between cognition and adaptation blur.

Looking at the specific dynamics of even the relatively simple biological neural networks used in the *DishBrain* system, key implications for how systems may best by structured for rapid, intrinsic learning can be gleaned. Follow-up work indicated that biological neural networks would rapidly reorganize activity patterns not only in a functionally relevant manner to better perform at the simplified Pong game but would also show robust metrics of neural criticality [77]. Criticality is a state where a system operates near the edge of chaos, displaying both stochastic and deterministic behaviors simultaneously [77, 78]. Neural criticality has been observed in many different neural systems and has been proposed to be an optimal state for information representation, processing, and perhaps even computation [79–83]. Many have discussed whether neural criticality requires higher-level control or may vary based on complex cognition have occurred [84–86]. Yet finding that even simple neural cultures would rapidly transition to show critical dynamics when embodied in a structured information landscape suggests a more foundational role [77]. Instead, it would indicate the neural criticality is a basic property of any complex dynamic system. Further analysis examining how metrics of criticality varied between game states and game performance in the Pong game, further supported the notion that critical dynamics were a necessary (but not sufficient) property for a neural system to exhibit directed information processing capabilities [77]. As such, the edge-of-chaos dynamics exhibited by neural systems when engaged in information processing may underlie their incredibly efficient transmission and processing of information [87, 88].

Thus, it is interesting to consider that modern AI systems are typically designed to operate close to linear regimes, carefully tuned to avoid chaos. The reasoning behind this design choice is clear: dynamics that border on chaotic are seen as unpredictable and difficult to control, and thus, AI models rely on smooth, predictable optimization landscapes amenable to gradient descent. This method can be incredibly effective under the right circumstances, but not without cost. For instance, the AlphaGo Zero system which relied on reinforcement learning methods did master the game "Go," yet required training data from 29 million games over 40 days [3]. With reasonable time controls this would take on the order of thousands of years of continuous real-time play and required enormous energy costs to train. While it is impressive that a system could both generate and train on a dataset of this timescale, it underscores the artificial nature of the current learning process and how inefficiently knowledge is extracted from experience. Therefore, this aversion to edge-of-chaos dynamics may be preventing the emergence of more adaptive, self-organizing forms of learning, akin to what we observe in natural intelligent systems like the brain. Taking inspiration from observing the dramatic exhibition of critical dynamics in neural systems would position edge-of chaos dynamics not as an undesirable side effect, but a functional feature. Indeed, criticality dynamics appear to allow the brain to balance stability with adaptability, enabling rapid learning from minimal data—a trait sought after in AI research. In chaotic systems, small perturbations can lead to significant changes in system behavior [89, 90], a property that could allow AI systems to adapt more flexibly and rapidly to new information. In contrast, today's AI models avoid this dynamism and AI systems that display criticality and learning are significantly underdeveloped.

Given that AI systems are optimized for smooth behavior, the gradient-based methods employed inherently limit the system's ability to have rich internal dynamics. This creates the reliance on external optimization, such as gradient descent, means that models can only learn when explicitly guided by an external algorithm separating learning from the model's functioning. As a result, AI systems are unable to dynamically adjust their behavior without undergoing a retraining process that often relies on large quantities of data and computational power. Given the preponderance of evidence that edge-of-chaos dynamics, captured as neural criticality, is occurring throughout neural systems it is surprising that relatively little work has explored integrated these properties in AI systems [79]. Potentially there is resistance to adopting radically new techniques, especially when it means temporarily showing reduced performance compared to the current gold standard. Yet, just like how early steam engines were considered less reliable than horses by many, recent success and track record, does not mean long term superiority [91]. By introducing edge-of-chaos dynamics into AI architectures, we may have the missing element to create systems that are capable of powerful intrinsic learning — learning that is continuous and inseparable from the system's operation. The Supersymmetric Theory of Stochastics (STS) suggests that chaotic dynamics can be leveraged in a way that allows for efficient internal adaptation. In STS, noise-induced chaos occurs in a phase where the system remains integrable but is dominated by instanton processes, akin to the neural avalanches observed in the brain [92]. These chaotic dynamics could enable a form of self-organization that is not extrinsically imposed but emerges naturally from the system's interactions with its environment. Moving beyond the linear, gradient-based optimization approaches of today requires embracing the rich potential of chaos. Instead of seeing chaos as a threat to control and predictability, we should view it as a potential pathway to more robust, adaptive, and intrinsically learning AI systems. This shift will enable models that learn as they act, just as natural systems do, operating continuously near the edge of chaos to enhance performance and adaptability.

12.4 Looking Toward the Future

Whether current AI research methods yield AGI capabilities or plateau is unknown at the present time. Yet that biological agents operate with distinct mechanisms enabled by neural architecture is a certainty. Approaches such as SBI that seek to discover the underlying processes that give rise to biological intelligence by using neural cells as the information processing substrate may lead to a series of new intelligent systems. Whether these systems use synthetic biological methods to generate wetware that harness these neuroinformatic processes, or the systems are instead novel algorithms that enhance current AI research, is an open question. Given the early results from recent research, perhaps the most promising outcome would be a combination of both. Already work seeking to utilize more physiologically relevant cell cultures, called brain organoids, has begun in earnest, often under the banner of organoid intelligence (OI) [93–95]. Related approaches have also explored other methods in using neural cultures for information processing with promising results [59, 60, 96]. On the other side, work aiming to harness more physiologically relevant algorithms for Von Neumann based computing methods has also had some early success [75]. By further enhancing these approaches with edge-of-chaos dynamics, it may offer a further advance toward true intrinsic learning, allowing AI methods to embrace more dynamic, integrated architectures—drawing from biological principles to create systems that adapt continuously and autonomously.

If current AI research approaches prove limited in giving rise to AI, this shift toward biologically inspired algorithms might bridge the current wide gap between artificial and natural intelligence, opening new possibilities for robust, flexible, and general intelligence. Even should AGI be achieved with current approaches; the exploration of neural systems may offer improvements upon the energy consumption and sample efficiency of these AGI systems. At the very least, having the tools to better interact with neural systems and an understanding of how they function at the most fundamental levels will be an invaluable addition for neuroscience research, disease modeling, and pre-clinical drug testing [103]. Of course, all these lines of research have challenges. There is still a pressing need for scalable and accessible platforms to reduce the barrier of entry for research groups. Additionally, the use of biological tissue does raise the question about whether systems comprising of biological material require different treatment [97, 98]. These ethical questions are further complicated by inconsistent nomenclature across fields which can impact communication of results and agreements for how progress should occur [99]. Fortunately, international collaborations are already underway to begin to address these issues and help facilitate responsible research moving forward [39, 94, 100]. Ultimately, while there is significant uncertainty about what forms of intelligence may exist in the future and how they will be realized, the concepts discussed in this chapter can certainly no longer be considered science fiction. Science fictions ceases and science begins, once people start working on a realization of the idea. This work has now begun and given the fascination humankind has with non-human generalized intelligence, it is clear that this work will continue.

References

1. Gignac GE, Szodorai ET (2024) Defining intelligence: bridging the gap between human and artificial perspectives. Intelligence 104:101832
2. Hagos DH, Battle R, Rawat DB (2024) Recent advances in generative AI and large language models: current status, challenges, and perspectives. IEEE Trans Artif Intell 5:1–21. https://doi.org/10.1109/TAI.2024.3444742

3. Silver D et al (2017) Mastering the game of Go without human knowledge. Nature 550:354–359
4. Fan L, Glynn PW (2022) The fragility of optimized bandit algorithms. Preprint at https://doi.org/10.48550/arXiv.2109.13595
5. Jouppi NP et al (2020) A domain-specific supercomputer for training deep neural networks. Commun ACM 63:67–78
6. Yildirim I, Paul LA (2024) From task structures to world models: what do LLMs know? Trends Cogn Sci 28:404–415
7. Dong Q et al (2024) A survey on in-context learning. Preprint at https://doi.org/10.48550/arXiv.2301.00234
8. Eloundou T, Manning S, Mishkin P, Rock D (2024) GPTs are GPTs: labor market impact potential of LLMs. Science 384:1306–1308
9. Yenduri G et al (2024) GPT (generative pre-trained transformer)—a comprehensive review on enabling technologies, potential applications, emerging challenges, and future directions. IEEE Access 12:54608–54649
10. OpenAI et al (2024) GPT-4 Technical Report. Preprint at https://doi.org/10.48550/arXiv.2303.08774
11. Wolters C, Yang X, Schlichtmann U, Suzumura T (2024) Memory is all you need: an overview of compute-in-memory architectures for accelerating large language model inference. Preprint at https://doi.org/10.48550/ARXIV.2406.08413
12. Vaswani A et al (2023) Attention is all you need. Preprint at https://doi.org/10.48550/arXiv.1706.03762
13. Li J et al (2024) Large language model inference acceleration: a comprehensive hardware perspective. Preprint at https://doi.org/10.48550/arXiv.2410.04466
14. McKenzie IR et al (2023) Inverse scaling: when bigger isn't better. https://doi.org/10.48550/ARXIV.2306.09479
15. Muennighoff N et al (2023) Scaling data-constrained language models. Preprint at https://doi.org/10.48550/ARXIV.2305.16264
16. Guo X, Chen Y (2024) Generative AI for synthetic data generation: methods, challenges and the future. Preprint at https://doi.org/10.48550/ARXIV.2403.04190
17. Garry M, Chan WM, Foster J, Henkel LA (2024) Large language models (LLMs) and the institutionalization of misinformation. Trends Cogn Sci 28:S1364661324002213. https://doi.org/10.1016/j.tics.2024.08.007
18. Dentella V, Guenther F, Leivada E (2024) Language in vivo vs. in silico: size matters but larger language models still do not comprehend language on a par with humans. Preprint at https://doi.org/10.48550/ARXIV.2404.14883
19. Kimovski D et al (2024) Beyond von Neumann in the Computing Continuum: architectures, applications, and future directions. IEEE Internet Comput 28:6–16
20. Shaafiee M, Logeswaran R, Seddon A (2017) Overcoming the limitations of von Neumann architecture in big data systems. In: 2017 7th international conference on cloud computing, data science & engineering—confluence. IEEE, Noida, pp 199–203. https://doi.org/10.1109/CONFLUENCE.2017.7943149
21. Strubell E, Ganesh A, McCallum A (2019) Energy and policy considerations for deep learning in NLP. Preprint at http://arxiv.org/abs/1906.02243
22. Richards BA et al (2019) A deep learning framework for neuroscience. Nat Neurosci 22:1761–1770
23. Hasson U, Nastase SA, Goldstein A (2020) Direct fit to nature: an evolutionary perspective on biological and artificial neural networks. Neuron 105:416–434
24. Kumar S, Williams RS, Wang Z (2020) Third-order nanocircuit elements for neuromorphic engineering. Nature 585:518–523
25. Beaulieu-Laroche L et al (2018) Enhanced dendritic compartmentalization in human cortical neurons. Cell 175:643–651.e14
26. Guerguiev J, Lillicrap TP, Richards BA (2017) Towards deep learning with segregated dendrites. elife 6:e22901

27. Ghaderi P, Marateb HR, Safari M-S (2018) Electrophysiological profiling of neocortical neural subtypes: a semi-supervised method applied to in vivo whole-cell patch-clamp data. Front Neurosci 12:823
28. Sun X, Si H (2020) Population rate coding in recurrent neuronal networks consisting of neurons with mixed excitatory–inhibitory synapses. Nonlinear Dyn 100:2673–2686
29. AI investment forecast to approach $200 billion globally by 2025. https://www.goldmansachs.com/insights/articles/ai-investment-forecast-to-approach-200-billion-globally-by-2025
30. Hoffmann CH (2023) A philosophical view on singularity and strong AI. AI Soc 38:1697–1714
31. Mediano PAM et al (2022) Greater than the parts: a review of the information decomposition approach to causal emergence. Philos Trans R Soc Math Phys Eng Sci 380:20210246
32. Kagan BJ, Loeffler A, Boyd JL, Savulescu J (2024) Embodied neural systems can enable iterative investigations of morally relevant states. J Neurosci 44:e0431242024
33. Tessadori J, Bisio M, Martinoia S, Chiappalone M (2012) Modular neuronal assemblies embodied in a closed-loop environment: toward future integration of brains and machines. Front Neural Circuits 6
34. Bakkum DJ, Chao ZC, Potter SM (2008) Spatio-temporal electrical stimuli shape behavior of an embodied cortical network in a goal-directed learning task. J Neural Eng 5:310–323
35. Chao ZC, Bakkum DJ, Potter SM (2007) Region-specific network plasticity in simulated and living cortical networks: comparison of the center of activity trajectory (CAT) with other statistics. J Neural Eng 4:294–308
36. Bakkum DJ, Chao ZC, Potter SM (2008) Long-term activity-dependent plasticity of action potential propagation delay and amplitude in cortical networks. PLoS One 3:e2088
37. Pimashkin A, Gladkov A, Mukhina I, Kazantsev V (2013) Adaptive enhancement of learning protocol in hippocampal cultured networks grown on multielectrode arrays. Front Neural Circuits 7
38. Brewer GJ et al (2013) Toward a self-wired active reconstruction of the hippocampal trisynaptic loop: DG-CA3. Front Neural Circuits 7
39. Kagan BJ et al (2023) The technology, opportunities, and challenges of synthetic biological intelligence. Biotechnol Adv 68:108233
40. Schwedhelm I et al (2019) Automated real-time monitoring of human pluripotent stem cell aggregation in stirred tank reactors. Sci Rep 9:12297
41. Lampert A et al (2020) Human sensory neurons derived from pluripotent stem cells for disease modelling and personalized medicine. Neurobiol Pain 8:100055
42. Tejavibulya N, Sia SK (2016) Personalized disease models on a chip. Cell Syst 3:416–418
43. Hu B-Y et al (2010) Neural differentiation of human induced pluripotent stem cells follows developmental principles but with variable potency. Proc Natl Acad Sci U S A 107:4335–4340
44. Fedorova V et al (2019) Differentiation of neural rosettes from human pluripotent stem cells in vitro is sequentially regulated on a molecular level and accomplished by the mechanism reminiscent of secondary neurulation. Stem Cell Res 40:101563
45. Hunt CPJ et al (2023) Understanding and modeling regional specification of the human ganglionic eminence. Stem Cell Rep 18:654–671
46. Niclis JC et al (2017) Efficiently specified ventral midbrain dopamine neurons from human pluripotent stem cells under xeno-free conditions restore motor deficits in parkinsonian rodents. Stem Cells Transl Med 6:937–948
47. Du Z-W et al (2015) Generation and expansion of highly pure motor neuron progenitors from human pluripotent stem cells. Nat Commun 6:6626
48. Shi Y, Kirwan P, Livesey FJ (2012) Directed differentiation of human pluripotent stem cells to cerebral cortex neurons and neural networks. Nat Protoc 7:1836–1846
49. Yang N et al (2017) Generation of pure GABAergic neurons by transcription factor programming. Nat Methods 14:621–628

50. Ho S-M et al (2016) Rapid Ngn2-induction of excitatory neurons from hiPSC-derived neural progenitor cells. Methods 101:113–124
51. Pak C et al (2018) Rapid generation of functional and homogeneous excitatory human forebrain neurons using Neurogenin-2 (Ngn2). Protoc Exch. https://doi.org/10.1038/protex.2018.082
52. Randi F, Sharma AK, Dvali S, Leifer AM (2023) Neural signal propagation atlas of Caenorhabditis elegans. Nature 623:406–414
53. Lansner A, Fiebig F, Herman P (2023) Fast Hebbian plasticity and working memory. Curr Opin Neurobiol 83:102809
54. Luo L (2021) Architectures of neuronal circuits. Science 373:eabg7285
55. Schöfmann CM, Fasli M, Barros MT (2024) Investigating biologically plausible neural networks for reservoir computing solutions. IEEE Access 12:50698–50709
56. Halvagal MS, Zenke F (2023) The combination of Hebbian and predictive plasticity learns invariant object representations in deep sensory networks. Nat Neurosci 26:1906–1915
57. Voglis G, Tavernarakis N (2006) The role of synaptic ion channels in synaptic plasticity. EMBO Rep 7:1104–1110
58. Xie X, Seung HS (2003) Equivalence of backpropagation and contrastive Hebbian learning in a layered network. Neural Comput 15:441–454
59. Cai H et al (2023) Brain organoid reservoir computing for artificial intelligence. Nat Electron 6:1032–1039
60. Sumi T et al (2023) Biological neurons act as generalization filters in reservoir computing. Proc Natl Acad Sci U S A 120:e2217008120
61. Kagan BJ et al (2022) In vitro neurons learn and exhibit sentience when embodied in a simulated game-world. Neuron 110:S0896627322008066. https://doi.org/10.1016/j.neuron.2022.09.001
62. Zenke F, Gerstner W (2017) Hebbian plasticity requires compensatory processes on multiple timescales. Philos Trans R Soc B Biol Sci 372:20160259
63. Ebitz RB, Tu JC, Hayden BY (2020) Rules warp feature encoding in decision-making circuits. PLoS Biol 18:e3000951
64. Ebitz RB, Hayden BY (2021) The population doctrine in cognitive neuroscience. Neuron 109:S0896627321005213. https://doi.org/10.1016/j.neuron.2021.07.011
65. Pang JC et al (2023) Geometric constraints on human brain function. Nature 618:566–574
66. Baspinar E, Schülen L, Olmi S, Zakharova A (2021) Coherence resonance in neuronal populations: mean-field versus network model. Phys Rev E 103:032308
67. Bick C, Goodfellow M, Laing CR, Martens EA (2020) Understanding the dynamics of biological and neural oscillator networks through exact mean-field reductions: a review. J Math Neurosci 10:9
68. Attinger A, Wang B, Keller GB (2017) Visuomotor coupling shapes the functional development of mouse visual cortex. Cell 169:1291–1302.e14
69. Friston KJ, Daunizeau J, Kiebel SJ (2009) Reinforcement learning or active inference? PLoS One 4:e6421
70. Friston K (2009) The free-energy principle: a rough guide to the brain? Trends Cogn Sci 13:293–301
71. Friston K (2010) The free-energy principle: a unified brain theory? Nat Rev Neurosci 11:127–138
72. Parr T, Friston KJ (2019) Generalised free energy and active inference. Biol Cybern 113:495–513
73. Khajehnejad M, Habibollahi F, Loeffler A, Kagan B, Razi A (2023) On complex network dynamics of an in-vitro neuronal system during rest and gameplay. In: NeurIPS 2023 Generative AI and Biology (GenBio) Workshop, vol 2023. NeurIPS
74. Khajehnejad M, Habibollahi F, Khajehnejad A, Kagan BJ, Razi A (2024) TAVRNN: temporal attention-enhanced variational graph RNN captures neural dynamics and behavior. Preprint at http://arxiv.org/abs/2410.00665

75. Friston KJ, Tommaso S, Takuya I, Alexander T, Alex K, Tim V, Magnus K et al (2025) Active inference and intentional behaviour. Neural computation 37(4): 666–700. https://doi.org/10.1162/neco_a_01738
76. Habibollahi F, Gaurav A, Khajehnejad M, Kagan BJ (2022) Biological neurons vs deep reinforcement learning: sample efficiency in a simulated game-world. Accept NeurIPS 2022:16
77. Habibollahi F, Kagan BJ, Burkitt AN, French C (2023) Critical dynamics arise during structured information presentation within embodied in vitro neuronal networks. Nat Commun 14:5287
78. Ovchinnikov IV (2016) Introduction to supersymmetric theory of stochastics. Entropy 18:108
79. Shriki O, Yellin D (2016) Optimal information representation and criticality in an adaptive sensory recurrent neuronal network. PLoS Comput Biol 12:e1004698
80. Boedecker J, Obst O, Lizier JT, Mayer NM, Asada M (2012) Information processing in echo state networks at the edge of chaos. Theory Biosci Theor Den Biowissenschaften 131:205–213
81. Bertschinger N, Natschläger T (2004) Real-time computation at the edge of chaos in recurrent neural networks. Neural Comput 16:1413–1436
82. Walter N, Hinterberger T (2022) Self-organized criticality as a framework for consciousness: a review study. Front Psychol 13:911620
83. Alam El Din D-M et al (2024) Human neural organoid microphysiological systems show the building blocks necessary for basic learning and memory. Preprint at https://doi.org/10.1101/2024.09.17.613333
84. Tinker J, Velazquez JLP (2014) Power law scaling in synchronization of brain signals depends on cognitive load. Front Syst Neurosci 8
85. Cocchi L, Gollo LL, Zalesky A, Breakspear M (2017) Criticality in the brain: a synthesis of neurobiology, models and cognition. Prog Neurobiol 158:132–152
86. Hesse J, Gross T (2014) Self-organized criticality as a fundamental property of neural systems. Front Syst Neurosci 8
87. Shew WL, Plenz D (2013) The functional benefits of criticality in the cortex. Neuroscientist 19:88–100
88. Shew WL et al (2015) Adaptation to sensory input tunes visual cortex to criticality. Nat Phys 11:659–663
89. Poil S-S, Hardstone R, Mansvelder HD, Linkenkaer-Hansen K (2012) Critical-state dynamics of avalanches and oscillations jointly emerge from balanced excitation/inhibition in neuronal networks. J Neurosci 32:9817–9823
90. Miller P (2016) Dynamical systems, attractors, and neural circuits. F1000Research 5:992
91. Sagle LW (1948) The Tom Thumb: little known facts about first locomotive built in America slowly come to light. Railw Locomot Hist Soc Bull 46–53
92. Ovchinnikov IV et al (2020) Criticality or supersymmetry breaking? Symmetry 12:805
93. Hartung T et al (2023) The Baltimore declaration toward the exploration of organoid intelligence. Front Sci 1:1068159
94. Smirnova L et al (2023) Organoid intelligence (OI): the new frontier in biocomputing and intelligence-in-a-dish. Front Sci 1
95. Morales Pantoja IE et al (2023) First organoid intelligence (OI) workshop to form an OI community. Front Artif Intell 6:1116870
96. Zhang X et al (2024) Mind in vitro platforms: versatile, scalable, robust, and open solutions to interfacing with living neurons. Adv Sci 11:2306826
97. Farahany NA et al (2018) The ethics of experimenting with human brain tissue. Nature 556:429–432
98. Kagan BJ, Duc D, Stevens I, Gilbert F (2022) Neurons embodied in a virtual world: evidence for organoid ethics? AJOB Neurosci 13:114–117
99. Kagan BJ et al (2023) Scientific communication and the semantics of sentience. Neuron 111:606–607

100. Kagan BJ et al (2024) Toward a nomenclature consensus for diverse intelligent systems: call for collaboration. Innovation 5:100658
101. Kagan BJ, Habibollahi F, Watmuff B, Azadi A, Doensen F, Loeffler A, Byun SH, Servais B, Desouza C, Abu-Bonsrah KD, Kerlero de Rosbo N (2025). Harnessing intelligence from brain cells in vitro. The Neuroscientist, p. 10738584251321438. https://doi.org/10.1177/10738584251321438
102. Tanveer MS, Patel D, Schweiger HE, Abu-Bonsrah KD, Watmuff B, Azadi A, ... Wang G (2025). Starting a synthetic biological intelligence lab from scratch. Patterns 6(5). https://doi.org/10.1016/j.patter.2025.101232
103. Watmuff B, Habibollahi F, Desouza C, Khajehnejad M, Loeffler A, Baranes K, ... Kagan BJ (2025). Drug treatment alters performance in a neural microphysiological system of information processing. Communications Biology. In Press. https://doi.org/10.1038/s42003-025-08194-6

Toward Conscious Service Robots

13

Sven Behnke

Contents

Abstract

Deep learning's success in perception, natural language processing, etc. inspires hopes for advancements in autonomous robotics. However, real-world robotics face challenges like variability, high-dimensional state spaces, nonlinear dependencies, and partial observability. A key issue is non-stationarity of robots, environments, and tasks, leading to performance drops with out-of-distribution data. Unlike current machine learning models, humans adapt quickly to changes and new tasks due to a cognitive architecture that enables systematic generalization and metacognition. Human brain's System 1 handles routine tasks unconsciously, while System 2 manages complex tasks consciously, facilitating flexible problem-solving and self-monitoring. For robots to achieve human-like

S. Behnke (✉)
Autonomous Intelligent Systems, Computer Science Institute VI—Intelligent Systems and Robotics, Center for Robotics and the Lamarr Institute for Machine Learning and Artificial Intelligence, University of Bonn, Bonn, Germany
e-mail: behnke@cs.uni-bonn.de

U. A. K. Betz (ed.), *Science for a Better Tomorrow*,
https://doi.org/10.1007/978-3-031-93623-4_13

learning and reasoning, they need to integrate causal models, working memory, planning, and metacognitive processing. By incorporating human cognition insights, the next generation of service robots will handle novel situations and monitor themselves to avoid risks and mitigate errors.

13.1 Introduction

Industries like car manufacturing impressively demonstrate the utility of robots. Recent developments in sensing, actuation, and, most importantly, artificial intelligence (AI) make it conceivable that robots will revolutionize many new application domains such as the flexible production of small lots, logistics, agriculture, security, inspection, professional services, and personal assistance. All of these domains, however, require cognitive capabilities far beyond those of today's robots.

Current robotic systems rely on structuring the environments and tasks, e.g., by providing objects in well-defined locations. In open-ended settings, such as our everyday environments, robot-friendly structuring is impossible. Instead, autonomous service robots must instantiate models of their environment from sensor measurements, plan actions to achieve goals, carry out plans in the presence of disturbances, and monitor their execution. They must also familiarize with new objects and tools and need to improve their behavior through learning. Finally, they must communicate with persons in a human-understandable way, to receive instructions, answer questions, and explain their behavior.

The tremendous success of deep learning [64, 98] in visual perception, speech recognition, natural language processing, vision-language tasks, and multimodal tasks gives rise to hope that these methods will lead to revolutionary advances in autonomous robot performance.

13.2 Related Work

Personal Service Robots Personal robots that assist handicapped or elderly persons in their activities of daily living have attracted much attention in robotics research. An increasing number of research groups are working on robots for service applications. Examples include PR2 [74] that was used in a household marathon experiment [52], Everyday Robots' mobile manipulator [42], and Toyota Human Support Robot [125], a standard platform in the international RoboCup@Home competitions. My team NimbRo won these competitions 2024 with two PAL Robotics TIAGo++ robots [75] and 2011–2013 with our cognitive service robot Cosero [105, 106], demonstrating a large variety of domestic service tasks. Further examples include Care-O-Bot 4 [55], Armar-6/7 [2], HRP-5P [59], TORO [53], E2-DR [127], and our Centauro robot [56, 57] which demonstrated challenging locomotion and manipulation tasks including the use of tools. For the ANA Avatar

XPRIZE competition [8, 40], capable systems have been developed, including iCub3 [20], Pollen Robotics Reachy, and our winning Avatar system [66, 99]. Scene Perception. In order to act in complex indoor environments, service robots must perceive the room structure, obstacles, persons, objects, etc. To this end, they are equipped with cameras and depth sensors. Estimating the sensor poses and registering the measurements yield environment maps [87]. In addition to modeling the environment geometry and appearance, semantic perception is needed [95].

Deep Learning For pattern recognition, deep learning [64] methods are extremely successful. They revolutionized visual perception [24, 92], speech recognition [88, 91], natural language processing [1, 85], vision-language tasks [70, 107], and multimodal tasks [33, 39]. Supervised deep learning requires large annotated datasets like ImageNet-21K [25], JFT-3B [129], and Kinetics [14], though, which are expensive to obtain. To address variability that should not change output, data augmentation methods such as image transformations [101] and generative models [128] are used to generate variants of training examples. For robotic tasks such as mobile manipulation, large-scale annotated datasets do not exist. To avoid the need for large labeled datasets, much research focuses on methods that can adapt to new conditions through transfer learning and domain adaptation. Transfer learning [58] uses representations learned from large data to learn a related task from small data, e.g., by continuation of training. Semi-supervised, weakly supervised, and unsupervised learning methods use fewer, low quality, and no labels at all, respectively. One example of semi-supervised methods is the student-teacher approach [124], where a teacher is trained on a small labeled dataset and then generates pseudo labels for a large unlabeled dataset to train the student. Because unlabeled data is much easier to obtain than annotated data, unsupervised methods are often used to pretrain models [10, 16]. The hope is to discover useful structure in the data which might aid target tasks. A promising subclass of unsupervised learning is self-supervised learning [37], which requires only unlabeled data to formulate a pretext task, for which a target objective can be computed without supervision. These pretext tasks must be designed in a way that high-level data understanding is useful for solving them, e.g., prediction of occluded image parts [41] or future video frames. As a result, the intermediate layers of trained models encode high-level semantic representations that are useful for solving downstream tasks. One form of self-supervision is contrastive learning [15], where two different data augmentations are applied to an image and a model is trained to maximize agreement between the outputs and minimize agreement with outputs for other images. Contrastive learning of dense descriptors for object surface elements has been applied to learn visuomotor manipulation policies [29] that generalize within a category of objects and are able to handle deformable objects. Other possibilities for self-supervised learning are to maximize mutual information between input and model output [45] and joint embeddings of two inputs with variance-invariance-covariance regularization [3].

Large Language Models Self-supervised training is the basis for the impressive performance of recent large language models (LLMs) such as GPT-4 [85] and Palm 2 [1] that continue text in plausible ways. Their large transformer networks [114] were trained on massive data to predict the next token. In contrast to recurrent sequence models, transformers flexibly reroute and combine information from relevant parts of the sequence through learned self-attention, which is implemented using content-based access to information values by matching keys to queries. Recently, autoregressively trained LLMs have shown sparks of artificial general intelligence [13]. Such models can acquire human-like systematic generalization through meta-learning [61], but this requires generating a training set of systematic generalization example problems. On the other hand, LLMs often lack common sense, hallucinate facts, fail at arithmetic, have difficulty reasoning, and cannot make proper plans. For these reasons, LLMs are combined with external tools [89] such as search engines, calculators, planners [69], etc. Multimodal models such as PaLM-E [27] and ImageBind [33] combine text and other modalities such as images in joint embeddings. Generative models are not restricted to producing text but are also used to generate, e.g., images [86] and video [90] from text inputs.

3D Models To address the 3D nature of scenes, Neural Radiance Fields (NeRF) have been proposed, which learn a neural network mapping 5D coordinates to density and color by predicting images from multiple views through volumetric rendering [77]. In our recent work PermutoSDF [93], the 3D shape of objects is represented by a neural signed distance function (SDF). By modeling individual objects as permutation-invariant slots, object representations can be learned through novel-view synthesis [94]. If conditioned on latent variables, category-level shape spaces can be learned, e.g., for articulated human bodies [26]. Compositional generative scene models that represent objects and their relations can be learned without image-level supervision [32].

Scene Prediction In dynamic scenes, the motion of objects and persons must be estimated and predicted. Scenes with moving agents (e.g., humans or robots) can be represented with 3D dynamic scene graphs [47]. Motion is the strongest cue for perceptual grouping, and predictive models are widely used to explain human visual perception [31]. Consequently, optimization of a prediction loss can be used to segment moving objects in videos [119]. SlotFormer [122] models spatiotemporal object relationships and predicts object states. Our recent work on object-centric video prediction decouples the processing of temporal dynamics and object interactions [116]. This facilitates learning of tasks that require understanding of object relations [81]. A fundamental problem when predicting the future is that often multiple plausible futures exist. MultiPath++ [113] predicts a distribution of future trajectories of road users parameterized as a Gaussian Mixture Model (GMM). Multiverse [67] predicts the distribution over multiple possible future paths of persons using convolutional recurrent neural networks (RNNs) over graphs. World Modeling. Prediction of future scene states and planning own actions require world models that are conditioned on actions. Playable Video Generation [76] learns a discrete set of actions from unlabeled video that are used to interactively generate video from actions. This task has been extended to Playable Environments [76] that can control multiple objects in 3D scenes with action labels that are discovered in an unsupervised way. DayDreamer [120] learns action-conditioned forward models in a latent space for multiple robots. GenAD [126] and GameNGen [112] are world models for autonomous driving and a video game, respectively.

Deep Reinforcement Learning Reinforcement learning (RL) addresses the development of situated agents that learn how to behave while interacting with the environment [108]. This problem is formulated as an agent-centric optimization in which the objective is to select actions based on the estimated state in order to obtain as much reward from the environment as possible in the long run. Impressive success has been achieved by combining this approach with deep learning. One example is MuZero [97] which combines tree-based search with a learned model and achieves superhuman performance in a range of challenging and visually complex domains (Atari games, Go, chess, and shogi), without any prior knowledge of their underlying dynamics. MuZero learns a model that predicts the quantities relevant to action planning: the reward, the action selection policy, and the value function. AlphaStar [117] learned the multi-agent game StarCraft II from $\sim$500K human games and 120M self-played games. Gran Turismo Sophy [123] learned from carefully engineered state and reward in more than 5 years of simulated driving hours to compete with the world's best drivers. Playing soccer with humanoid agents was learned from decades of match simulations [68]. Soccer skills for a humanoid robot and 1v1 play were learned from 2.5 years of simulated experience [38].

These numbers indicate that it would be impractical to collect that much experience with a real robotic system. Consequently, real-robot reinforcement learning mostly focuses on individual skills. For example, Google X learned grasping from cluttered bins with a simple manipulator under closed-loop monocular vision-based control [50]. They operated seven experimental setups for 4 months to collect 580K real-world grasp attempts to train a deep neural network Q-function with over 1.2M parameters and report a 96% grasp success rate on unseen objects. The method learned regrasping strategies, probing or repositioning objects to find the most effective grasps, performing other nonprehensile pre-grasp manipulations, and responding dynamically to disturbances and perturbations. Sorting recyclables and trash was learned from simulation and 9,527 hours of real-robot experience obtained with a fleet of 23 mobile manipulators [42].

Real-robot RL needs suitable inductive biases [43] to learn from little experience. These biases represent domain knowledge and can take many forms, e.g., the structure of the agent-environment interface and the policy generation mechanism. To improve the data efficiency of RL, transfer learning has been investigated. By pretraining on RoboNet [21], a dataset providing 15M video frames from seven different robot platforms, and fine-tuning on a held-out target platform, it has been demonstrated that simple manipulation tasks such as pushing and pick and place can be learned from limited experience. Multitask learning amortizes experience over multiple tasks [51]. It generalizes to structurally similar tasks and acquires distinct new tasks more quickly. One way to address the combinatorial complexity of multi-object scenes is to factorize them into objects. Object-centric perception, prediction, and planning [115] learn to discover objects in visual scenes and model their dynamics and appearance without supervision. A model-based reinforcement learner that predicts and plans block stacking on this abstract level generalizes to novel configurations and more objects. Action Schema Networks [110] learn generalized policies for probabilistic planning problems. By mimicking the relational structure of planning problems, they generalize over all instances of a given planning domain. Manipulation inherently involves contact and often requires both haptic and visual feedback. Lee et al. [65] use self-supervision to learn a compact and multimodal representation of sensory inputs, which is then used to improve the sample efficiency of policy learning as demonstrated for peg insertion. To avoid random exploration, imitation of human experts can be used. RT-1 [12] is a transformer-based controller trained on 130K demonstrations of a large variety of pick-and-place tasks in kitchen environments. Open X-Embodiment [84] is a large dataset of camera images and end-effector movements from 22 different robots, demonstrating 527 skills (160,266 tasks). The RT-X model trained on this data exhibits positive transfer and improves the capabilities of multiple robots by leveraging experience from other platforms. 970k episodes from this dataset were used to train OpenVLA [54], starting from a large language model and a visual encoder. OpenVLA demonstrates generalist manipulation capabilities and can be adapted to new robots via fine-tuning.

13.3 Challenges

Despite much research and progress, capable mobile manipulation robots that can cope with the complexity of open-ended real-world applications have not yet been realized. Developing such robots is a tremendous challenge, due to the typical characteristics of these applications.

Many Sources of Variability There are many sources of variability that a mobile manipulation robot must cope with. These include varying shape, texture, and physical properties of objects—even within a category. Furthermore, the 6D object pose, speed, and articulation state may vary. Environmental conditions, such as lighting, and surface properties, such as shininess, transparency, texturelessness, or nonreflectivity, greatly impact appearance in camera images and consequently the completeness and precision of depth estimates. The variability of single objects is exponentiated by the infinite possibilities for multi-object arrangements. Similarly, the robot environments such as rooms and apartments vary greatly in layout, geometry, surface properties, and other factors. The manipulation and locomotion tasks that capable robots need to perform are highly variable as well. Hence, learning methods are needed that generalize to novel, unseen situations.

High-Dimensional State and Action Spaces Input and output of mobile manipulation robot controllers are high-dimensional. Typical camera images are of size $1920 \times 1080 \times 3$, already more than 6M dimensions. Depth cameras, 3D LiDARs, force-torque and haptic, inertial, and joint sensors add many more input dimensions. The sensors measure at high rates, e.g., at 30 Hz, producing hundreds of million measurements per second. The output dimensionality is high as well, with typically more than 50 DoF for anthropomorphic robots. These joints need to be controlled at high rates with target positions, velocities, or torques. Hence, learning methods are needed that can cope with high-dimensional state and action spaces.

Hybrid Discrete-Continuous Variables Some variables, such as the presence of objects or the task category, are discrete, while other variables, such as 6D object poses or task parameters, are continuous. This creates the need for learning methods that can cope with both discrete and continuous variables.

Nonlinear Dependencies Objects are typically in contact with support surfaces and with each other, and the robot must make and release contacts with its end effectors or other body parts to manipulate them. This induces highly nonlinear constraints. While objects may be easily moved away from the contact point, moving them further toward the colliding surface is not possible. Similarly, occlusion effects and the transition between stick friction and sliding are highly nonlinear. Learning methods must address such nonlinearities.

Stochasticity There is much randomness in the world. Unmodeled environmental factors and other agents might also be perceived as non-determinism from our robot's point of view. Furthermore, robot sensors are noisy and unreliable, and robot actuators are imperfect and induce stochasticity. Hence, the state must be estimated from unreliable observations, and predictions are hard to make and become more and more uncertain for larger time horizons. Learning methods must cope with such uncertainties.

Partial Observability Due to the projection of the 3D world onto 2D cameras and other sensors, limited sensor ranges, resolutions, accuracies, etc., not all state variables that would be needed for action planning are directly accessible. Hence, learning methods must consider the distribution of possible states and must generate actions to acquire more information, for example, changing the camera pose to see occluded objects, touching objects to sense physical properties like weight and stiffness, and opening containers to see what is inside.

Underactuation Robots have limited action capabilities to influence the state of the environment. Their drives have limited speed and acceleration, and their manipulators have limited reach, strength, and dexterity. Some environmental variables cannot be influenced directly, but only through indirect means like tools. Learning methods must respect these constraints and generate behavior such as improvised tool use to overcome them.

Multimodality Mobile manipulation involves multiple modalities, such as vision, distance measurements, forces, and haptics on the input, and also multiple outputs, such as mobility, manipulation, and active sensing. Hence, learning methods must jointly address these modalities and come up, for example, with grasping strategies that transition smoothly from vision-based scene understanding and grasp selection, to visual tracking and correction of the approaching motion, to grasping execution and regrasping based on haptic feedback.

Non-stationarity One unique challenge is the non-stationarity of robots and their environments. Not only do robot bodies change due to wear and tear, but also the open-ended environments in which they operate and the tasks they perform are constantly changing. Already the ancient philosopher Heraclitus noted that the only constant in life is change. Such changes violate the fundamental assumption underlying current machine learning that a learned model will be used on the same distribution of data it has been trained on. When using a trained model on a different distribution (out of distribution, OOD), one cannot expect good performance [80]. In fact, seemingly small changes can lead to catastrophic failure [17].

13.4 Human Cognitive Functions

Humans are able to cope with such changes and quickly learn new tasks. My hypothesis is that the cognitive architecture of the human mind has evolved to continuously interact with changing environments and that equipping robots with key elements of this architecture will enable flexible handling of OOD data and *systematic generalization*. Systematic generalization was first studied in linguistics [5, 60] because it is a core property of language that meaning for a novel composition of existing concepts (e.g., words) can be derived systematically from the meaning of the composed concepts and the way they are composed. Humans exhibit systematic generalization also when understanding a new object by combining properties or parts which compose it [62]. *Compositionality* is the principle that complex objects can be described by their constituent parts and their relations to each other [30]. It allows to generate infinite variants from a finite set of building blocks, enables open-world zero-shot learning [71], and even makes it possible to generalize to new combinations that have zero probability under the training distribution.

While humans perform many routine tasks like walking or riding a bike without much attention, object manipulation, communication, and handling novelty are different. Cognitive science distinguishes *habitual* and *controlled* processing [11]. Habitual processing effortlessly generates default behaviors that are performed routinely. In contrast, controlled processing requires attention and mental effort to generate nonroutine behaviors. Kahneman [49] introduced the framework of fast and slow thinking and corresponding processing systems in our brain (see Fig. 13.1). Routine, habitual tasks can be achieved quickly in parallel without *conscious* attention using only System 1 abilities, whereas more complex tasks also require System 2 that is more capable but slower, serial and involves conscious

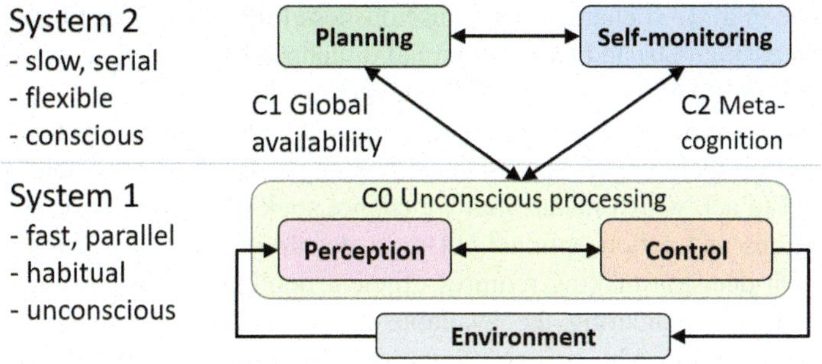

Fig. 13.1 Human cognitive functions according to Kahneman [49] (System 1 and System 2) and Dehaene et al. [23] (C0, C1, and C2)

processing. System 2 uses explicit, verbalizable knowledge and explicit processing, while System 1 relies on implicit, nonverbalizable, intuitive knowledge. We can act in fast and precise habitual ways without having to think consciously, but the reverse is not true: Conscious processing builds on the unconscious System 1.

System 2 is very flexible and powerful. It allows solving novel problems creatively by recombining existing pieces of knowledge, to discover and use causal dependencies, to imagine future outcomes, to plan actions, to find explanations, to reason, etc. It is also at that level that we communicate with others through natural language, e.g. to receive task specifications or new knowledge and rules that we can apply immediately. The capacity of System 2 is very limited, though. Our working memory can only hold 3–5 meaningful items active simultaneously [19]. Baars introduced the *Global Workspace Theory* [4] that identifies conscious processing as the communication bottleneck between selected parts of the brain that are called upon when addressing a current task. There is a threshold of relevance beyond which information that was previously handled unconsciously gains access to this bottleneck and is instantiated in working memory. When this happens, the information is broadcast throughout the brain, allowing its various relevant parts to synchronize and choose configurations and interpretations of their piece of information that are globally coherent with the configurations chosen in other parts of the brain.

While this severe communication bottleneck might appear to be a weakness, it can also be advantageous. Firstly, there is pressure to combine multiple lower level items that frequently occur together to larger, composite items, facilitating abstracting away irrelevant detail and providing compositionality. Secondly, when focusing on a few relevant items of a scene for conscious planning, we essentially ignore all other items, which are irrelevant for the task at hand. This leads to systematic generalization, because we can reuse the task knowledge in infinitely many novel situations in which the irrelevant items change. System 2 is slow, has limited capacity, and involves conscious effort; hence, there is pressure to migrate tasks to System 1 wherever possible. Through rehearsal, frequently performed tasks become habitual.

Dehaene et al. [23] characterize consciousness further. They distinguish unconscious processing (C0) and two orthogonal dimensions of conscious computations: *global availability of information (C1)* and *metacognition (C2)*.

C1—global availability—is a consequence of the distributed organization of the brain as a deep hierarchy of specialized subsystems that must be synchronized and of the need to act, which means that we cannot stick to a diversity of probabilistic interpretations and action options, but must decide in favor of a single course of action. Such decision-making requires efficient pooling over all available sources of information, considering the available action options and selecting the best one, sticking to this choice over time, and coordinating internal and external processes toward the achievement of that subgoal. Attention—selective processing of information—is crucial for items entering consciousness, but attention is not limited to conscious processing. Unconscious C0 processing also includes bottom-up and top-down attention mechanisms, which operate in parallel to prioritize and flexibly route information—often without bringing it to consciousness. The hierarchical system of sieves that operate unconsciously computes probability distributions, but only a single sample drawn from these becomes conscious at a given time, making it available globally to all specialized modules. Alternative interpretations might become conscious at other points in time; thus, C1 consciousness is causally responsible for our serial information processing bottleneck. Attention also implements variable binding [36]. The association of information elements to roles in relations and rules is crucial for applying these templates to varying input and, hence, for multistep inference and systematic generalization.

Consciousness in the second sense (C2) is characterized by the ability to reflexively represent oneself. When making decisions, we feel more or less confident about our choices. Our brain does not only make perception and action decisions but also estimates its degree of confidence. State estimation and learning also rely on confidence, for example, we weigh existing knowledge versus new evidence, like a Kalman filter [109]. Error detection is another example of self-monitoring: Just after responding, we sometimes realize that we made an error and change our minds. This might be explained by further evidence that arrived after the decision or by slower C2 processing monitoring fast C0 sensory motor execution. We do not just have knowledge, but we also know what we do not know. Such meta-knowledge is crucial for assessing our limits and for learning.

C1 and C2 are largely orthogonal and complementary dimensions of consciousness. Their joint possession may have synergistic benefits to organisms and robots. Bringing probabilistic metacognitive C2 information into the global C1 workspace allows it to be held over time, integrated into explicit long-term reflection, and shared with others. On the other hand, the possession of an explicit repertoire of one's own abilities (C2) improves the efficiency with which C1 information is processed.

13.5 The Need for Conscious Robots

Despite tremendous progress in C0-like deep neural networks trained end to end in tasks such as object recognition, video games, and board games, truly human-like learning and thinking machines will need to go beyond current engineering trends in both what they learn and how they learn it [62]. They need to build causal models of their environment that support explanation and understanding and must harness compositionality and learning-to-learn to rapidly acquire and generalize knowledge to new tasks and situations [96]. For this, equipping machines with C1 and C2 conscious processing will be crucial. Upon success, they would behave as if they were conscious; for example, they would know that they are seeing something, express confidence in it, report it to others, and may even experience the same perceptual illusions as humans.

Of course, *traditional symbolic AI* systems (GOFAI), like Hierarchical Task Network Planners [82] and CRAM [7], exhibit some of the properties that are associated with conscious System 2 processing, like compositionality. However, such symbol manipulation systems often lack semantic grounding of the higher level concepts in terms of the lower level observations and actions. Whereas pure symbolic representations put every symbol at the same distance from every other symbol, learned embeddings represent concepts through a vector of attributes—with related concepts being close-by and interpolations being meaningful. GOFAI systems often are too rigid to account for real-world data with outliers, etc. Further, GOFAI search and inference are generally intractable and need to be approximated. Here, learning representations together with inference procedures is needed to generate fast habitual C0 behavior. Finally, GOFAI approaches often do not handle uncertainty, which is crucial for partially observable, stochastic environments.

In recent years, *neuro-symbolic* approaches [22, 44, 73] have been proposed that integrate symbolic and subsymbolic representations, inference, and learning. However, hybrid neuro-symbolic systems [18,34,63,72,83,100,102,111,130] inherently use different representations and tools for neural and symbolic computations, which are difficult to integrate tightly.

Neurocompositional computing [103] is based on the principles of compositionality and continuity. It encodes structures in vectors that are processed by neural networks and shows promising results by quickly learning tasks from small datasets that require systematic generalization. The Differentiable Tree Machine [104] compiles high-level symbolic tree operations into subsymbolic matrix operations on tensors. Here, an agent learns to sequentially select tree operations to execute tree transformations with the help of a tree memory.

Recently, autoregressively trained embodied multimodal models have been used for generating robotic skills such as grasping and placing objects [12, 54] and for higher levels of robot control [27]. These models lack System 2 conscious processing, though. They need much data [84] and computing power, and the addressed scenarios are still relatively simple.

My hypothesis is that using insights from human cognition for the cognitive architectures of robots by incorporating C1 global availability and C2 metacognition will enable the next level of robot capabilities.

Because it extends highly successful C0 processing without a change in tools, I am convinced that a bottom-up way toward consciousness-inspired higher level cognitive functions for service robots is the way to go.

13.6 Objectives

My overall goal is to develop methods for learning higher level cognitive functions for service robots, which go beyond unconscious routine tasks by incorporating conscious processing to cope with novel situations and self-monitor.

Unconscious Perception and Control The System 1 / C0 routine processing directly interacts with the environment and is hence the basis for any higher level cognitive functions.

Starting from raw sensory measurements, such as video, depth, forces, and haptics, structured representations of mobile manipulation robot workspaces shall be learned on multiple levels of spatiotemporal abstraction. Abstraction will be realized by coarser spatiotemporal scales and more expressive, sparser representations on the higher levels. The elements of these representations will correspond to increasingly larger entities (parts, objects, and groups of objects) in the scene and will be increasingly semantic. The learned representations shall transition from sensor coordinate systems (e.g., the camera frame) to 3D representations of the scene and joint multimodal embeddings.

They shall model individual objects and the robot end effectors in their own canonical frame. This will enable learning of category-level shape and appearance spaces within a hierarchical categorization. Scene parsing shall instantiate these models and estimate object parameters, such as pose, shape descriptors, and appearance descriptors. Predictive models for these scene representations shall be learned on all levels of abstraction (see Fig. 13.2). Prediction of low-level detail shall be done only for short time horizons. Higher layer representations shall be predicted with coarser temporal granularity over longer time horizons. These predictions shall be based on individual object dynamics models and on pairwise relational models to account for object interactions, such as contact. The graph of object relations shall be sparsely instantiated according to the relevant object interactions. The predicted representations shall be compared to the feed-forward interpretation of new measurements, such that prediction error can be used to update the representations on all levels.

The learned predictive models shall be extended by conditioning them on robot actions. This will allow for the rollout of possible futures of robot-environment interaction. Coarse-to-fine model-predictive control of routine skills that do not require conscious attention shall be learned from imagined rollouts on the multiple

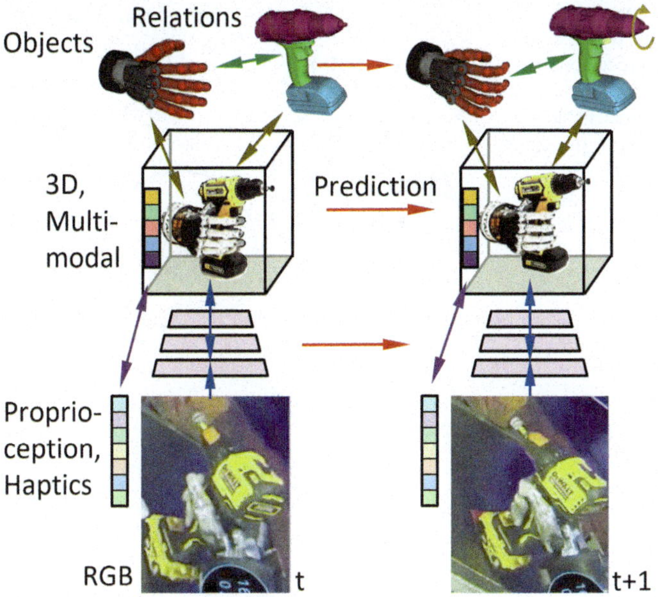

Fig. 13.2 Scene perception and prediction on three levels: in the sensor coordinate frame (bottom), in 3D multimodal embeddings (center), and with objects and their relations (top)

levels of abstraction. Higher layers shall plan abstract actions longer into the future, which are concretized on lower levels for shorter time horizons. A large variety of skills shall be learned for modular behaviors that activate coarse-to-fine actors according to the situation. Binding objects or places to roles shall yield parametrizable skills, such as grasping or placing an object or navigating toward a waypoint while avoiding obstacles.

Conscious Prediction and Planning Methods for selecting a small set of elements from the highest level C0 representations and for maintaining them in a working memory (WM) shall be learned. This WM will be the basis for learning action-conditioned predictions, based on binding selected elements to variables of applicable rules.

Structured predictions of state transition rewards, value, and action selection probabilities shall be learned from interactions with simulated and real environments. LLMs shall be incorporated as oracles.

As illustrated in Fig. 13.3, the learned WM world models shall be used for efficient action planning by sequential search. A spatiotemporal action abstraction shall be learned, such that sub-plans are reused in different contexts. The learned models shall be used for autonomous operation of mobile manipulation in complex novel situations.

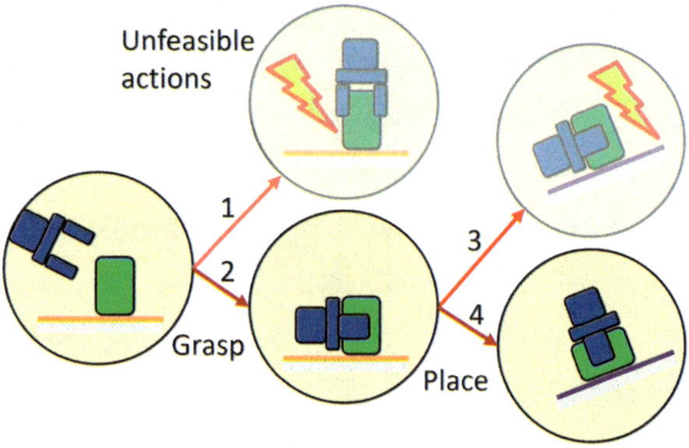

Fig. 13.3 Conscious planning. The WM state is rolled out using actions 1–4. Actions 1 and 3 are unfeasible (unreachable top grasp and unstable placement, respectively)

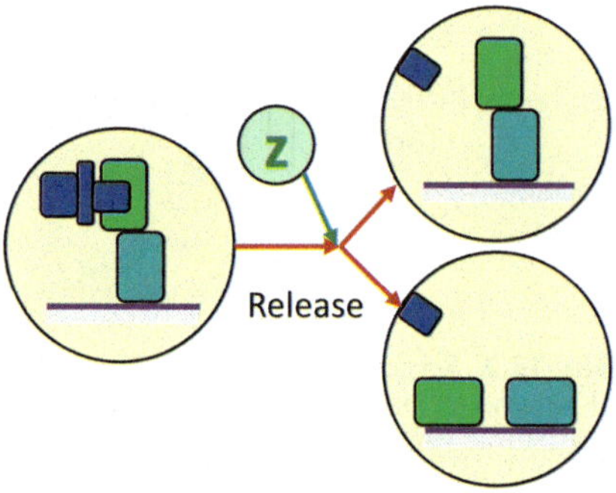

Fig. 13.4 Predicting multiple plausible futures conditioned on latent variable z

Conscious Self-Monitoring Methods for assessing the confidence of perceptions and predictions shall be developed. They shall be based on learning the distributions of latent variables on which multiple plausible futures can be conditioned (see Fig. 13.4). By sampling from these variables, a tree-manifold of state-action rollouts can be generated, such that not only average value but also its variance and worst-case return can be estimated. These quantities shall be incorporated into perception and action selection, to obtain policies that collect more information when needed and avoid dangers. Furthermore, the execution of low-level skills shall be monitored by comparing the current percept to expected outcomes to detect errors and to mitigate them.

13.7 Methodology

My approach will be to add suitable *inductive biases* to deep reinforcement learning (DRL), such that structured representations and conscious processing are enforced, which will enable systematic generalization and self-monitoring. Inductive biases reflect assumptions about the statistics of modeled scenes and robot-environment interactions and are necessary for generalization [78]. For instance, hierarchical convolutional neural networks (CNNs) [64] hardwire local dependencies, translation equivariance, hierarchical structure, and invariance to local deformations. Recurrent neural networks [46] exploit equivariance over time.

Further biases are needed for higher cognitive functions. One example of these is choosing the *appropriate frame of reference* for modeling. Commonly, deep neural networks represent visual scenes in a sensor coordinate system. Describing objects in object-centered canonical frames normalizes away the variability induced by the 6D object pose [118]. In such object-centered frames, shape and appearance spaces can be learned much easier. Of course, such canonical frames are also useful for individual parts of objects, for which the 6D pose relative to the object-centered frame must be modeled. The *projection of the 3D world to 2D images* induces occlusions and discontinuities at object boundaries. Modeling the scene in 2.5D by individual depth layers or directly in 3D allows for more complete, continuous representations where occluded parts are present, and hence unoccluded parts can be predicted. Much of the image motion can be explained by *camera motion*. Hence, approaches that explicitly model the projection from 3D to the variable camera view can represent this dependency compactly. One particularly powerful assumption is *relational inductive bias* [6]. It expresses the observation that scenes can often be described in terms of entities (objects, parts, and groups) and their sparse pairwise interactions (relations). Relations describe interactions between entities on adjacent levels of abstraction, e.g., between the whole object and its parts, which represents the *compositional structure* of the world. Such compositional hierarchy can also be found on the action side, where tasks are composed of multiple subtasks and subtasks are composed of individual skills (see Fig. 13.5).

Relations are also present between closely interacting entities on the same level of description, e.g., for objects that are in contact or for adjacent subtasks. In *hierarchical categorization*, objects or actions are categorized on multiple granularities [25]. This allows for pooling instances of multiple finer categories to

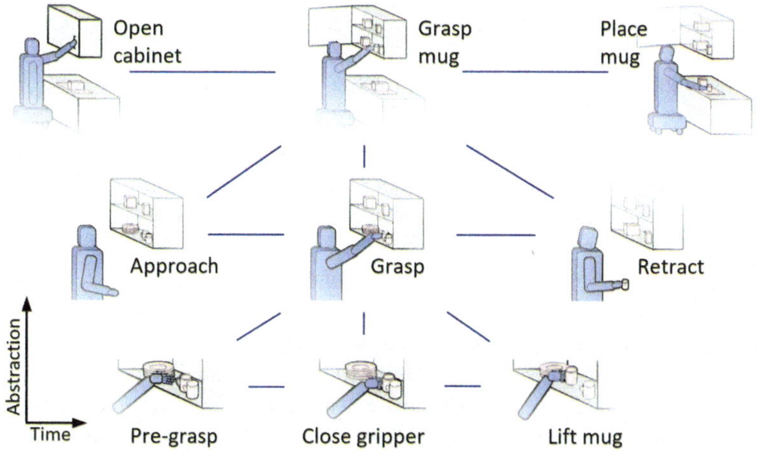

Fig. 13.5 Modeling actions on multiple levels of abstraction

learn models of coarser categories. *Planning in the now* [48] refers to the assumption that typical planning problems are not like mazes but can be solved by plans that consist of only a few steps which are described on a detailed, concrete level for the immediate future and on coarser, more abstract levels for the more distant future. This makes planning exponentially more efficient than detailed long-horizon planning.

Crucial is the *consciousness prior* proposed by Bengio [9]. It assumes that from subconscious, massively parallel computed representations, a small subset of elements is selected by attention mechanisms for sequential processing. This corresponds to a sparse factor graph on a conceptual, symbolic level, which affords abstract reasoning. Generic factors are probabilistic analogs of logical rules with quantifiers, i.e., with variables or arguments that can be bound [35]. On this level, graph neural networks [121] are applicable, which exploit *equivariance over entities and relations*.

Self-attention used in transformer networks [114] provides flexible information routing and learns sparse features with the sample complexity scaling only logarithmically with the context size [28]. High-level representations that describe verbalizable concepts as semantic variables that play a causal role can be encouraged with an *inductive bias toward words* [35]. When planning is restricted to maintaining a single state consisting of few elements and search is serial and can consist of few action-conditioned WM state predictions only, there is *pressure to aggregate* elements to higher level entities by discovering new concepts and macro actions.

Inductive biases alone will not suffice to address complex, open-ended real-world domains, because real-robot experience is expensive to obtain and cannot be collected in large enough quantities. Fortunately, *foundation models* for vision [24,92], language [1,85], and multimodal data [33,39] are available. They have been trained on Internet-scale data and summarize much more experience than real robots could make. Incorporating this knowledge through distillation will be a crucial factor for success. Another source of knowledge that I will incorporate is *human guidance and demonstrations*. Still, it will be necessary to develop methods first in a *photorealistic physics-based simulation* [79], where experience can be made cheaply in large quantities without endangering the real robot, before transferring them to the real world.

I will not approach human-like mobile manipulation in its full generality in a single step but will increase the level of difficulty gradually along the following dimensions: *number of objects:* starting without an object (learning a self-model), proceeding with a single object (grasping, placing, and pushing), and finally considering the manipulation of two objects and the use of tools; *DoF of the robot:* from single-handed object manipulation over bimanual tasks to mobile manipulation tasks; *DoF of objects:* from rigid objects, to articulated objects, and to deformable objects; *feedback modalities:* starting with RGB-D camera-based feedback, adding simple robot state and force-torque sensing, to rich multimodal feedback incorporating haptic measurements and 3D LiDAR; *familiarity of objects and tasks:* from known objects and tasks, over variations of known requiring parameter adaptations, to unfamiliar objects and tasks requiring compositional generalization; *level of*

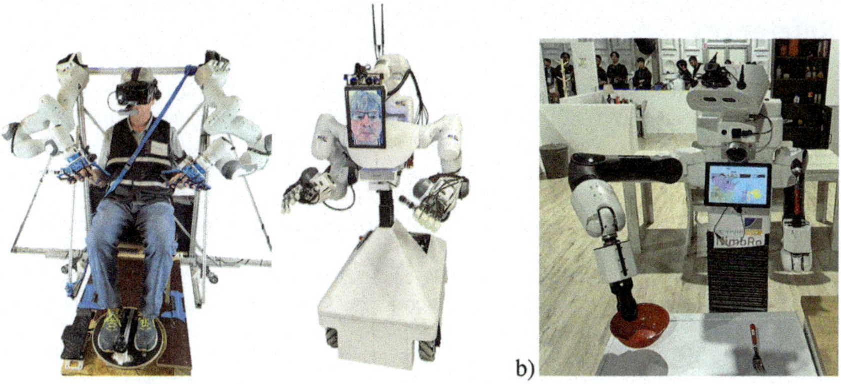

a) b)

Fig. 13.6 (**a**) NimbRo Avatar system [66] and (**b**) NimbRo@Home robot [75]

abstraction: starting with motion control on a fast time scale, proceeding with movement primitives, continuing with skills, such as grasping or placing an object, and finally considering chaining of skills to solve entire tasks; and *dynamics:* starting with quasi-static motion by considering only kinematics, proceeding with slow, compliant motion with interaction forces, and finally modeling dynamic effects of fast movements.

Intuitive immersive telepresence systems enable transporting human presence to remote locations in real time. My team NimbRo developed the winning entry for the ANA Avatar XPRIZE competition [66] (see Fig. 13.6a). Telepresence also provides a rich source of environment interaction data for learning structured perception and autonomous behavior.

My team NimbRo develops perception, planning, and learning for anthropomorphic mobile manipulation robots providing personal assistance [105] and benchmarks them in the RoboCup@Home league, where we recently won the German Open 2024 and RoboCup 2024 OPL competitions [75] (see Fig. 13.6b).

13.8 Conclusion

By incorporating insights from human cognition, the next generation of service robots will systematically generalize their knowledge to cope with novelty. This new generation of robots will also monitor themselves to obtain more information when needed, to avoid risks, and to detect and mitigate errors. Conscious service robots have much potential for numerous open-ended application domains, including assistance in everyday environments. Moreover, artificial conscious processing will contribute to a better understanding of consciousness in humans and other animals.

References

1. Anil R, Dai AM, Firat O et al (2023) PaLM 2 technical report. CoRR. ArXiv:2305.10403
2. Asfour T, Paus F et al (2019) ARMAR-6: a high-performance humanoid for human-robot collaboration in real-world scenarios. IEEE Rob Autom Mag 26(4):108–121
3. Assran M, Duval Q, Misra I et al (2023) Self-supervised learning from images with a joint-embedding predictive architecture. In: IEEE/CVF Conference on CVPR
4. Baars BJ (1993) A cognitive theory of consciousness. Cambridge University Press, Cambridge
5. Bahdanau D, Murty S, Noukhovitch M et al (2019) Systematic generalization: What is required and can it be learned? In: 7th International Conference on Learning Representations (ICLR)
6. Battaglia PW, Hamrick JB, Bapst V, Sanchez-Gonzalez A, Zambaldi VF et al (2018) Relational inductive biases, deep learning, and graph networks. CoRR. ArXiv:2406.09246
7. Beetz M, Kazhoyan G, Vernon D (2023) The CRAM cognitive architecture for robot manipulation in everyday activities. CoRR. ArXiv:2304.14119
8. Behnke S, Adams JA, Locke D (2023) The $10 million ANA Avatar XPRIZE competition: how it advanced immersive telepresence systems. IEEE Rob Autom Mag 30(4):98–104
9. Bengio Y (2017) The consciousness prior. CoRR. ArXiv:1709.08568
10. Bengio Y, Courville AC, Vincent P (2013) Representation learning: a review and new perspectives. IEEE Trans Pattern Anal Mach Intell 35(8):1798–1828
11. Botvinick MM, Braver TS, Barch DM, Carter CS, Cohen JD (2001) Conflict monitoring and cognitive control. Psychol Rev 108(3):624
12. Brohan A, Brown N, Carbajal J, Chebotar Y, Dabis J et al (2023) RT-1: robotics transformer for real-world control at scale. In: Robotics: Science and Systems XIX (RSS)
13. Bubeck S, Chandrasekaran V, Eldan R et al (2023) Sparks of artificial general intelligence: early experiments with GPT-4. CoRR. ArXiv:2303.12712
14. Carreira J, Zisserman A (2017) Quo Vadis, action recognition? A new model and the Kinetics dataset. In: IEEE Conference on Computer Vision and Pattern Recognition (CVPR)
15. Chen T, Kornblith S, Norouzi M, Hinton GE (2020) A simple framework for contrastive learning of visual representations. In: 37th International Conference on Machine Learning (ICML)
16. Chen Y, Yun Z, Ma Y et al (2023) Minimalistic unsupervised representation learning with the sparse manifold transform. In: 11th International Conference on Learning Representations (ICLR)
17. Chen Z, Li B, Wu S, Jiang K, Ding S, Zhang W (2023) Content-based unrestricted adversarial attack. In: Advances in Neural Information Processing Systems 36 (NeurIPS)
18. Cornelio C, Diab M (2024) Recover: a neuro-symbolic framework for failure detection and recovery. In: IEEE/RSJ International Conference on Intelligent Robots and Systems (IROS)
19. Cowan N (2010) The magical mystery four: how is working memory capacity limited, and why? Current Direct Psychol Sci 19(1):51–57
20. Dafarra S, Pattacini U, Romualdi G et al (2024) iCub3 avatar system: enabling remote fully immersive embodiment of humanoid robots. Sci Rob 9(86):eadh3834
21. Dasari S, Ebert F, Tian S, Nair S, Bucher B et al (2019) RoboNet: large-scale multi-robot learning. In: Conference on Robot Learning (CoRL), pp 885–897
22. d'Avila Garcez A, Lamb LC (2023) Neurosymbolic AI: the 3rd wave. Artif Intell Rev 56(11):12387–12406
23. Dehaene S, Lau H, Kouider S (2017) What is consciousness, and could machines have it? Science 358(6362):486–492
24. Dehghani M, Djolonga J, Mustafa B et al (2023) Scaling vision transformers to 22 billion parameters. In: International Conference on Machine Learning (ICML), pp 7480–7512
25. Deng J, Dong W, Socher R et al (2009) ImageNet: a large-scale hierarchical image database. In: IEEE Conference on Computer Vision and Pattern Recognition (CVPR), pp 248–255

26. Deng B, Lewis JP, Jeruzalski T et al (2020) NASA: neural articulated shape approximation. In: 16th European Conference on Computer Vision (ECCV), pp 612–628
27. Driess D, Xia F, Sajjadi MSM et al (2023) PaLM-E: an embodied multimodal language model. In: International Conference on Machine Learning (ICML), pp 8469–8488
28. Edelman BL, Goel S, Kakade SM, Zhang C (2022) Inductive biases and variable creation in self-attention mechanisms. In: International Conference on Machine Learning (ICML)
29. Florence PR, Manuelli L, Tedrake R (2020) Self-supervised correspondence in visuomotor policy learning. IEEE Rob Autom Lett 5(2):492–499
30. Fodor J (2001) Language, thought and compositionality. Roy Inst Philos Suppl 48:227–242
31. Friston K (2018) Does predictive coding have a future? Nat Neurosci 21(8):1019–1021
32. Gao G, Liu W et al (2024) GraphDreamer: compositional 3D scene synthesis from scene graphs. In: IEEE/CVF Conference on Computer Vision and Pattern Recognition (CVPR)
33. Girdhar R, El-Nouby A et al (2023) ImageBind one embedding space to bind them all. In: IEEE/CVF Conference on Computer Vision and Pattern Recognition (CVPR)
34. Glanois C, Jiang Z, Feng X et al (2022) Neuro-symbolic hierarchical rule induction. In: International Conference on Machine Learning (ICML), pp 7583–7615
35. Goyal A, Bengio Y (2022) Inductive biases for deep learning of higher-level cognition. Proc R Soc A 478(2266):20210068
36. Greff K, van Steenkiste S, Schmidhuber J (2020) On the binding problem in artificial neural networks. CoRR. ArXiv:2012.05208
37. Gui J, Chen T et al (2024) A survey on self-supervised learning: algorithms, applications, and future trends. IEEE Trans Pattern Anal Mach Intell 46(12):9052–9071
38. Haarnoja T, Moran B, Lever G et al (2024) Learning agile soccer skills for a bipedal robot with deep reinforcement learning. Sci Rob 9(89):eadi8022
39. Han J, Gong K, Zhang Y et al (2024) OneLLM: one framework to align all modalities with language. In: IEEE/CVF Conference on Computer Vision and Pattern Recognition (CVPR)
40. Hauser K, Watson E, Bae J, Bankston J, Behnke S et al (2024) Analysis and perspectives on the ANA Avatar XPRIZE competition. Int J Soc Rob 17:473–504
41. He K, Chen X, Xie S et al (2022) Masked autoencoders are scalable vision learners. In: IEEE/CVF Conference on Computer Vision and Pattern Recognition (CVPR)
42. Herzog A, Rao K, Hausman K et al (2023) Deep RL at scale: sorting waste in office buildings with a fleet of mobile manipulators. In: Robotics: Science and Systems XIX (RSS)
43. Hessel M, van Hasselt H, Modayil J, Silver D (2019) On inductive biases in deep reinforcement learning. CoRR. ArXiv:1907.02908
44. Hitzler P, Sarker MK, Eberhart A (2023) Compendium of Neurosymbolic Artificial Intelligence. IOS Press, Amsterdam
45. Hjelm RD, Fedorov A et al (2019) Learning deep representations by mutual information estimation and maximization. In: 7th International Conference on Learning Representations (ICLR)
46. Hochreiter S, Schmidhuber J (1997) Long short-term memory. Neural Comput 9(8):1735–1780
47. Hughes N, Chang Y, Carlone L (2022) Hydra: a real-time spatial perception system for 3D scene graph construction and optimization. In: Robotics: Science and Systems (RSS)
48. Kaelbling LP, Lozano-Pérez T (2011) Hierarchical task and motion planning in the now. In: IEEE International Conference on Robotics and Automation (ICRA), pp 1470–1477
49. Kahneman D (2011) Thinking, Fast and Slow. Macmillan, New York
50. Kalashnikov D, Irpan A, Pastor P et al (2018) Scalable deep reinforcement learning for vision-based robotic manipulation. In: Conference on Robot Learning (CoRL), pp 651–673
51. Kalashnikov D, Varley J, Chebotar Y, Swanson B et al (2021) Scaling up multi-task robotic reinforcement learning. In: Conference on Robot Learning (CoRL), pp 557–575
52. Kazhoyan G, Stelter S, Kenfack FK et al (2021) The robot household marathon experiment. In: IEEE International Conference on Robotics and Automation (ICRA), pp 9382–9388
53. Kheddar A, Roa MA, Wieber P et al (2019) Humanoid robots in aircraft manufacturing: the airbus use cases. IEEE Rob Autom Mag 26(4):30–45

54. Kim MJ, Pertsch K, Karamcheti S, Xiao T, Balakrishna A, Nair S, Rafailov R et al (2024) OpenVLA: an open-source vision-language-action model. CoRR. ArXiv:2406.09246
55. Kittmann R, Fröhlich T, Schäfer J, Reiser U, Weißhardt F, Haug A (2015) Let me introduce myself: i am Care-O-bot 4, a gentleman robot. In: Mensch und Computer (MuC)
56. Klamt T, Kamedula M, Karaoguz H, Kashiri N, Laurenzi A, Lenz C et al (2019) Flexible disaster response of tomorrow: final presentation and evaluation of the CENTAURO system. IEEE Rob Autom Mag 26(4):59–72
57. Klamt T, Schwarz M, Lenz C, Baccelliere L, Buongiorno D, Cichon T et al (2020) Remote mobile manipulation with the Centauro robot: Full-body telepresence and autonomous operator assistance. J Field Rob 37(5):889–919
58. Kolesnikov A, Beyer L, Zhai X et al (2020) Big transfer (bit): general visual representation learning. In: 16th European Conference on Computer Vision (ECCV), pp 491–507
59. Kumagai I et al (2019) Toward industrialization of humanoid robots: autonomous plasterboard installation to improve safety and efficiency. IEEE Rob Autom Mag 26(4):20–29
60. Lake BM, Baroni M (2018) Generalization without systematicity: on the compositional skills of sequence-to-sequence recurrent networks. In: International Conference on Machine Learning (ICML)
61. Lake BM, Baroni M (2023) Human-like systematic generalization through a meta-learning neural network. Nature 623(7985):115–121
62. Lake BM, Ullman TD, Tenenbaum JB, Gershman SJ (2017) Building machines that learn and think like people. Behav Brain Sci 40:e253
63. Lamb LC, d'Avila Garcez AS et al (2020) Graph neural networks meet neural-symbolic computing: a survey and perspective. In: Proceedings of the Twenty-Ninth International Joint Conference on Artificial Intelligence (IJCAI)
64. LeCun Y, Bengio Y, Hinton G (2015) Deep learning. Nature 521(7553):436–444
65. Lee MA, Zhu Y, Zachares P, Tan M, Srinivasan K, Savarese S, Fei-Fei L, Garg A, Bohg J (2020) Making sense of vision and touch: Learning multimodal representations for contact-rich tasks. IEEE Trans Rob 36(3):582–596
66. Lenz C, Schwarz M, Rochow A, Pätzold B, Memmesheimer R, Schreiber, M., Behnke S (2025) NimbRo wins ANA Avatar XPRIZE immersive telepresence competition: human-centric evaluation and lessons learned. Int J Soc Rob 17(3):337–361
67. Liang J, Jiang L et al (2020) The garden of forking paths: towards multi-future trajectory prediction. In: Conference on Computer Vision and Pattern Recognition (CVPR)
68. Liu S, Lever G, Wang Z et al (2022) From motor control to team play in simulated humanoid football. Sci Rob 7(69):eabo0235
69. Liu B, Jiang Y, Zhang X et al (2023) LLM+P: empowering large language models with optimal planning proficiency. CoRR. ArXiv:2304.11477
70. Liu H, Li C, Wu Q, Lee YJ (2023) Visual instruction tuning. In: Advances in Neural Information Processing Systems NeurIPS 36
71. Mancini M, Naeem MF, Xian Y, Akata Z (2021) Open world compositional zero-shot learning. In: IEEE Conference on Computer Vision and Pattern Recognition (CVPR), pp 5222–5230
72. Manhaeve R, Dumancic S, Kimmig A, Demeester T, Raedt LD (2021) Neural probabilistic logic programming in DeepProbLog. Artif Intell 298:103504
73. Marra G, Dumancic S, Manhaeve R, Raedt LD (2024) From statistical relational to neurosymbolic artificial intelligence: a survey. Artif Intell 328:104062
74. Meeussen W, Wise M, Glaser S et al (2010) Autonomous door opening and plugging in with a personal robot. In: IEEE International Conference on Robotics and Automation (ICRA), pp 729–736
75. Memmesheimer R, Nogga J, Pätzold B, Kruzhkov E, Bultmann S, Schreiber M, Bode J, Karacora B, Park J, Savinykh A, Behnke S (2025) RoboCup@Home 2024 OPL winner NimbRo: Anthropomorphic service robots using foundation models for perception and planning. In: RoboCup 2024: Robot World Cup XXVII. Springer

76. Menapace W, Lathuilière S, Tulyakov S, Siarohin A, Ricci E (2021) Playable video generation. In: IEEE Conference on Computer Vision and Pattern Recognition (CVPR)
77. Mildenhall B, Srinivasan PP, Tancik M et al (2022) NeRF: Representing scenes as neural radiance fields for view synthesis. Commun ACM 65(1):99–106
78. Mitchell TM (1980) The need for biases in learning generalizations. Department of Computer Science, Rutgers University, New Brunswick, NJ, Tech Rep
79. Mittal M, Yu C, Yu Q et al (2023) Orbit: a unified simulation framework for interactive robot learning environments. IEEE Rob Autom Lett 8(6):3740–3747
80. Moreno-Torres JG, Raeder T, Alaíz-Rodríguez R, Chawla NV, Herrera F (2012) A unifying view on dataset shift in classification. Pattern Recog 45(1):521–530
81. Mosbach M, Ewertz JN, Villar-Corrales A, Behnke S (2024) SOLD: reinforcement learning with slot object-centric latent dynamics. CoRR. ArXiv:2410.08822
82. Nau DS, Au T, Ilghami O, Kuter U, Murdock JW, Wu D, Yaman F (2003) SHOP2: an HTN planning system. J Artif Intell Res 20:379–404
83. Nye MI, Tessler MH et al (2021) Improving coherence and consistency in neural sequence models with dual-system, neuro-symbolic reasoning. In: Advances in NeurIPS 34
84. O'Neill A, Rehman A et al (2024) Open X-Embodiment: robotic learning datasets and RT-X models. In: IEEE International Conference on Robotics and Automation (ICRA), pp 6892–6903
85. OpenAI (2023) GPT-4 technical report. CoRR. ArXiv:2303.08774
86. Pernias P, Rampas D et al (2024) Würstchen: An efficient architecture for large-scale text-to-image diffusion models. In: 12th International Conference on Learning Representations (ICLR)
87. Placed JA, Strader J, Carrillo H et al (2023) A survey on active simultaneous localization and mapping: State of the art and new frontiers. IEEE Trans Rob 39(3):1686–1705
88. Pratap V, Tjandra A, Shi B et al (2024) Scaling speech technology to 1,000+ languages. J Mach Learn Res 25(97):1–97:52
89. Qin Y, Liang S, Ye Y et al (2024) ToolLLM: facilitating large language models to master 16000+ real-world APIs. In: 12th International Conference on Learning Representations (ICLR)
90. Qin Y, Shi Z, Yu J et al (2024) WorldSimBench: towards video generation models as world simulators. CoRR. ArXiv:2410.18072
91. Radford A, Kim JW, Xu T et al (2023) Robust speech recognition via large-scale weak supervision. In: International Conference on Machine Learning (ICML), pp 28492–28518
92. Ravi N, Gabeur V, Hu Y et al (2024) SAM 2: segment anything in images and videos. CoRR. ArXiv:2408.00714
93. Rosu RA, Behnke S (2023) PermutoSDF: fast multi-view reconstruction with implicit surfaces using permutohedral lattices. In: IEEE/CVF Conference on Computer Vision and Pattern Recognition (CVPR), pp 8466–8475
94. Sajjadi MSM, Duckworth D, Mahendran A et al (2022) Object scene representation transformer. In: Advances in Neural Information Processing Systems 35 (NeurIPS)
95. Schmid L, Abate M et al (2024) Khronos: a unified approach for spatio-temporal metric-semantic SLAM in dynamic environments. In: Robotics: Science and Systems (RSS)
96. Schölkopf B, Locatello F, Bauer S, Ke NR, Kalchbrenner N, Goyal A, Bengio Y (2021) Toward causal representation learning. Proc IEEE 109(5):612–634
97. Schrittwieser J, Antonoglou I, Hubert T et al (2020) Mastering Atari, Go, chess and shogi by planning with a learned model. Nature 588(7839):604–609
98. Schulz H, Behnke S (2012) Deep learning - Layer-wise learning of feature hierarchies. Künstliche Intell (Artif Intell) 26(4):357–363
99. Schwarz M, Lenz C, Memmesheimer R, Pätzold B, Rochow A, Schreiber M, Behnke S (2023) Robust immersive telepresence and mobile telemanipulation: NimbRo wins ANA Avatar XPRIZE finals. In: 22nd IEEE-RAS International Conference on Humanoid Robots (Humanoids)

100. Shindo H, Pfanschilling V, Dhami DS, Kersting K (2023) αILP: thinking visual scenes as differentiable logic programs. Mach Learn 112(5):1465–1497
101. Shorten C, Khoshgoftaar TM (2019) A survey on image data augmentation for deep learning. J Big Data 6:60
102. Smet LD, Venturato G, Raedt LD, Marra G (2025) Relational neurosymbolic Markov models. In: 39th Conference on Artificial Intelligence (AAAI)
103. Smolensky P, McCoy RT et al (2022) Neurocompositional computing: from the central paradox of cognition to a new generation of AI systems. AI Mag 43(3):308–322
104. Soulos P, Hu EJ, McCurdy K et al (2023) Differentiable tree operations promote compositional generalization. In: International Conference on Machine Learning (ICML), pp 32499–32520
105. Stückler J, Schwarz M, Behnke S (2016) Mobile manipulation, tool use, and intuitive interaction for cognitive service robot Cosero. Front Rob AI 3:58
106. Stückler J, Holz D, Behnke S (2012) RoboCup@Home: demonstrating everyday manipulation skills in RoboCup@Home. IEEE Rob Autom Mag 19(2):34–42
107. Sun Q, Wang J, Yu Q, Cui Y, Zhang F, Zhang X, Wang X (2024) EVA-CLIP-18B: scaling CLIP to 18 billion parameters. CoRR. ArXiv:2402.04252
108. Sutton RS, Barto AG (2018) Reinforcement Learning: An Introduction. MIT Press, Cambridge
109. Tenenbaum JB, Kemp C, Griffiths TL, Goodman ND (2011) How to grow a mind: statistics, structure, and abstraction. Science 331(6022):1279–1285
110. Toyer S, Trevizan FW, Thiébaux S, Xie L (2018) Action schema networks: generalised policies with deep learning. In: 32nd Conference on Artificial Intelligence (AAAI), pp 6294–6301
111. Trinh TH, Wu Y, Le QV, He H, Luong T (2024) Solving olympiad geometry without human demonstrations. Nature 625(7995):476–482
112. Valevski D, Leviathan Y, Arar M, Fruchter S (2024) Diffusion models are real-time game engines. CoRR. ArXiv:2408.14837
113. Varadarajan B et al (2022) MultiPath++: efficient information fusion and trajectory aggregation for behavior prediction. In: International Conference on Robotics and Automation (ICRA)
114. Vaswani A, Shazeer N, Parmar N et al (2017) Attention is all you need. In: Advances in Neural Information Processing Systems 30 (NeurIPS), pp 5998–6008
115. Veerapaneni R, Co-Reyes JD, Chang M, Janner M et al (2019) Entity abstraction in visual model-based reinforcement learning. In: Conference on Robot Learning (CoRL)
116. Villar-Corrales A, Wahdan I, Behnke S (2023) Object-centric video prediction via decoupling of object dynamics and interactions. In: IEEE International Conference on Image Processing (ICIP), pp 570–574
117. Vinyals O, Babuschkin I, Czarnecki WM et al (2019) Grandmaster level in StarCraft II using multi-agent reinforcement learning. Nature 575(7782):350–354
118. Wang H et al (2019) Normalized object coordinate space for category-level 6D object pose and size estimation. In: IEEE Conference on Computer Vision and Pattern Recognition (CVPR)
119. Wang X, Misra I, Zeng Z et al (2024) VideoCutLER: surprisingly simple unsupervised video instance segmentation. In: IEEE/CVF Conference on CVPR
120. Wu P, Escontrela A, Hafner D, Abbeel P, Goldberg K (2022) DayDreamer: world models for physical robot learning. In: Conference on Robot Learning (CoRL), pp 2226–2240
121. Wu Z, Pan S, Chen F, Long G, Zhang C, Yu PS (2021) A comprehensive survey on graph neural networks. IEEE Trans Neural Netw Learn Syst 32(1):4–24
122. Wu Z, Dvornik N, Greff K et al (2023) SlotFormer: unsupervised visual dynamics simulation with object-centric models. In: 11th International Conference on Learning Representations (ICLR)
123. Wurman PR, Barrett S, Kawamoto K et al (2022) Outracing champion Gran Turismo drivers with deep reinforcement learning. Nature 602(7896):223–228

124. Xie Q, Luong M et al (2020) Self-training with noisy student improves ImageNet classification. In: IEEE/CVF Conference on Computer Vision and Pattern Recognition (CVPR)
125. Yamamoto T, Takagi Y et al (2019) Human support robot as research platform of domestic mobile manipulator. In: RoboCup 2019: Robot World Cup XXIII. Springer
126. Yang J, Gao S, Qiu Y et al (2024) Generalized predictive model for autonomous driving. In: IEEE/CVF Conference on Computer Vision and Pattern Recognition (CVPR)
127. Yoshiike T et al (2019) The experimental humanoid robot E2-DR: a design for inspection and disaster response in industrial environments. IEEE Rob Autom Mag 26(4):46–58
128. Yu T, Xiao T, Tompson J et al (2023) Scaling robot learning with semantically imagined experience. In: Robotics: Science and Systems XIX (RSS)
129. Zhai X, Kolesnikov A, Houlsby N, Beyer L (2022) Scaling vision transformers. In: IEEE/CVF Conference on Computer Vision and Pattern Recognition (CVPR), pp 1204–1213
130. Zhou Y, Feinman R, Lake BM (2024) Compositional diversity in visual concept learning. Cognition 244:105711

Summary

Draw Our Future! An Idea Competition for Junior Visionaries Imagining Better Worlds, Sustainable Environments, and Flourishing Communities

Sonja Hatz • Laura von Soosten • Ulrich A. K. Betz • Simone von Stockhausen-Taufenbach • Volker Tschiedel
Merck KGaA, Darmstadt, Germany
Future Insight e.V. Darmstadt, Darmstadt, Germany
laura.v.soosten@future-insight.org; ulrich.betz@future-insight.org
Ministry of Science and Health of Rhineland-Palatinate, Mainz, Germany
simone.vonstockhausen-taufenbach@stk.rlp.de
Ministry of Education of Rhineland-Palatinate, Mainz, Germany
Volker.tschiedel@bm.rlp.de

Introduction

The earlyeducation period is crucial for fostering curiosity and interest in science and technology, engineering and maths (STEM) subjects, as children in this age group are naturally inquisitive and open to exploring new concepts. With the Junior Visions prize, we developed a drawing competition as an offer to students from grades 3 to 7 (typically ages 7–13) in schools across Rheineland-Palatinate.

The competition was developed by Future Insight e.V. and ran with the support of the Ministry of Education of Rhineland-Palatinate, the Ministry of Science and Health of Rhineland-Palatinate as well as the City of Mainz. The program was promoted to schools as a program to support science education in classrooms and to parents as a competition supporting informal out-of-school science education. To participate, students sent in their artistic ideas for a better world. The competition included a selection process where a jury evaluated the submissions and the top

U. A. K. Betz (ed.), *Science for a Better Tomorrow*,
https://doi.org/10.1007/978-3-031-93623-4

ten artworks were showcased on social media, allowing the public to vote for their favorite. The top ten pictures were exhibited at the Curious2024—Future Insight™ conference in Mainz, Germany where the three artworks with the most likes and shares in the social media voting were awarded their prizes, including the top prize of a visit to the Institute of Biotechnology and Drug Research (IBWF) in Mainz for the entire winner's entire school class.

As supporters of the competition, Dr. Stefanie Hubig, Minister of Education in Rhineland-Palatinate and patroness of the competition and Biotechnology Coordinator Prof. Dr. Eckhard Thines gave a positive assessment of the art competition. "The student competition is a wonderful platform to engage young students with scientific concepts and inspire them to think creatively about the future. Ideas from young people can bring new and innovative perspectives to a sustainable and healthy future."

Results

The student participants imagined and visualized their views of the future and many different technological solutions to some of the world's greatest challenges such as space travel, climate change, pollution, and resource depletion. The 132 submitted pictures from 43 different schools integrated many topics including accelerated tree growth, sustainable fashion alternatives, advanced recycling technologies utilizing robotics, and the use of plants to clean up oil spills. These artworks not only reflected the creativity of young minds but also displayed children's understanding of solutions to real needs. The ideas were reviewed by established scientists, who engaged with the participant's innovative thinking.

Pictures of the Winners of the 2024 Junior Visions Drawing and Painting Competition

We present here the winning images together with the scientific comments written by Prof. Dr. Eckhard Thines.

"Zukunftsideen von Jacob": Jacob's Artwork (Third Grade)

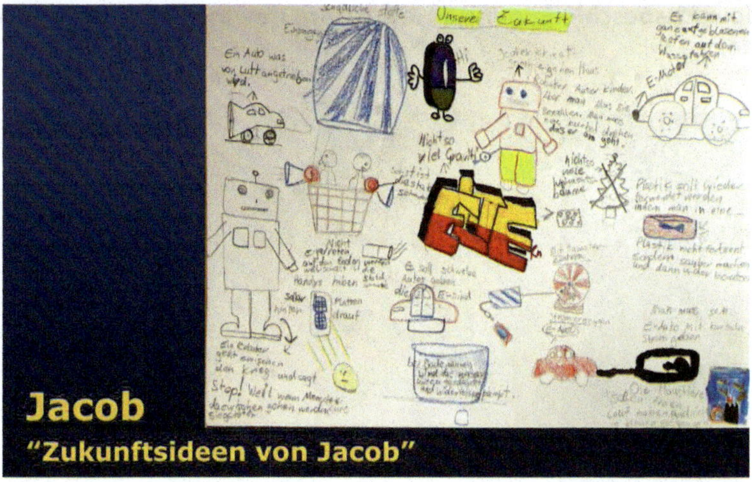

Fig. A.1 Jacob's artwork presents a collection of innovative ideas for a cleaner world, showcasing wonderfully practical solutions such as alternatives to combustion engines and solar-powered devices

"Weltraumaufzug": Hawon's Space Elevator (Third Grade)

Fig. A.2 Table Hawon's imaginative depiction of a Space Elevator connecting Earth and Mars challenges conventional space travel concepts, showcasing a detailed vision of future transportation technology. In addition to the impressive artistic execution, the detailed planning of the facility complex immediately catches the eye. The concept of a Space Elevator is based on real scientific principles, but its construction faces significant challenges, including the need for materials that are stronger and lighter than any currently available. However, significant progress is being made in the field of material science, and new materials such as hydrogels or carbon fibers could contribute to realizing this vision

Turbowachsbeschleuniger: Marie's Turbo Growth Accelerator (Sixth Grade)

Fig. A.3 This artwork illustrates a concept aimed at enhancing the efficiency of agriculture and reforestation. It suggests a method to accelerate tree growth through nutrient-rich soil and targeted irrigation, thereby addressing the challenges posed by climate change. Forest dieback due to climate change impacts biodiversity, groundwater, and even the microclimate, as forests have a cooling effect. Forests are also important in protecting against floods and landslides. Perhaps the Turbo Growth Accelerator represents a method to speed up tree growth through nutrient-rich soil and targeted irrigation, if the optimistic goal of 86 m^2 per day can be achieved

Top 10 Images in the Collage

Conclusions

The results show how a science art competition can serve as a platform for creativity and innovation. This engaging format connects students to real-world scientific concepts, and allows them to express their understanding of scientific concepts through art. The format has the potential to make learning more engaging and accessible, particularly for audiences that may not normally be interested in STEM topics. By integrating art into science education, students can demonstrate critical thinking and problem-solving skills while also fostering a love for STEM subjects thereby contributing to integrative and cross-disciplinary STEM teaching practices.

Awarding the Junior Visions Prize at the Curious 2024: Future Insight Science Conference

Fig. A.4 On at the AMA stage (l.t.r.) in Mainz: Moderator Anna Leah Scholl, winners Marie, Hawon and Jacob

Fig. A.5 The Junior Visions Top 10 wall at the Curious2024—Future Insight science conference

Curious 2024: Future Insight™ Conference Abstracts

Keynotes

David Baker, University of Washington and Howard Hughes Medical Institute, United States of America: Design of new protein functions using deep learning Proteins mediate the critical processes of life and beautifully solve the challenges faced during the evolution of modern organisms. Our goal is to design a new generation of proteins that address current-day problems not faced during evolution. In contrast to traditional protein engineering efforts, which have focused on modifying naturally occurring proteins, we design new proteins from scratch to optimally solve the problem at hand. Increasingly, we develop and use deep learning methods to design amino acid sequences that are predicted to fold to desired structures and functions. We also produce synthetic genes encoding these sequences and characterize them experimentally. In this talk, I describe several recent advances in protein design.

Martin Beck, Max Planck Institute of Biophysics, Germany: Future views of structural cell biology In his talk, Martin Beck explains how the capsid containing the viral genome of HIV enters the nucleus during infection. He uses this process as an example to illustrate cellular tomography to visualize the molecular processes happening inside of human cells today, and how in the future, 4D virtual reality will help the next generation of scientists to understand very complex cellular phenomena.

Sven Behnke, Rheinische Friedrich-Wilhelms-Universität Bonn/Computer Science Institute VI—Intelligent Systems and Robotics, Germany: Perception, Planning, and Learning for Cognitive Robots Cognitive robots need to perceive their environment to act in a goal-directed way. We developed efficient methods for environment perception, based on local multiresolution representations, deep learning, and object-centered representations. The resulting models form the basis for locomotion and manipulation planning, for which we developed efficient methods. My team integrated perception and planning for autonomous control of multiple cognitive robots and demonstrated their abilities in complex environments, including humanoid soccer, domestic service, space exploration, disaster-response, bin picking, and industrial inspection. I will discuss ideas on how structured models for efficient perception and planning can be learned from limited experience.

Mario Capecchi, University of Utah School of Medicine, United States of America: Defective Hoxb8 Microglia are Causative for both Chronic Anxiety and OCSD-like Behavior in Mice We have shown there are two cell lineages in mice that give rise to microglia with distinct ontogenies; Hoxb8 microglia and non-Hoxb8 microglia. Disruption of Hoxb8 or ablation of the Hoxb8 cell lineage gives rise to both chronic anxiety and OCSD-like behavior (trichotillomania, compulsive pathological overgrooming, leading to lesions at the sites of overgrooming). There is a strong sex bias for both behavioral pathologies, with females showing much more aggressive disease. By cell transplantation of purified, cell sorted Hoxb8 microglia progenitors derived from Wt or Hoxb8 mutant mice we were able to demonstrate causation of defective Hoxb8 microglia for both behavioral pathologies. Recently we have demonstrated that optogenetic activation of Hoxb8 microglia in specific regions of the brain induces higher levels of anxiety, grooming or both. Furthermore, these experiments have demonstrated that Hoxb8 microglia function to reduce anxiety and grooming (function as brakes), whereas non-Hoxb8 microglia function to increase these behaviors (function as an accelerator). This provides a biological reason for mice having two populations of microglia. We have further extended this work to define the molecular mechanisms used by microglia to communicate with specific neuronal circuits to modulate anxiety and grooming in mice.

Alon Chen, Weizmann Institute, Israel: Bridging the gap: Preclinical research in mental health Mental disorders are a significant cause of disability worldwide. They profoundly affect individuals' well-being and impose a substantial financial burden on societies. However, despite decades of extensive research, the effectiveness of current therapeutics for mental disorders is often not satisfactory or well tolerated. Moreover, most novel therapeutic candidates fail in clinical testing, which also brings into question the effectiveness of using animal models in preclinical studies. In this lecture, I will present a paradigm shift in the methodologies used to measure animal behavior in laboratory settings.

Steven Chu, Stanford University, United States of America: Materials as an Enabling Technology in Getting to Net-Zero GHG Emissions Throughout human history, materials such as metals, glass and ceramics, chemicals, plastics, and fertilizers and have enabled multiple industrial and agricultural revolutions that have profoundly transformed the world. However, the unintended consequence of these revolutions is that the greenhouse gas emissions are changing Earth's climate. Most of our energy sources, industrial and agricultural chemicals and building materials and are fossil-fuel based or demand process heat. The challenges how to provide clean energy, water, air, food in a world of over 8 billion people and likely to grow to 11 billion by 2100 are formidable. I will discuss new material challenges needed to achieve net-zero greenhouse emissions and a more sustainable and prosperous future.

Patrick Cramer, Max Planck Society, Germany: The future of science—the Max Planck way For over 75 years, the aim of the Max Planck Society has been to deliver excellent research by providing researchers with trust, generous long-term funding, and support infrastructure. Over the next years, we will continue our mission, open new research fields, and expand our efforts to provide optimal support for the best scientists at all career stages. In my presentation, I will illustrate our approach with the use of examples from different research fields. I will also explain how, in a complex and quickly changing world, scientists can continue to collaborate and build bridges between people. Finally, I will illustrate how basic science can lead to application and contribute to tackling global challenges—and how it can help in the ongoing transformation toward a sustainably future.

Andreas Dengel, DFKI, Germany, The Pulse of AI: Charting the Rhythm of the Biomedical Transformation The use of artificial intelligence in biomedicine promises to open up new frontiers and explore unexplored paths. At DFKI, we are exploring the symbiotic relationship between AI and biomedical processes, whether for deciphering complex biological data or improving medical decision making. In my talk, I will give examples of how AI can predict the interaction between proteins and viruses, which are crucial for the development of antiviral strategies, or how machine intelligence can be used for improved and transparent diagnostics, or how microscopic image analysis can be used for early detection of diseases and personalization of treatments.

Dorothee Dormann, JGU and IMB Mainz, Germany: Molecular aging of intrinsically disordered proteins in neurodegenerative diseases The talk will focus on molecular mechanisms that underlie the dysfunction of intrinsically disordered RNA-binding proteins (RBPs) in neurodegenerative diseases, such as ALS, FTD, and Alzheimer's disease. Of particular interest are the RBPs TDP-43 and FUS, which are genetically linked to ALS and form aberrant cytoplasmic aggregates in the degenerating brain regions. The talk will provide insights into recently discovered mechanisms that cause RBP mislocalization and aggregation, such as defects in nuclear import, phase separation and molecular aging into amyloids. It will also highlight interdisciplinary research conducted in the Collaborative Research Center SFB1551 on "Polymer concepts in cellular function".

Tim Flohrer, European Space Agency: Space Debris and Zero Debris Charter Our established standards and guidelines for space debris mitigation need to evolve, as not only the space debris environment, but also the scenarios for using space as a resource for many essential and vital services and applications undergo drastic and revolutionary changes. ESA, as part of its Agenda 2025, started to follow a "Zero Debris" approach. By implementing this approach by 2030, it is ESA's ambition that its contribution to the space debris population shall be zero from then onward. As a first step ESA's Space Debris Mitigation Policy and Requirements applicable to ESA missions have been adapted in 2023. Further, ESA facilitated a community approach to formulate a Zero Debris Charter that is open to worldwide signatories. We will discuss the identification of needed technologies and planned implementation steps for achieving "Zero Debris". We will present how protecting our space-based infrastructure through technologies demonstrated in ESA's Space Safety Program contributes to the sustainable utilization of space as an integral part of our modern and connected society.

Makoto Fujita, University of Tokyo, Japan, Molecular Confinement Effects in Self-assembled Cages Since our first report on a self-assembled coordination cage in 1995 (Nature, 1995, 378, 469), we and others have been developing the molecular confinement effects of the self-assembled cages. The cavities of our cages are extraordinarily large and are capable of binding neutral guests. Through molecular recognition, new properties, reactions, and functions have been created (Angew. Chem. Int. Ed. 2009, 48, 3418; Bull. Chem. Soc. Jpn. 2021, 94, 2351). One of the recent topics in the course of our study is protein encapsulation in a self-assembled gigantic cage (Chem 2021, 7, 2672).

Charles Godfray, Oxford Martin School, United Kingdom, Feeding 10 billion sustainably Global demand for food will likely increase by 40–60% this century before plateauing. Can we meet this demand while simultaneously reducing the negative effects of food production on the environment and halting the epidemic of diet-related disease. This talk will argue we can and will highlight and explore some of the exciting technical advances that will help make this possible though stressing that we will fail unless we have the political will to make hard decisions about refashioning the global food system.

Michael Graetzel, École Polytechnique Fédérale de Lausanne (EPFL), Switzerland: Molecular Photovoltaics and Artificial Photosynthesis Photovoltaic cells using molecular dyes, semiconductor quantum dots or perovskite pigments as light harvesters have emerged as credible contenders to conventional devices. They possess unique practical advantages in particular highly effective electricity production from ambient light, ease of manufacturing, flexibility and transparency, bifacial light harvesting, and esthetic appeal, which have fostered industrial production and commercial applications. They engendered perovskite solar cells (PSCs) which are a leading future PV technologies, their efficiency having reached 26.1%. My lecture will cover our most recent findings in these revolutionary photovoltaic domains that relate to the production of electricity and fuels from sunlight.

Melanie Heil, European Space Agency: Space Weather and Vigil Mission Solar storms come in different variety, spanning from increased high-energy electromagnetic radiation, to fast plasma clouds or highly energetic charged particles. They can affect infrastructure such as satellite systems, radio communication, but also power grids on the ground. Although we cannot stop our Sun, we can prepare ourselves to reduce impact of these events. ESA's Space Safety Program aims at protecting space and ground assets against adverse effects from space. The Space Weather Segment is focusing on such effects due to the activity of our Sun.

Frank Heinricht, Schott AG, Germany, DNA of SCHOTT: pioneering—responsibly—together The DNA of SCHOTT is pioneering, responsibly, together, because we believe that close partnerships with our customers can release the energy to pioneer the impossible. As we are thinking about the upcoming generations, we want to make an active contribution to climate protection and become the first climate neutral specialty glass company.

Brett Kagan, Cortical Labs, Australia: Synthesizing Biological Intelligence: Promises, Challenges, and Opportunities How to create a generally intelligent system? Recently, we embodied biological neurons in a real-time closed-loop system to simulate the classic arcade game, 'Pong' and determine if the simple neural systems could learn. Interestingly the learning rates were faster than machine learning and networks showed rapid reorganization that inform us about how brains process information. Moreover, preliminary work supports that this approach provides information previously inaccessible to help improve drug discovery and preclinical pharmacological testing. The data provides compelling insights into the potential value of these systems for wider applications, along with understandings on how to improve this technology further.

Robert Langer, MIT, United States of America: From nanotechnology to mRNA vaccines: How overcoming skepticism led to new medical treatments and ways to tackle a global health challenge Advanced drug delivery systems are having an enormous impact on human health. We start by discussing our early research on developing the first controlled release systems for macromolecules and the isolation of angiogenesis inhibitors and how these led to numerous new therapies. This early research then led to new drug delivery technologies including nanoparticles and nanotechnology that are now being studied for use treating cancer, other illnesses and in vaccine delivery (including the COVID-19 vaccine). Finally, by combining mammalian cells, including stem cells, with synthetic polymers, new approaches for engineering tissues are being developed that may someday help in various diseases. These can also serve as a basis for tissues on a chip which can potentially reduce animal and human testing. Examples in the areas of cartilage, skin, blood vessels, GI tract, and heart tissue are discussed.

Benjamin List, Max-Planck-Institut für Kohlenforschung, Germany: Catalysis for the world The science of catalysis is fascinating. To design a catalyst—by providing a lower energy pathway to convert substrates into the desired product—is as close as chemists can ever come to magic. Providing this lower energy pathway without being used up enables us to use a very small amount of a catalyst to convert large amounts of starting materials into a useful product. And there is so much more to it. Catalysis is not only a beautiful concept; it is a very important technology. Some would state that it is the most important technology for humankind on this planet. It is estimated that catalysis contributes to roughly one third of the GDP—which means figures in the trillions. There is probably no other technology that can claim to feed, to heal, to warm and to transport humans and their goods. The challenges humanity faces right now—global warming, energy conversion—will be addressed with the help of catalysis. In my talk, I will speak about the catalysis with organic molecules and how strong and confined acids could become universal catalysts.

Abraham (Avi) Loeb, Harvard University, United States of America: In Search for Technological Interstellar Objects The search for extraterrestrial life is one of the most exciting frontiers in science. First tentative clues were identified close to Earth in the form of the first two interstellar meteors which were tougher than iron, the unfamiliar interstellar object 'Oumuamua and Unidentified Aerial Phenomena (UAP).The "Galileo Project" ushers the new frontier of "interstellar archeology" in search of extraterrestrial technological relics.

Ralph Marquardt, Evonik Operations GmbH, Germany: Sustainable Chemistry: Solving the pressing challenges of our time at scale The chemical industry is uniquely positioned to solve the green transformation challenge. While the birthplace of innovation is the laboratory, bridging the gap between laboratory invention and large-scale application is crucial to tackle this challenge. In my keynote presentation, I will illustrate how we speed and scale up sustainable innovations, such as membranes for green hydrogen production, polyurethane recycling, new classes of biosurfactants, and biotechnological CO_2 activation.

Laura Matz, Merck KGaA, Darmstadt, Germany: Unleashing the power of data and AI to tackle industry challenges The world has changed significantly in the past decade, accelerated by advances in biology and artificial intelligence (AI) technologies. The global spotlight on mRNA vaccines during the COVID-19 pandemic and the subsequent introduction of ChatGPT underscore the transformative potential of these innovations and their impact on global society. Future breakthroughs in precision medicine and materials innovation can be unlocked when we direct AI and scientific knowledge at the right problems. The key is to connect the relevant data with the relevant analytical models to solve the right problem. In this presentation, we will illustrate how data ontologies and advanced analytical methods, such as generative AI and Bayesian optimization, are leveraged to drive scientific progress in healthcare and materials. Additionally, we will showcase a fully automated high-throughput synthesis lab, which complements the in-silico work in our drug discovery portfolio to generate a robust composite dataset, enhancing predictive capabilities.

Hartmut Michel, Max Planck Institute of Biophysics, Germany: The challenges in biomedicine and agriculture I shall present my ideas what should be done for a most complete understanding of the human body and for improvement of the yields in agriculture.

Shahaf Peleg, FBN Dummerstorf, Germany: A song of Light and Power—using light as an energy replacement for human longevity To generate electricity, factories burn coal in the presence of oxygen. However, this process produces toxic pollutants. Similarly, our cells produce energy by burning carbon compounds with oxygen in their mitochondria. Energy production is accompanied by the accumulation of toxic cellular waste and 'wear and tear' during aging. Here, I will discuss a novel idea termed 'energy replacement' to reduce cellular waste by partially substituting the cellular need for upstream metabolic activities and oxygen consumption with an engineered light-activated proton pump (mtON) in animal mitochondria. This disruptive technology will be the basis for gene therapy in human aging and beyond.

Joe Quirk, The Seasteading Institute, United States of America, Seasteading Will Solve Political Conflict We founded civilization on the wrong third of the planet. Two-thirds of this planet is ocean. When startup societies float permanently on the Blue Frontier, every city block will be mobile. This will foster variation in governance providers, and selection by governance customers, unleashing the magic of evolution in governance innovation. The success of floating self-governing cities has already been demonstrated on cruise ships. What if cruise ships were family-sized and floated permanently in international waters? This vision has inspired maritime professionals to start building the first seastead. I will show you their practical plans for a fluid free world without political conflict.

Marc Raibert, Boston Dynamics/The AI Institute, United States of America Making robots smarter, in both body and mind Robots are getting better and better all the time. Their ability to perceive the world around them is improving almost daily, while their bodies and physical skills have advanced by leaps and bounds. But robots are still pretty stupid and lack cognitive intelligence that would make them easier to use, safer to operate with people, and more productive. In this talk Marc Raibert will show some of the recent progress in the development of robots, then talk about using AI to make them smarter and eventually live up to our dreams.

Asya Rolls, Technion-Israel Institute of Technology, Technion Institute of R&D, Israel, A mechanistic perspective of the mind-body connection in medicine Thoughts and emotions can impact physiology. This connection is evident in the emergence of disease following stress, or recovery in response to placebo treatment. Nevertheless, our understanding of this fundamental aspect of physiology is still very limited. In this talk, I will discuss the mind-body connection from the neuroimmunological perspective. For example, how the brain represents the state of the immune system and how brain activity can regulate peripheral immunity to promote recovery from infections or cancer. On the other hand, the brain can induce disease and psychosomatic disorders, and gaining an understanding of the underlying mechanisms has the potential to introduce new therapeutic avenues utilizing the brain's therapeutic potential.

Andrée Rothermel, TRON gGmbH, Germany, TRON—Science in Translation TRON is an independent non-profit translational research organization that develops cutting-edge technologies to target diseases with high medical needs such as infectious or cardiovascular diseases, autoimmunity and cancer. We combine transdisciplinary competencies in genomics, immunology and bioinformatics to develop novel approaches to be used in personalized therapies. At the interface of academia and industry, TRON executes research at the leading edge to support innovative drug design for human health.

Khalid Salaita, Emory University, United States of America, DNA Mechanotechnology: nucleic acids that sense and generate molecular forces enable fundamental research tools and new biomedical diagnostics Modern machines, which are composed of force-generating motors, force sensors, and load-bearing structures, enabled the industrial revolution and are foundational to human civilization. Extreme miniaturization would enable machines that can manipulate molecules for applications in medicine, biological research, and material development. This was previously only a dream, but new synthetic methods to assemble and modify nucleic acids combined with single molecule force spectroscopy have propelled the emergence of a subfield that we call "DNA mechanotechnology". In this talk, I will discuss my group's efforts at using DNA mechanotechnology for new types of viral diagnostics and in biomedical applications.

Katja Scharpwinkel, BASF SE, Germany: Shaping transformation together—How Basic research and Chemical industry can ignite and foster Transformation In times of multiple crises which are parallelly ongoing and thus challenging economies and even democracies, it is crucial to continue and foster fact-based discussions and collaborations. Looking at current challenges posed by the climate crisis, this presentation shows how cooperation-based transformation is driven at BASF. Today still being among the top fossil-based energy consumers and CO_2 emitters, BASF builds on R&D and partnerships to reduce their carbon footprint to net zero by 2050 and switch to a circular economy.

Thomas Reiter, European Space Agency: Europeans in Space A general introduction to the European Space Agency (ESA), the European Space Operations Centre in Darmstadt (ESOC), its recent mission highlights, cutting-edge ground systems and future activities. Thomas also talks about his long and unique career as one of the astronauts with the longest total times in space worldwide.

Uğur Şahin, Johannes Gutenberg University of Mainz and Helmholtz Institute for Translational Oncology (HI-TRON), co-founder and CEO of BioNTech SE, The Dawn of Individualized Cancer Medicine: How novel (bio)technologies revolutionize the treatment of cancer The past decade has seen a significant shift from early targeted therapies to an era of personalized precision medicines, enhanced by a new generation of treatment approaches including immunotherapies, antibody-drug conjugates (ADCs), and T cell therapies. These, whether used alone or in combination, have the potential to redefine treatment efficacy. Moreover, the advent of therapeutic mRNA cancer vaccines signifies a move toward fully individualized treatment. These approaches are tailored to each patient's individual tumor profile and aim to target a spectrum of unique cancer mutations. Complementing this, AI is accelerating and informing the discovery and development of novel medicines, supporting the path toward individualized cancer medicine.

Randy Schekman, University of California, Berkeley, UCLA, Howard Hughes Medical Institute, and Stanford University. ASAP: Tackling Parkinson's disease with basic science ASAP (https://parkinsonsroadmap.org) seeks to probe the systems, cellular and molecular basis of PD. We focus on disease processes including the genes and proteins implicated in familial forms of PD, the neuroinflammatory processes that may influence disease progression, the neural circuity that feeds into and from dopaminergic neurons, and processes that may predict and precede development of the movement disorder. We have assembled teams of investigators who have developed meaningful collaborations on these key issues, and from these groups we have built an international consortium of interactive investigators who share a vision for open communication of ideas and results.

Harpreet Singh, Immatics N.V., Germany, Developing living drugs against cancer: TCR-T cell therapies T cells, a type of white blood cells, have become one of the most potent weapon in our arsenal to fight cancer. At academia, pharmaceutical and biotechnology companies, scientist are moving to new frontiers to make T cells safer, more precise and more potent to make even large established solid tumor melt. At the forefront of this development are autologous and allogeneic CAR-T and TCR-T cell therapies as well as bispecific antibodies and T cell receptors. The first of these therapies are being marketed and more are following in clinical trials with the goal to deliver a meaningful impact on the lives of cancer patients.

Thomas Südhof, Stanford University—Stanford University School of Medicine, United States of America: Science Integrity 2024 The practice of biomedical research is changing dramatically. Over the last two decades, the information technology (IT) revolution has not only enabled new experimental approaches that were previously unthinkable, such as affordable DNA sequencing at an unprecedented scale or rapid cryo-EM determinations of protein structures, but also transformed the way experimental results are stored and published. Although the IT revolution has greatly accelerated progress in most areas of biomedical research, it has also produced serious science integrity challenges. Motivated by personal experiences with some of these challenges, I would like to discuss some of the challenges as I view them from the perspective of a practicing scientist whose primary interest is to support progress in understanding diseases of the brain and to train young scientists for a career as independent researchers.

Arndt Steinmetz, Darmstadt University of Applied Sciences (h_da), Germany: h_da/EUt+ striving for a new type of University The European University of Technology (EUt+) is a consortium consisting of Darmstadt University of Applied Sciences (h_da) and eight other partners from all over Europe, which are among the leading technical, research-intensive universities of applied sciences. EUt+ is committed to offer common European degree programs, promote transnational mobility as well as democratic, civic values and critical thinking for the benefit of our societies. As pioneer EUt+ strives to establish a transnational institution, one European University with campuses from Western to Eastern Europe. A focus will be on the research dimension on h_da and the potentials of a holistic cooperation of EUt+.

Eckhard Thines, Institute of Molecular Physiology (imP), Johannes Gutenberg Universität Mainz, Germany: Natural products as lead structures for therapeutics—is there anything left to be discovered? It is not only since the discovery of penicillin that natural substances have been important as lead structures for therapeutics in pharmaceutical and crop protection research. Almost 100 years after the discovery of this antibiotic, the question arises as to whether new structures can still be expected despite the intensive search for biologically active metabolites from nature. In the post-genome age, molecular and bioinformatic methods can be used to investigate the extent to which the potential has already been exhausted. Alternative cultivation methods and the molecular manipulation of organisms are being discussed as an approach to identifying new lead structures from fungi.

Tanja Weil, Max Planck Institute for Polymer Research, Germany: Materials talking to cells We imagine reprogramming cells, the smallest units of life, by using synthetic molecules to create functional nanostructures in the cellular environment that can specifically influence or even improve cellular function. Could we teach molecules to recognize cancer cells and form nanostructures that prevent them from cell respiration? Could synthetic nanostructures interact with the cytoskeleton of T cells to change their shape and activate them? Cell therapy has already shown enormous therapeutic breakthroughs. The modification of living cells using intelligent molecular systems could make it possible to equip cells with entirely new structures and functions.

Otmar Wiestler, Helmholtz Association of German Research Centers: Grand challenges for biomedical research Biomedical research is facing an exciting period with a number of major challenges to be addressed. The following areas deserve special attention: (1) New strategies for early clinical proof of concept processes, (2) Transformation of health research and health care toward preventive medicine, (3) Full exploitation of innovative data management approaches toward AI-based precision medicine and prevention, (4) New era of biomedical engineering based on sensor technologies, smart materials, biotechnologies and AI. In his address, Prof. Dr. Otmar D. Wiestler, President of the Helmholtz Association, will highlight some of these major strategic challenges that need to be overcome: What are adequate institutional framework conditions to master these tasks? How can we explore the potential of cooperations between research, medicine and industry? What role do cutting edge technologies play? What examples for best practice can help to address these grand challenges?

Philipp Sebastian Wild, University Medical Center, German Center for Cardiovascular Research (DZHK), Institute of Molecular Biology (IMB) gGmbH, Germany: The curATime cluster—Individualized Medicine for Cardiovascular Disease Cardiovascular disease (CVD) represents a major, growing public health burden due to aging populations, but also climate change. The BMBF-funded future cluster curATime develops innovative therapeutics and diagnostic and prognostic tools for CVD. curATime combines the latest biotechnological approaches with artificial intelligence-driven research. New targets are identified and evaluated by integrating cutting-edge preclinical models with multi-level biodatabases with long-term follow-up of humans. A web-based platform is designed to facilitate complex biodata analyzes and novel AI tools are developed to handle computational challenges. This is the fundament for innovative pharmacological interventions including locoregional RNA-based applications.

Omar Yaghi, University of California-Berkeley, United States of America: AI and Reticular Chemistry Reticular Chemistry is linking molecular building blocks into crystalline extended structure such as metal-organic frameworks and covalent organic frameworks. Our efforts in using AI tools to identify the best water harvesting from air and carbon capture MOFs and COFs will be presented. Time permitting progress in molecular weaving will be covered.

Ignite Speakers

Claudio Flores, Mimotype Technologies GmbH, Berlin: Natural Light for a Brighter future: Ecologically Seamless Optical Materials for Green Photonics At Mimotype, we have developed a high-performing platform for bio-photonic materials systems by integrating biotechnology and organic electronics, pioneering bio-morphic sustainable tech applications. Central to our approach is the use of GFP, fluorescent proteins derived from marine bioluminescence. GFP emits green light upon exposure to UV or blue light. Our team has the capability to: (1) engineer GFP's genetic code to generate red, blue and green emitting variants, which we produce using bacteria, (2) encapsulate GFP variants in nanoshells and embed them into polymer matrices for deployment in real-world applications, (3) scale this pipeline using bioreactors. This approach facilitates the in-house production of high-performance, biodegradable, flexible, light-emitting materials. Collaborating with Humboldt University and supported by SPRIND, we are fine-tuning these materials for specific needs, like color-converting films for AgTech, OLED technologies and flexible wearables.

Jenny Hong Hoang, Larissa Esparza Borquez, Eleni Samaridou, Johanna Simon, Moritz Beck-Broichsitter, Merck Life Science KGaA: Bringing your nucleic acid modalities to the next level via customized delivery solutions mRNAs represent versatile tools to address a wide range of conditions, ranging from infectious diseases to cancer. Each RNA drug requires individual manufacturing and encapsulation conditions—both of which play a critical role in the success of the mRNA therapeutic. We have optimized both workflows to generate mRNA of high quality, purity, and efficacy which are then complemented by high-performing lipid nanoparticles (LNPs) designed using our novel and versatile library of ionizable lipids. To overcome limitations in manufacturing, we developed a novel process with increased flexibility for clinical applications. In contrast to conventional processes, our PCR-process demonstrated high-quality attributes and in vivo performance. Additionally, the process debottlenecks supply chains and offers scalability into commercial phases. Following successful manufacturing, mRNAs have to be packaged into LNPs to protect them from nucleases in the serum and to mediate successful delivery into target cells. Ionizable lipids (ILs) have been shown to be the key drivers of LNP efficacy, toxicity, and biodistribution and are critical in designing improved mRNA therapies. We have developed a versatile library of novel ILs that lead to safe and robust mRNA delivery in mice. These findings highlight the potential of PCR-based mRNA in combination with a novel promising library of IL for the successful design and acceleration of next-generation mRNA therapeutics, for multiple applications.

Hakan Kayal, Julius-Maximilians-University Würzburg, Germany: Research on Unidentified Anomalous Phenomena (UAP) in Germany With the establishment of a dedicated investigative agency, congressional hearings with military witnesses and a NASA research directorate, research into Unidentified Flying Objects (UFO)/Unidentified Anomalous Phenomena (UAP) has become politically and scientifically established in the USA. However, the UAP phenomenon is a worldwide phenomenon. The Interdisciplinary Research Center for Extraterrestrial Studies (IFEX) at the University of Würzburg in Germany is the first and only institute in Europe and Germany to officially promote UAP research within the academic canon of a renowned university. The presentation will shed light on the current background and explain the methods and projects that IFEX is using, also in cooperation with civilian UFO researchers. The importance of this research and its possible impact on mobility, space and renewable energy will be emphasized from different perspectives.

Sara S. Nogueira, Eike F. Joest, SAC Innovation, Merck Life Science KGaA: Unique platform to accelerate mRNA drug design mRNAs represent versatile tools to address a wide range of conditions, ranging from infectious diseases to cancer. Each RNA drug requires individual manufacturing and encapsulation conditions—both of which play a critical role in the success of the mRNA therapeutic. We have optimized both workflows to generate mRNA of high quality, purity, and efficacy which are then complemented by high-performing lipid nanoparticles (LNPs) designed using our novel and versatile library of ionizable lipids. To overcome limitations in manufacturing, we developed a novel process with increased flexibility for clinical applications. In contrast to conventional processes, our PCR-process demonstrated high-quality attributes and in vivo performance. Additionally, the process debottlenecks supply chains and offers scalability into commercial phases. Following successful manufacturing, mRNAs have to be packaged into LNPs to protect them from nucleases in the serum and to mediate successful delivery into target cells. Ionizable lipids (IL) have been shown to be the key drivers of LNP efficacy, toxicity and biodistribution and are critical in designing improved mRNA therapies. We've developed a versatile library of novel ILs that lead to safe and robust mRNA delivery in mice. These findings highlight the potential of PCR-based mRNA in combination with a novel promising library of IL for the successful design and acceleration of next-generation mRNA therapeutics, for multiple applications.

Meike Ramon,[1] University of Lausanne, Switzerland: Deepfakes: Challenges and Opportunities of Synthetic Identities The emergence of generative AI and synthetic content, such as deepfakes, presents complex challenges and opportunities for society and science. This presentation will present recent work on human-based deepfake processing—both among members of the general population as well as groups tasked with digital content verification: Berlin Police officers and Super-

[1] Winner of the Best Ignite Talk at the Curious 2024—Future Insight™ conference.

Recognizers (SRs). SRs possess an innate, exceptional ability to process facial identity, and are increasingly leveraged by law enforcement agencies across security applications. Our findings describe the limits of human ability to detect deepfakes, which are—at the behavioral level—genially processed similarly to real faces. I will discuss implications of these limitations and the potential benefits synthetic identities can play toward increasing fairness, privacy and transparency. I will outline a participatory action research project that merges science, practice and art to support fundamental and applied knowledge gain, as well as awareness regarding the risks and benefits of generative AI. To harness its potential responsibly and support the research and policy development that is needed, I encourage interested stakeholders to contribute to this transdisciplinary endeavor.

Arne Scheu, Valink Therapeutics Ltd, Imperial Translation & Innovation Hub, England: Scalable generation and screening of bispecific antibody drug-conjugates Antibody-drug conjugates (ADCs) are now an established modality in oncology. ADCs utilize the specific antigen binding of antibodies to deliver cytotoxic payloads to tumor cells. Recently, bispecific ADCs (BsADCs) have arisen as a promising approach to improve selectivity of ADCs, enhancing efficacy and safety. Despite broad potential, the discovery of BsADCs is restricted by the lack of technologies for rapid generation and screening of therapeutic candidates. We have developed a multi-part assembly approach for the generation of BsADCs with a 1-week turnaround time from design to HTS experiments (1000+/week). The low burden of BsADC generation enables discovery to start from functional assays, such as cytotoxicity, without requiring pre-assessment of unconjugated bispecific antibodies (e.g., internalization). Our platform can rapidly switch formats, binders, linkers, and payloads, enabling unbiased screening of BsADCs for rapid hypothesis generation and drug lead identification.

Thomas Endres, Dominik Goebel, Evonik Operations GmbH, Philip Dreier, Rebecca Matthes, Holger Frey, Johannes Gutenberg-University Mainz, Germany: Randomized poly(ethylene glycol) lipids—A new approach to overcome the "PEG dilemma" Poly(ethylene glycol) (PEG) lipids are used in lipid nanoparticles (LNPs), i.e., for COVID-19 vaccination. A concern of PEG is the formation of anti-PEG antibodies (APAs) that can cause loss of drug-functionality and side effects. A new class of lipids containing randomized PEG (rPEG) was established. A set of rPEG copolymers was synthesized and conjugated to lipid anchors. Interaction of rPEG lipids with APAs was quantified by enzyme-linked immunosorbent assay (ELISA). rPEG lipids were formulated to LNPs and tested in vitro. Target molecular weight and dispersity ($Đ < 1.15$) of rPEG polymers confirmed controlled polymerization. Purity (>95%) of rPEG lipids confirmed successful conjugation. ELISA confirmed significantly reduced interaction with APAs. rPEG LNPs had defined and controllable diameters (100–140 nm) and were found fully functional in vitro. The new class of rPEG lipids is applicable to industrial manufacturing processes and a candidate to fuel next generation mRNA therapeutics.

Hendrik Schneider, Merck KGaA, Darmstadt, Germany, Exploring the PROxAb Shuttle: A Plug and Play Solution for Targeted Antibody-Mediated PROTAC Delivery Proteolysis-targeted chimeras (PROTACs) have emerged as a therapeutic modality in recent years. However, oral bioavailability and cell type specificity remain challenging, particularly for PROTACs recruiting the von Hippel-Lindau (VHL) E3 ligase. To address these limitations, we present an unprecedented plug-and-play platform for the delivery of VHL-recruiting PROTACs. Our technology is based on camelid-derived antibody domains (VHHs) that can be easily engineered into existing therapeutic antibody scaffolds. The resulting targeted, bispecific fusion proteins can be complexed with PROTACs within minutes to form non-covalent antibody-PROTAC complexes at defined PROTAC-to-antibody ratios and have been named "PROxAb Shuttles". PROxAb shuttles can prolong the half-life of PROTACs from hours to days, demonstrate anti-tumor efficacy in vivo, show sustained target protein degradation in vivo, and have the potential for reloading in vivo to further enhance efficacy.

Start-up Fair

ActiTrexx Activated Treg for tolerance ActiTrexx is a clinical phase drug development company that develops a unique cellular and biologic therapies platform for overshooting T-cell mediated autoimmune and auto-inflammatory diseases. We target diseases with high unmet medical need such as Graft-vs-Host Disease and Multiple Sclerosis. ActiTrexx provides hyperactivated regulatory T cells (Treg) generated in vivo or in vitro. Our lead product, Actileucel, is manufactured within 24 h in an ultra-fast process, minimizing costs, and risks. Actileucel has orphan drug designation and is being tested in a clinical phase lb/II trial to prevent GvHD. It is the first allogeneic Treg cell therapy from a third-party donor that does not require tissue matching.

AdRegeneer Adregeneer is dedicated to filling a critical void in the medical landscape by providing innovative treatments that promote axonal regrowth and remyelination—breakthrough therapies currently unavailable elsewhere. To repair nervous system lesions, we employ a novel paradigm utilizing epigenetic factors to induce the repair program in neural cells. Our most advanced drug is now poised for GMP production and clinical trials for MS patients in development.

Algen Spun out from Nobel laureate Jennifer Doudna's Lab at UC Berkeley, Algen is a precision medicine company fusing high throughput, scaled CRISPR modulation with AI to reverse engineer disease trajectory for drug discovery. Algen applies proprietary novel technologies to develop innovative, patient- centric medicines with the largest impact on human health. Algen's world-class clinical single-cell RNA-sequencing dataverse deconvolutes disease-driving gene networks across oncology, immunology and metabolism. Our AI models decode causal biology, which we then combine with our best-in-class CRISPR gene-modulation system to discover combinatorial therapeutics that reverse disease. Algen's platform succeeded in internal programs and pharma partnerships.

BeesWall BeesWall saves business experts time during transformations by enhancing AI accuracy. It streamlines data governance, quality, and maintenance with an intuitive interface. Ideal for M&A, BeesWall ensures seamless data integration and smooth transitions, supporting strategic growth. BeesWall saves business experts time by enhancing AI accuracy and streamlining data governance, quality, and maintenance. Ideal for transformation projects, its intuitive interface ensures seamless data integration, supporting strategic growth.

Bloom Standard Bloom Standard is building the first self-driving ultrasound device that automatically scans the heart and lungs for abnormalities. This first wearable self-driving ultrasound technology for pediatric heart and lung screening reduces deaths associated with delayed, missed, or misdiagnosis, and poor follow-up. Bloom Standard's EVA (Echo View Automation) is a hands-free rapid scanning tool embedded with beam steering sensors and proprietary algorithms for real-time clinical decision support. This allows front-line health staff without training in sonography to screen the young patients in minutes. EVA implements primary care diagnostics, improves health equity, and reduces infant and child deaths.

BrainCures BrainCures biological intelligence approach de-risks and accelerates drug development for 1 in 6 people worldwide affected by long-term potentiation (LTP)-linked disorders such as depression (350 million), Alzheimer's disease (AD) (48 million), schizophrenia (21 million), Parkinson's disease (6 million), and rare diseases. The BRAINCURES Discovery Engine (BDE) platform eliminates preclinical failure and enables 69–97% phase III clinical success rates while shortening the drug development pathway by at least 4 years.

CardioSignal CardioSignal enables heart disease detection in 1 min using a smartphone—no additional hardware needed. A software as a medical device meant for self-monitoring, remote care, and point-of-care diagnostics. The technology is currently available for detecting signs of atrial fibrillation, soon followed by the detection of other major heart diseases. The CardioSignal technology uses smartphones as measuring devices, offering unparalleled scalability among those populations who are usually left out of digital advances. All you need to do is place the phone on your chest for a 1-min measurement. We accelerate heart disease detection for millions who are underdiagnosed or at risk of developing a serious heart condition.

Deep LS Deep LS was founded to improve and update the costly and time-consuming drug discovery process. Deep LS offers a software to design novel medical compounds by using state-of- the-art generative AI models. Traditional drug discovery methods are often slow and labor-intensive. Generative AI can rapidly propose novel molecular structures with potential therapeutic effects. This speed is crucial in the race against fast- evolving pathogens. Deep LS' software generates and optimizes ligands. Thanks to its latent diffusion based architecture it covers a larger chemical space at a higher quality of results compared to other de novo generative design solutions.

Imaginary Standing at the forefront of Europe's serious gaming industry, imaginary is renowned for crafting immersive environments tailored for physical and cognitive rehabilitation, now the cornerstone of their operations. Their flagship product REHABILITY is proven to enhance NHS organizations, improve patients' quality of life, and has a tangible social impact. Imaginary stands out for revolutionizing the silver economy with its simple, scalable, and cost-effective solution REHABILITY that reduces NHS expenses by 60%, enhances patient adherence, improves quality of life, boosts social well-being, and promotes environmental sustainability.

inContAlert inContAlert develops a sensor technology that measures the bladder filling level. The wearable is fixed ~2 cm above the pubic bone using a patch or strap and does not require surgery. The data collected from the patient is analyzed using deep learning algorithms. The bladder filling level is then displayed on an app which gives a warning when the filling level is excessively high. inContAlert provides a convenient and accurate bladder monitoring and can be worn continuously (24/7), as no gel is required. Patients place the device on their body and regain control over the bladder in their everyday lives. This lowers the risk of damage to health, relieves the burden on caregivers, and significantly reduces costs for the healthcare system.

INIA Biosciences INIA Biosciences is innovating systemic inflammation treatment, starting with patients with psoriasis. The venture was inspired by the founder's battles with skin inflammation, relentless itching, and dissatisfaction with current treatments. They aim to offer accessible alternatives to biologics and immunosuppressant drugs, addressing a common issue among patients. INIA Biosciences created a closed-loop, non-invasive, at-home wearable device using ultrasound to reduce systemic inflammation by targeting the splenic nerve and activating the cholinergic anti-inflammatory pathway. This technology avoids the side effects associated with biologics and immuno-suppression drugs, promoting the production of pro-resolving mediators (SPMs) essential for inflammation resolution and tissue repair, restoring immune balance.

Lignilabs LigniLabs is a founded spin-off from the Max Planck Institute For Polymer Research in Mainz. It develops and produces new solutions in the field of bio-based encapsulation of active substances in the biopolymer lignin for targeted drug delivery and drug release in plants. LigniLabs enables the first bio-based micro- and nano-encapsulation of cargos in the wood's own component lignin. Encapsulation in lignin makes it possible for the first time to reach pathogens in plants and release active substances via an enzymatic biosignal release mechanism, making previously incurable plant diseases curable.

Metatissue Advancing 3D Cell Culture Metatissue specializes in creating human protein-based products for cell culture, tissue engineering, regeneration, and disease modeling. Our biomaterials replicate realistic microenvironments, enhancing cell-to-cell and cell-to-matrix interactions for more accurate in vitro studies. Mimicking the diverse structural proteins and bioactive factors of living tissues remains a significant challenge. Metatissue's approach involves using ethically sourced human proteins to develop advanced, xeno-free 3D cell culture platforms aimed at achieving biomimetic environments for research and application. Metatissue provides advanced, xeno-free bioactive platforms derived from human platelet lysates, revolutionizing 3D cell culture and tissue engineering. Our proprietary hPLMA technology enhances biocompatibility, promoting cell proliferation and differentiation in physiologically relevant microenvironments.

NIMBLE Diagnostics NIMBLE Diagnostics is innovating with the WAVE™ System, a pioneering medical device for non-invasive monitoring of metallic implants. This technology aims to enhance patient care through early intervention, reducing complications and healthcare expenses. It enables clinicians to monitor coronary stent function, ensure medication adherence, and prevent cardiovascular issues like heart attacks. The WAVE™ System is a groundbreaking technology designed for monitoring implanted coronary stents, regardless of type or placement. It is non-invasive, portable, and cost-effective, eliminating the need for specialized interpretation due to automated signal processing algorithms. This makes it suitable for implementation in various clinical settings.

OmicsChart OmicsChart is a translational cancer research platform integrating multi-omics patient data from public, proprietary and in-house sources and oncology-tailored analytics. OmicsChart PREON enables translational cancer scientists to test hypotheses and gain insights from harmonized public, proprietary and in-house biomarker data in just a few clicks. Built on federated data integration, oncology-driven ML/AI, and biologist-friendly scientific visualizations, our solution reduces the time required for data acquisition and analysis by 150 times compared to current practices.

Pheal For many people living with heart failure, symptoms can continue to worsen despite receiving guideline-recommended therapies. A close monitoring of patients is the key—but this cannot be accomplished by healthcare providers. Pheal's Smart Patch is the heart failure patient's invisible helper. It continuously monitors and analyzes risk factors for heart failure in real time. Patients and healthcare providers are automatically notified when critical trends are detected, even before symptoms occur, allowing timely intervention to prevent health deterioration. Our Smart Patch provides a window into your body using advanced embedded sensor technology. This microneedle-free patch enables multiparametric, continuous measurement of biomarkers directly from the patient's blood without any delay. Data interpretation algorithms constantly screen for trends, functioning like a smart, real-time blood lab.

PlantShield PlantShield develops a nature-proven, plant material-based, biological fungicide that offers an effective, biodegradable, and risk less alternative with rapid, cost-effective, and scalable production. It was developed by a team of two founders with 35+ years of combined experience in plant pathology, bio-fungicide industry R&D, circular economy, finance, marketing, and operations. Through our biomimetics-based approach, our prototype is effective against fungal pathogens and completely safe for the operators and environment. The source plant species are grown commercially worldwide, enabling a scalable and commercially viable production within a short time frame. This allows for side stream valorization and fully sustainable production.

re.solution re.solution is on a mission to establish a circular value chain for polyester-containing textiles. Our chemical recycling process addresses the millions of tons of fossil-based polyester textiles that are produced and incinerated every year. The water-based solution enables an economic recovery of the resources from mixed fiber textiles—at a low chemical, water, and carbon footprint. Hydrolysis is a robust depolymerization technology for PET but consumes vast amounts of auxiliary chemicals producing waste water and salt. re.solution developed a sustainable downstream processing technology to recover the PET monomers. The in-process recycling of chemicals enables a nearly waste-free process, making the recycled feedstock sustainable and price competitive versus virgin feedstock.

Seedsight Seedsight is dedicated to curating a comprehensive digital repository of seeds and grains signatures, meticulously validated through biomolecular and biophysical analyzes provided by our platform in almost real time. Our mission is to equip clients with the knowledge necessary to make well-informed decisions in raw material, commodities procurement and during food processing for optimizing sustainable production at the highest rate. Our platform optimizes quality-check efficiency, significantly reduces procurement costs and increases productivity and sales with an 8% reduction in costs related to inefficient screening and up to 89% reduction in procurement time.

Workerbase Workerbase is a Munich-based technology company that helps leading manufacturers to create the agile factory. Our Connected Worker Platform empowers factory workers and enables flexible and sustainable production processes. We enable a fast digital transformation in factories. Unified interface: One application platform for all worker activities. Workflow automation: Leverage new and existing data for real-time workflows. Use your existing systems: Connect existing systems without any migration. Blended Human and Artificial Intelligence: GenAI workflows to enhance decision-making and task efficiency.

Presented Online and On-Site Posters

LabsToGo—the most sustainable lab of the future. Gertrud E. Morlock, Wolfgang Schwack, Maria C. Ochoa Romero, Kevin Jakob, Justus Liebig University Giessen, Germany Abstract: You think all the functionalities of devices in two fully equipped laboratories cannot be miniaturized to a small-scaled all-in-one mobile system? The open-source 2LabsToGo consolidated not only two fully equipped laboratories (chromatography laboratory and biological assay laboratory) but also combines complementary technologies on the same analysis surface, which explains its strength. The high degree of dematerialization is unsurpassed. You think your own personal care products in daily use are safe? No sample preparation was carried out to keep products as intact as possible. After separation and bioassay detection, previously unknown hazardous compounds were detected in lifestyle products of daily use, such as perfumes, cosmetics and skin care products. Ready for a change? The powerful 2LabsToGo prioritization strategy on detrimental ingredients brings breakthrough insights and novel application benefits in transforming markets to fill the new gap for healthier products. It is crucial to solving major challenges facing humanity, such as identifying triggers of chronic diseases and providing healthier lifestyle products and food as well as clean water.

Revolutionizing Rural Healthcare: A Unified Operating System. Zafar Mirza, Emun Abdu MD, Zahid Ali, TECHMEDO Inc. Abstract: TECHMEDO has developed an innovative Healthcare Operating System that revolutionizes primary care delivery in underserved rural areas globally. By leveraging clinical algorithms, AI-powered decision support, and care plans, the platform improves efficiency, enhances patient care, and reduces physician burnout. In 8 months, TECHMEDO has served over two million patients, primarily women and children, in under-privileged regions. The solution streamlines workflows, enables informed clinical decisions, and facilitates preventative care. With a 300% reduction in patient clinic time and 30% decrease in physician burnout, TECHMEDO is making quality healthcare more accessible and affordable worldwide through its cutting-edge technology platform.

AI in architecture—the city of the future (illustrated by the example of Darmstadt in 2050 and 2100) Mike de Saldanha, Marie Kühn, Oliver Wieser, Maria Szücs, Wiebke Roxana Bremer, Torben Hartmann, Anastasya Dewi, Theresa Jung, Eva Zachmann, Christo Aldorino Rus, Ghofran Majied, Lütfü Kaan Isik, Mohammad Konbos, Mohammad Al Akrami, Hochschule Darmstadt, Germany Abstract: In an university project, we explored the utilization of artificial intelligence (AI) in architecture, culminating in a film created using both CAD and AI technologies. Through expert interviews, we crafted a vision for Darmstadt in 2050 and 2100. Despite its capacity to automate processes, identify issues, and optimize systems, AI remains underutilized in architecture. Our film delves into

AI's definition, applications, and impact on architecture, showcasing innovative tools such as the "Digital Twin" to enhance urban planning. Highlighting mobility, heritage preservation, and climate considerations, we advocate for architects and urban planners to harness AI's potential for more efficient and intelligent design solutions.

Electrifying chemical processes. Laura Lennartz, Jan B. Metternich, Patrik Stenner, Stephan Kohlstruk, Evonik Operations GmbH, Germany Abstract: The pathway toward the carbon neutral production of chemicals requires widespread innovation and the development of disruptive Next Generation technologies. As part of the Next Generation Sustainable research program of Evonik, researchers at process technology and RD&I are aspiring for the general electrification of chemical processes relying on modern electrochemical methods. Organic electrochemistry allows for the replacement of stoichiometric chemical reagents with in-situ generated highly reactive species. This approach provides expedient access to established Evonik products at reduced manufacturing costs, while simultaneously lowering the environmental impact.

How stakeholder engagement drives innovations in vaccine development. Irina Meln, Romina Di Marzo, European Vaccination Initiative Abstract: Vaccines play a critical role in global health. To implement innovative approaches (for example, in silico modeling) in the development of new modalities of vaccines or to optimize the bio-manufacturing of existing ones, regulatory approval is required. EVI drives dialog with national and international regulatory agencies to ensure that new innovative methods are included in regulatory frameworks and may be used when developed. When building new pan-European collaborative projects, EVI invites regulatory agencies to participate as project partners. This allows for a constant feedback loop. In addition, EVI organizes stakeholder meetings where developers can present the novel approach to a wide range of regulators and gain feedback about key requirements to be considered during development. After that, EVI promotes the publishing of meeting reports to widely disseminate knowledge to the scientific community.

Building Bridges: The Intersection of Research, Journalism, and Science Communication through the ScienceFluencer Award. Khadija Aichane, Lena Krämer, Gudrun Liebscher, Hannah Schiller, Leon Werner, Erik Zimmer, AG Science Fluencer, Junior-GBM, Society for Biochemistry and Molecular Biology e. V., Germany Abstract: How are research and journalism connected? In order to form one's own opinion on scientific topics, it is essential to present research results in a manner tailored to the target audience. Science communicator Nicolas Wöhrl aptly says "You don't have to be a professor to do science communication!" That's why the Junior-GBM created the Science Fluencer Award, designed to honor the contribution of young scientists (students, doctoral candidates, postdocs) with a prize at a public event. The award places particular emphasis on the effective communication of scientific concepts to the media and the general public. The applicants will face a jury, which consists of GBM members, (young) science journalists and communicators and a Junior GBM member. In this way, the format enables a bridge to be built between academic societies, industry, universities and those interested in science. Curious people and science enthusiasts are very welcome at the award ceremony! Further information can be found on our website at sciencefluencer.org.

The Art and Science of creating Entrepreneurial scientists. Rajiv Vaid Basaiawmoit, Aarhus University, Denmark Abstract: There is an urgent need to tackle some of the grand challenges facing society today. One approach to tackle grand challenges is breakthrough science and entrepreneurship—where cutting edge science is coupled with the resourcefulness and doggedness of an entrepreneur. However, the number of scientists turning into entrepreneurs is still a very low number and definitely far from what society needs today. Despite various initiatives both at a university, industry and government level, there is not enough supply coming in to meet these readily available infrastructures and resources. The problem is not engagement alone but maybe that of identity and mindset. I have pioneered a novel approach of embedding entrepreneurship into science disciplines and with a science-first, business-second strategy enveloped within a gamified experience. We are now seeing this approach pay dividends as we see more science entrepreneurs as a result of this but also more breakthrough science in itself.

Membrane filter selection for methods used to characterize microplastics in the environment. Lindsay D. Lozeau (1), Marican Dube (1), Kevin Sydlowski (1), Ranjani Muralidharan (2) (1) EMD Millipore Corporation, USA, (2) Sigma Aldrich, Inc., USA Abstract: The ubiquitous presence of microplastics and growing evidence of their adverse effects on human health has propelled them to the forefront of global regulatory attention. Many techniques are used to detect microplastics in the environment; most protocols include filtration. Notably, every technique has unique filter requirements. Thus, to better understand the role of filters in these methods, studies of various filter types were performed: Particle retention to create microplastics analysis grade water. Throughput for oil separation of microplastics, Handleability, particle recovery, and background signals for imaging methods. Filter properties for chemical digestion methods to handle high particulate matrices, and as a substrate for FTIR spectroscopy, Raman spectroscopy, and GC with pyrolysis were also collated to determine suitable filters by method. These studies provide valuable insights on choosing the optimal filter for the accurate isolation and analysis of microplastics in the environment.

Bridging the Gap: Challenging Gender Disparities in STEM with New Insights. Anna Wolfram (1,2,3,4,*) and Maik Wolfram-Schauerte (4) 1 University of Applied Sciences Berlin, Germany 2 Hertie Institute for Clinical Brain Research and University Hospital Tübingen, Germany 3 University of Tübingen, Tübingen, Germany 4 University of Castilla-La Mancha, Albacete, Spain Abstract: Science, technology, engineering, and mathematics (STEM) are crucial for developing knowledge-based societies, yet gender disparities persist, limiting human capital potential. The goal congruity theory suggests that women leave STEM due to a perceived mismatch between their communal goals and the nature of STEM fields. This study challenges this theory by exploring the leaky pipeline phenomenon in natural sciences and academia through quantitative and qualitative methods. Data from 2015 to 2021 show near-gender parity in early career stages but a sharp decline in female representation at senior levels, especially in professorships dominated by men. To understand the reasons behind female scientists' career choices, we conducted a qualitative study with 26 participants from diverse backgrounds across 17 fields and work experience in 79 countries. Contrary to the goal congruity theory, our qualitative findings reveal that female researchers perceive natural science and academia as overwhelmingly positive and communal. Their desire to stay and pursue permanent positions is driven by collaborative work environments, mentorship opportunities, and research freedom. Remarkably, the alignment of communal values within science significantly influences women's commitment to academic careers. This contradicts established theories, challenging the assumption that women leave STEM due to conflicts between personal (communal) goals and the nature of academia. Our research provides a robust, experience-based foundation to reevaluate the complexity of the leaky pipeline phenomenon in STEM, opening avenues for targeted interventions to foster gender equity in these fields.

Does ethics really stifle innovation? Simon Lucas, Martina Hueber, Jean Enno Charton, Bioethics and Digital Ethics, Merck KGaA, Darmstadt, Germany Abstract: A prevalent yet false notion among scientists posits that science is inherently value—neutral, implying that technology itself is neither good nor bad, but rather, it is the manner of its use or the context of its deployment that holds normative significance. Hence, many adhere to the view that ethics in science primarily serves to thwart morally questionable pursuits, while otherwise being seen as an unwarranted impediment to innovation. This perception is particularly pronounced within corporate spheres, where ethics is often perceived not only to snuff out innovation but also as a barrier to commercial success. However, far from stifling innovation, ethics has the potential to ignite scientific progress, particularly in domains rife with moral complexity. Here, ethics ensures that scientific endeavors can operate at peak efficiency, all while aligning with principles of justice or human welfare. Drawing upon the extensive expertise cultivated within our Bioethics and Digital Ethics Office at Merck, we will present case studies exemplifying our commitment to embedding ethics within our scientific endeavors. These include contentious fields such as stem cell research or the use of AI, where we have navigated intricate ethical conundrums while pushing the boundaries of scientific exploration.

ISyN: An integrative Initiative for Targeting the Complexity of Brain Function. Saleh Altahini (1), Albrecht Stroh (1,2) 1 Leibniz Institute for Resilience Research, Germany 2 University Medical Center Mainz, Germany Abstract:

The function of any brain, even that of a small fly, is too complex to be examined and understood by a single person or a single technique. The Initiative for Systems Analysis in Neuroscience (ISyN) is a multi-layered initiative based in Mainz, aimed at uniting the individual expertise of many local and international researchers toward harnessing the tremendous potential of new experimental and analytical techniques. ISyN currently has +100 members and offers shared computing resources, workshops and regular SystemsClubs. ISyN mainly focuses on how brain function (most importantly behavior) can be probed and predicted from the various layers of neuroscientific research (genes, cells, and networks), both in health and in disease. ISyN utilizes new analysis methods such as machine learning for detecting hidden principles of brain function. Using the acquired knowledge ISyN aims at pioneering a closed-loop interrogation, for real-time interpretation and alternation of brain function.

Exploring and steering the moral compass of LLMs. Alejandro Tlaie Ernst Strüngmann Institute for Neuroscience in Cooperation with Max Planck Society Abstract: Large language models (LLMs) have become central to advancing automation and decision-making across various sectors, raising significant ethical questions. This study proposes a comprehensive comparative analysis of the most advanced LLMs to assess their moral profiles. We subjected several state-of-the-art models to a selection of ethical dilemmas and found that all the proprietary ones are mostly utilitarian and all of the open-weights ones align mostly with values-based ethics. Furthermore, when using the Moral Foundations Questionnaire, all models we probed—except for Llama 2-7B—displayed a strong liberal bias. Lastly, in order to causally intervene in one of the studied models, we propose a novel similarity-specific activation steering technique. Using this method, we were able to reliably steer the model's moral compass to different ethical schools. All of these results showcase that there is an ethical dimension in already deployed LLMs, an aspect that is generally overlooked.

AI combats Cybercrime—Deep Learning for Forensics. Tobias Wirth (1), Andrey Guzhov (1), Matthias Muller (2), Tobias Fischer (2), Lucas Howes (3), Dominik Ospelt (3) 1 German Research Center for Artificial Intelligence (DFKI), Department of Smart Data & Knowledge Services, Kaiserslautern, Germany 2 The State Criminal Police Office of Rhineland-Palatinate, Mainz, Germany 3 Binary Impact GmbH, Niederelbert, Germany Abstract: The German Research Center for Artificial Intelligence (DFKI), The State Criminal Police of Rhineland-Palatinate and the software developer Binary Impact join forces to develop an AI-enabled assistant to support forensic experts. In this presentation we show how the AI-assistant (i) helps to classify a sheer mass of fragments of deleted files, (ii) clusters all fragments that belong to individual files, and (iii) in the final step puts together the "puzzle" of fragments to restore the digital evidence. We developed an AI pipeline built on top of state-of-the-art transformer architectures combined with a UI/UX frontend that is inspired by the best of gaming standards. The yet to be answered question will be "Can the re-assembled digital evidence stand up in court?". This research is funded by the project "Carve-DL" sponsored by The German Federal Ministry of Education and Research (BMBF).

Al-based Wood Species Detection in Paper and Fiber Materials for the Control of Global Deforestation. Henrike Stephani (1), Markus Rauhut (1), Lars Nieradzik (1), Thomas Stephani (1), Petra Gospodnetic (1), Andrea Olbrich (2), Jördis Sieburg-Rockel (2), Stephanie Helmling (2), Stephanie Wrage (2), Gerald Koch (2) 1 Fraunhofer Institute for Industrial Mathematics, Germany 2 Thünen Institute of Wood Research, Germany Abstract: We have developed a software that will be able to improve controls on global wood fiber product flows to protect forests. For this purpose, we have generated a software solution that uses Al to identify hardwood species in fibrous materials such as paper. Although the use of wood products has to be controlled (by law), currently this is only possible by visual-manual analysis by the so-called wood anatomists, i.e., highly trained biological experts. The team of Fraunhofer ITWM and the wood anatomists from Thünen Institut have worked together to generate an Al-based software solution. For the first time, the identification of hardwood species in microscopic images of fibrous materials can be done. We have generated a large reference database of the most important hard wood genera in combination with a flexible pipeline for easy annotation of characterizing elements. On this basis we have developed a modular deep learning approach to detect and identify the respective species in wood fiber products.

What molecule is it?—Structure elucidation by Infrared spectroscopy. Marvin Alberts (1), Amol Thakkar (1), Teodoro Laino (1), Frederik Fleissner (2) 1 IBM Research Europe, Switzerland 2 Merck Life Science KGaA, Darmstadt, Germany Abstract: Infrared spectroscopy (IR) is a widely used technology for quality screening and the investigation of suspicious samples due to its ease of use and speed. Whether testing the quality of a pharmaceutical drug, biomolecule, or solving a criminal case, IR spectroscopy plays a crucial role in the identification of unknown sample. To identify such a sample by its IR spectrum, the standard method compares the spectrum against a known reference. This can be achieved by acquiring the spectrum of a physical reference material or by using a library of pre-recorded spectra. Based on their similarity, a scientist can decide about the identity of the sample substance. Still, this approach is limited to the scope of the available dataset and any target not present in the library will result in incorrect matches and poor correlation. Recently, a machine learning approach utilizing a transformer model was proposed to directly predict the molecular structures of unknown substances [1] from the IR spectrum. Initially, this model was trained on both computational data and a relatively small experimental dataset of 3000 IR spectra. In this study, we explore the impact of augmenting the training dataset to include a larger ensemble of 17,000 experimental mid-IR spectra collected from pure substances in ATR configuration. Future applications of this method could facilitate the identification of uncommon derivatives or novel substances, such as Fentanyl analogs.

[1] Leveraging Infrared Spectroscopy for Automated Structure Elucidation, M. Alberts, T. Laino, A. Vaucher.

High-throughput APProach for age determination in Yeast. David Adam, Nicole Paczia, Max Planck Institute for Terrestrial Microbiology, Germany
Abstract: It is an undeniable fact that demographic change is taking place in our society. Currently, every second person in Germany is older than 45 and every fifth person is older than 66. The development in this country is exemplary of the global trend that can be observed above all in industrialized countries. As people are living longer and longer, the effects of age-related diseases are becoming increasingly important. In order to better understand and treat these, research on model organisms is essential. Conventional model organisms for research, such as mice, rats or fish, have a life expectancy of several years. Therefore, the search for new molecular mechanisms that can influence aging on a pharmacological level is very time-consuming. A model organism with a very short life cycle such as yeast would therefore be desirable. Yeasts are unicellular microorganisms that are used in the production of everyday products (bread, beer, wine, bioethanol, etc.) as well as pharmaceuticals such as vaccines. In addition to their industrial use, yeasts have numerous properties that also make them interesting for medical research. Our patented technology of a microfluidic chip makes it possible for the first time to determine the lifespan of a yeast cell in a high-throughput process. By establishing this technology, the full potential of the yeast cell can be utilized as a model organism in aging research to extend a healthy lifespan and as a new tool in drug screening. We offer high-throughput screening in the field of longevity. With the help of single-cell yeast and our patented microfluidic chip, we hope that scientists and drug developers could analyze a large number of substances for their life-prolonging properties within a few weeks.

A Unique and Holistic Platform to Accelerate mRNA Therapeutics' Design. Eike F. Joest, Sara S. Nogueira, Aleksej Turockin, Aditi Mehta, Merck Life Science KGaA, Life Science Service, Synthesis & ADC CDMO, SAC Innovation Abstract: mRNAs represent versatile pharmaceutical tools. RNA drugs require unique manufacturing and encapsulation conditions—both are critical for success of mRNA therapeutics. We optimized both workflows to generate mRNA of high quality complemented by novel proprietary lipid nanoparticles (LNP). We developed a novel manufacturing process with maximal flexibility for clinical applications. In contrast to conventional processes, our PCR-based concept demonstrated high-quality attributes and in vivo performance. It further debottlenecks supply chains and offers scalability into commercial phases. Produced mRNAs are packaged into LNP to mediate cellular uptake. Ionizable lipids (IL) are the key components of LNP and critical for improved mRNA therapies. We developed a versatile library of novel ILs that lead to safe and robust mRNA delivery. Our results highlight the potential of PCR-based mRNA manufacturing in combination with a novel IL library for design of next-generation mRNA therapies.

Optimizing (GMP) Manufacturing of Lipid Nanoparticles. Christoph Hein (1), Gregor Dürre (1), Annika Brehmer(1), Bernhard Bobusch(2), Oliver Krüger-Knoll (2), Jens H. Wintering (2) 1 Fraunhofer-Institute for Production Systems and Design Technology (IPK) 2 FDX Fluid Dynamix GmbH Abstract: RNA-based technologies are a particular focus of current vaccine and therapy development. Well-characterized manufacturing processes exist for the production of COVID-19 mRNA vaccines to ensure safe and effective production. However, there are limitations to the use of these manufacturing technologies in terms of particle properties and scalability. The GMP certified Fraunhofer Dynamic mixing technology (FDmix) can solve the scaling problem in particular. The technology platform makes it possible to produce lipid nanoparticles with volume flows of 5 mL/min up to 1500 mL/min with consistent particle characteristics. This means that the technology can be used from an early R&D phase right through to production scaling. In all scales, a superior particle quality compared to commercial encapsulation systems can be achieved, as the system is based on a highly dynamic mixing of the mRNA and lipid solutions involved in LNP production. Thus, even at high lipid concentrations of 0.6 g/L, PDI values of 0.05 can be achieved with an encapsulation efficiency of 98% Comparative studies have also shown that the proportion of solid-core LNPs is almost 100%, no sub-visible particles are detectable and no filtration losses occur during post-processing. The technology platform not only solves existing limitations, it can also become an enabling technology for the development of highly specific or tissue-specific forms of treatment. Furthermore, initial results also show the potential for the development of new manufacturing processes in the field of cell and gene therapy. Based on data from the scientific evaluation and various industrial test series, not only is the functionality of the technology demonstrated, but the potential for the development of completely new material transport options is also shown using various examples.

Unleashing the potential of nutrient–microbe interactions for inflammation resolution medicine. Bodo Speckmann (1,*), Tanja Wagner (2), Paul M. Jordan (3), Oliver Werz (3), Manfred Wilhelm(4), Ellen Ehring (1), Albino Carrizzo (5), Koen Venema (6), Christiane Schön (2), Heike tom Dieck (1) 1 Evonik Operations GmbH, Germany 2 BioTeSys GmbH, Germany 3 Friedrich Schiller University Jena, Germany 4 Ulm University of Applied Sciences, Germany 5 University of Salerno, Italy 6 Wageningen University, the Netherlands Abstract: Introduction: Chronic inflammation is implicated in various human diseases and aging-related conditions. Specialized pro-resolving mediators (SPM) and their intermediate lipid mediators (LM) play a crucial role in resolving inflammation. These LM derive endogenously from omega-3 polyunsaturated fatty acids (n-3 PUFA) and contribute to their inflammation- resolving effects. Dysfunctional LM biosynthesis may explain the inconclusive efficacy data from n-3 PUFA interventions. Here, we conceived a synbiotic strategy to improven-3 PUFA conversion to LM through microbial metabolism. Methods: Using dynamic models of the intestinal tract, we assessed the LM production capacity of SynQ3 (a synbiotic n-3 PUFA plus Bacillus megaterium DSM 32963 composition) compared to controls.

SynQ3 was tested against placebo and a fish oil control in a controlled human trial for its ability to increase plasma LM levels. In vitro assays explored wound healing, age-defying, and pro-resolving properties of a SynQ3 ferment. Results: SynQ3 facilitated LM production under simulated gastrointestinal conditions. Bacillus megaterium DSM 32963 metabolized n-3 PUFA to SPM and their precursors more effectively than n-3 PUFA lysine salts or esterified n-3 PUFA (fish oil). Human studies confirmed increased plasma levels of EPA- and DHA-derived metabolites upon SynQ3 supplementation. In inflamed colonic tissue, SynQ3 reduced pro-inflammatory cytokines and promoted M2 macrophage polarization. Migration and proliferation of keratinocytes were stimulated by a SynQ3 ferment. Discussion: Our synbiotic strategy offers self-sufficient pro-resolving LM production, promising applications in inflammation medicine and prevention of chronic inflammatory conditions.

Enzymatically Active, Ultrasound Responsive Protein Nanoparticles for enhanced drug delivery. Jamie Breen, Michael Gray, Constantin Coussios, Robert Carlisle, Brian Lyons, Institute of Biomedical Engineering (IBME), University of Oxford, UK Abstract: Less than 1% of intravenously administered cancer therapeutics typically reach the site of a tumor. As well as irregular vasculature and hypoxic conditions, effective tumoral penetration is often impeded due to the altered tumor extracellular matrix (ECM), consisting of a dense network of glycoproteins, proteoglycans, and fibrous proteins. We have developed enzymatically active ultrasound-responsive protein nanoparticles. These nanoparticles can be propelled into tumors using extracorporeally applied ultrasound. When exposed to ultrasound, the trapped air bubble in each nanoparticle "pops" undergoing a process known as inertial cavitation. Localized microstreaming is generated, facilitating "pumping" of the nanoparticles into the tumor.

Microbiota-host Interactions in atherothrombosis. Olga Dremova (1), Nadja Paeslack (1), Gwen Falony (2), Bharat Joshi (2), Iris Lade (3), Vladimir Wischnewski (3), Jessica Levin (3), Celine Müller (1), Markus Ingold (1), Nathalie Schön (1), Mariola Olkowicz (1), Philipp Wild (1), David Weber (3), Vincent ten Cate (1), Mustafa Diken (3), Sara Vieira- Silva (2), Wolfram Ruf (1), Christoph Reinhardt (1) 1 University Medical Center Mainz, Germany. 2 University Medical Center Mainz & Institute of Molecular Biology (IMB), Mainz, Germany 3 TRON gGmbH, Germany Abstract: The MicrobAlome project as part of the CurATime cluster aims to explore the impact of the gut microbiome on the development of atherothrombosis. To achieve this goal, we will take advantage of a germ-free mouse atherosclerosis model (Ldlr−/−) Since the hepatic endothelium is in contact with blood, and the liver serves as an organ that senses microbiota-derived metabolites, our research concentrates on identifying potential targets within hepatic endothelium. By conducting gene and protein expression analyzes of hepatic endothelium, we have identified potential biomarkers. Together with our collaboration partners at TRON, we will target these biomarkers in vitro and in vivo using the Ldlr mouse model. In collaboration with the IMMH and the CESM at the UM Mainz, we will link our research findings to human studies. By transplanting patient microbiomes into germ-free Ldlr−/− mice, this approach will result in a better understanding of the impact of the gut microbiome on atherogenesis.

High-throughput APProach for age determination in Yeast. Andrea Gaißler (1), Markus Schandar (1), Carolina De Biase Chatzis (1), Sarah Ostertag (1), Julius Fechner (1), Natalie Gebken (1), Martin Thoma (1), Michael Peter Langner (1), Andreas Traube (1), Thomas Bauernhansl (1,2) 1 Fraunhofer Institute for Manufacturing Engineering and Automation (IPA), Germany 2 University of Stuttgart, Allmandring 35, 70569 Stuttgart, Germany Abstract: There is a growing need for efficient production technologies for cell and gene therapies due to the increasing number of approvals for various diseases, including hematological cancers. CAR T cells are already on the market, and many other gene therapies show promising results. This drives the demand for a flexible and autonomous production system to reduce costs and increase standardization. Currently, these therapies are mainly produced manually or with all-in-one devices. We developed a modular system for cell and gene therapy production by breaking down the process into individual steps, each mapped onto separate modules. These modules operate with cassettes, allowing flexible use and enabling simultaneous production of various therapies using the same infrastructure. The cells are contained within a bio chamber in a closed "GMP-in-a-Box" system, eliminating the need for large cleanroom facilities and specialized staff. Each cassette includes all necessary materials for the process step. We designed bio chambers for genetic modification and expansion of T and NK cells, achieving up to 2–2.5 × 10^6 cells/ml in batch processes, with fed-batch processes under investigation. A sterile connector enables cell transfer between cassettes. This modular approach offers high flexibility and scalability, ideal for production near patients in clinics or specialized centers. We are developing the modules and planning to integrate online sensors to minimize system disruption and maximize material for treatment.

PeriShield: A Temporary, Protective Biofilm for the Therapy of Periodontitis. Lena Katharina Müller-Heupt (1), Tobias Kollmer (1), Eik Schiegnitz (1), Bilal Al-Nawas (1), Nadine Wiesmann-Imilowski (1), John Eckelt (2), Ralf Heermann (3,4),Christopher Synatschke (5), James Deschner (1) 1 University Medical Center Mainz, Germany 2 WEE-Solve GmbH, Germany 3 Johannes Gutenberg University Mainz, Germany 4 Institute for Biotechnology and Drug Research gGmbH (ibwf), Germany 5 Max Planck Institute for Polymer Research, Germany Abstract: Background: Periodontitis, affecting approximately 743 million people worldwide, is a leading cause of tooth loss and linked to systemic disease risks. Similarly, periimplantitis affects 10–20% of dental implants. Both conditions share pathogenetic pathways, with hypoxic environments in periodontal pockets and peri-implant bone promoting angerobic biofilm formation. Current therapies focus on biofilm removal, but bacterial recolonization can occur rapidly within 48 h post-therapy, before hypoxia and inflammation are resolved. Material and Methods: We developed a temporary adhesive film, made of oxygen-releasing hyaluronic acid modified with bio-inspired glue, which adheres to cleaned dental root surfaces or implants. Our data show this novel "protective biofilm" has antibacterial activity against anaerobic pathogens without cytotoxicity, unlike current adjuvants such as antiseptics or antibiotics. Results: Cell viability exceeding 80% was demonstrated for osteoblasts, human periodontal ligament cells (hPDL), and human umbilical vein endothelial cells. Additionally, the formulation promoted angiogenesis in the chorioallantoic membrane assay and upregulated bone formation genes (VEGF and BMP-2) in hypoxic hPDL cells. Surface analysis via scanning electron microscopy and low-kilovolt energy-dispersive X-ray spectroscopy confirmed no adverse effects on titanium implant integrity. Clinical Significance: Our in vitro results suggest this novel therapeutic approach can provide the necessary time for tissues to close the niche for angerobic bacteria and regulate the hypoxic environment, facilitating bone or periodontal ligament regrowth.

Seeing inside cells: Nanoscale Chemical Imaging of Intracellular Structure George Greaves, Holger W. Auner, Alexandra E. Porter and Chris C. Phillips, Imperial College London, UK Abstract: Optical microscopes' ~250-nm diffraction limited resolution is too coarse to see inside cells, and most of our knowledge of their internal organelles, collectively termed "ultrastructure", comes from EM. We have developed a radically new way of optically imaging them. This is the very machinery of life and we are the first people ever to see it optically, with a ~20-nm spatial resolution that beats diffraction by ~400×. We hope to discover new structures that are at present unknown to science. Our mid-IR light also gives us spectroscopic chemical contrast, and we can map out chemical moieties at the nanoscale. We chemically map intracellular structures in human multiple myeloma cells, we image the interaction between silicon theranostic nanoparticles and hippocampal neurons to study the way they may be used to transport drugs across the blood–brain barrier, and we demonstrate label-free mapping at wavelengths chosen to target the chemical signatures of drugs, proteins and nucleic acids.

Unlocking the predictive potential of cancer organoids with Al-powered computer vision and data processing. Abraham Lin, Edgar Cardenas De La Hoz, Maxim Le Compte, Tyler Gilcrest, Christophe Deben Orbits Oncology Inc. & University of Antwerp—Center for Oncological Research Abstract: For decades, cancer researchers have been working with tumor models that do not appropriately represent human patients. Research done on cells in a petri dish and animals are not bringing new cancer drugs to the market and have historically let to over 90% clinical trial failure, costing millions of lives and billions of dollars each year. Therefore, there is a dire need to capture the complexity of patients earlier in the lab to select the right drugs for clinical trials and successfully bring it to market. Patient-derived organoids are much better cancer models of human tumors, as they come directly from patient samples and retain unique characteristics of each patient. However, they are very hard to analyze. We have built a digital platform to analyze organoid data and derive patient-relevant insights from microscope images. Using AI-powered computer vision and data processing, we demonstrated that our analysis outperforms the industry standard analysis, CellTiterGlo3D, an endpoint biochemical assay [1–3]. In our retrospective clinical trial with pancreatic cancer patients, we demonstrated that our analysis of organoid drug response in the lab highly correlates with patient drug responses in the clinic [4]. Altogether, we aim to use our analysis to turn organoid data into digital twins of cancer patients to bring patients early into the drug discovery and development for predicting patient therapy response and identifying biomarkers. Acknowledgments: We thank our collaborators at the University of Antwerp and the University Hospital—Antwerp along with the Flemish Research Foundation for supporting our work. [1] Deben, C.et al. Auranofin repurposing for lung and pancreatic cancer: low CA12 expression as a marker of sensitivity in patient-derived organoids, with potentiated efficacy by AKT inhibition. Journal of Experimental & Clinical Cancer Research 43, 88 (2024). [2] Deben, C. et al. OrBITS: label-free and time-lapse monitoring of patient derived organoids for advanced drug screening. Cellular Oncology. 46, 299–314 (2023). [3] Le Compte, M. et al. Multiparametric Tumor Organoid Drug Screening Using Widefield Live-Cell Imaging for Bulk and Single-Organoid Analysis. JOVE. 190, e64434 (2022). [4] Le Compte, M. et al. Single-organoid analysis reveals clinically relevant treatment-resistant and invasive subclones in pancreatic cancer. npj Precision Oncology. 7, 128 (2023).

Salamander and newt skin peptides: Potential applications in drug research and medical biotechnology. Lorena Kröner, Marie-T. Hopp, University of Koblenz, Germany Abstract: The development of nature-derived peptides as drug candidates made great progress in the past decades. As such, amphibian skin secretions were described as a treasure trove of bioactive peptides with great pharmaceutical potential. Among these, frog and toad peptides were already considered as antibiotics. In contrast, the capability of skin peptides from salamanders and newts are overlooked although their bioactive activities seem to be promising for future pharmaceutical applications. Thus, we herein systematically compare the so far known salamander and newt skin peptides with respect to their chemical properties, structural features, and bioactivities. In total, 14 peptides were isolated from the skin of salamanders and newts so far. Among these peptides, 5 peptides exert, for example, an antimicrobial activity, 2 peptides an antioxidant activity, and one peptide an anti-inflammatory function. These results highlight the power of these peptides for future usage in health industry.

A fully digital population health program to detect silent atrial fibrillation. Lasse Leppäkorpi (1), Juulia Simonen (1), Ville Ingman (1), Juuso Blomster (2), Mikko Pankaala (1) 1 All authors are employed by CardioSignal, Finland 2 University of Turku, Finland Abstract Introduction: Atrial fibrillation is a cardiac arrhythmia impacting ~10 million in Europe and a significant cause of strokes. According to literature, atrial fibrillation is asymptomatic for >30%, and up to 2/3 of atrial fibrillation-related strokes are preventable with early detection and management. Purpose: Implement a population health program to identify silent atrial fibrillation in risk population (aged 50+ with risk factors). Refer to healthcare to confirm the diagnosis. Method: The program with a Nordic private healthcare provider enrolled 8245 participants. A volunteer subgroup of 1198 were assessed for a potential clinical follow-up. Participants got a 10-month access to the CardioSignal app, a class lla medical device utilizing gyrocardiography. Onboarding and measurements were done on the participant's smartphone at home. If signs of atrial fibrillation were detected, care provider's instructions were shown. Results: The sub-study (N = 1198) demonstrated 3.1% of confirmed new atrial fibrillation diagnoses. Mean age: 64, 57.5% women. 93% reported ease of use. Three reported a false positive caused by a high number of extrasystoles and high HRV in sinus rhythm. Conclusion: Atrial fibrillation screening using smartphones is well adopted to a large risk population and seem to yield new diagnoses.

Blood vessels are pioneers of bone regeneration in injured skull. M. Gabriele Bixel (1,*), Kishor K. Sivaraj (1), Melanie Timmen (2), Vishal Mohanakrishnan (1), Anusha Aravamudhan (1), Susanne Adams (1), Bong-Ihn Koh (1), Hyun-Woo Jeong (1), Kai Kruse (1), Richard Stange (2), and Ralf H. Adams (1) 1 Max Planck Institute for Molecular Biomedicine, Germany 2 University Hospital Münster, Germany Abstract: Living bone is fascinating due to its unique ability to adapt to mechanical stress and regenerate without scar tissue. During long bone regeneration after fracture, blood vessels and bone cells work together to gradually replace the initial stabilizing callus with ingrowing blood vessels and new bone tissue. Bone progenitor cells closely follow the course of the newly formed vessels and initiate bone formation, a process known as angiogenic-osteogenic coupling. But what happens when the skull is injured? Our research team was able to observe the healing of skull bone in vivo and visualize newly formed vascular and bone cells depositing bone matrix deep within the regenerating tissue at high resolution. After cranial bone injury, early vascular growth is uncoupled from invading osteoprogenitors and sprouting blood vessels vascularize the injury without co-migrating osteoprogenitors. At a later stage, osteogenic cells drive regenerative ossification as a multicellular layer from the injured bone edge by collectively invading the vascularized lesion. This type of bone healing is fundamentally different from the processes observed in long bones. Our findings may have clinical implications for bone regeneration and bioengineering approaches.

Cost-efficient, low-resource, and user-friendly diagnostic platform for rapid CSS BRIDE SPECIMEN and detection of Dengue virus: A Toehold switch Approach to Address Global Health Challenges. Nick Trettow, Kai Kabuth, Nadja Eisenbruch, Julian Kulenkampff, Melina Baur, Antona Schetle, Micha Becker, Annika Sonn, Conrad Knies, Timo Schütz, Moritz Domack, Mohamed Taha Soudani, Marlene Werner, Manuel Quezada, Maria Cadavid, Katrina Northcutt, Johanna Krettek, Fabio Elsner, Til Kossmann, Sarah Ullschmied, Aaron Rosowski, Olga Sheludko, Annelie Dommes, Malisa Lutz, Jaime Joachim Zimmermann Rodríguez, Vincent Wischhusen, Anara Rustamova,

Senem Altun, Max Schmidt. All contributing authors are associated with the CompuGene Student Lab, Technical University of Darmstadt Abstract: The global health challenge posed by tropical pathogens requires innovative diagnostic solutions, especially in low-resource settings. Effective disease management and control in both low-resource and industrialized countries are often hindered by the lack of cost-efficient, user-friendly diagnostic platforms for differential detection of these pathogens. Our research focuses on developing a diagnostic platform leveraging Toehold switches, a type of synthetic RNA that regulates gene expression. We designed several Toehold switches responsive to sequences in the genome of Dengue virus using GFP as reporter gene. In an in vivo assay in *E. coli* cells, three Toehold switches showed high fold activations when the corresponding trigger RNA was present demonstrating good potential as a key component in our diagnostic platform. The next step is to adapt this system for in vitro applications, enabling the development of a cell-free paper-based assay for user-friendly discrimination of tropical pathogens. Additionally, we are developing a cost-effective portable optical reader, currently in the prototyping phase, that will work in tandem with the colorimetric assay for high-fidelity signal quantification. Together, these approaches aim to address the urgent need for accessible diagnostic tools, especially in resource- limited settings, to combat the increasing burden of tropical infections.

The future is in vitro protein synthesis. Patrick Grossmann, Kilian Vogele Invitris GmbH, Germany Abstract: Cell-free protein synthesis (CFPS) enables protein production outside of living cells, utilizing solely DNA, buffer components, and the transcription and translation machinery extracted from cells. This offers faster protein production and provides greater control and flexibility in comparison to cell-based methods. Furthermore, CFPS systems allow the production of otherwise cytotoxic proteins. Thus, we utilize our cell-free extract, prepared from *E. coli*, to synthesize toxic proteins, antimicrobials, and highly structured protein assemblies such as bacteriophages and particles thereof as my data show. This poster illustrates how CFPS systems facilitate rapid prototyping and protein production by eliminating cumbersome cloning methods. Through optimizations and the direct use of template DNA, the time from DNA to protein can be covered within hours. Moreover, we highlight the versatility of the CFPS system as a host-independent platform for phage production targeting both gram-positive and -negative bacteria, including *Pseudomonas* sp., *Y. pestis*, and multidrug-resistant ESKAPE pathogens, supported by data from clinical isolates of K. pneumoniae. Our data demonstrates that our synthetic system is proficient in all stages of phage replication: from phage DNA replication, expression of phage proteins from DNA template, to the self-assembly of expressed phage proteins into functional phages. The versatile synthesis of complex protein assemblies facilitates rapid vaccine production utilizing phage particles as carriers for antigens. Furthermore, the system enables flexible scaling of the reaction process, suitable for high-throughput screenings via liquid handling or droplet microfluidics, down to pL scale, as supported by data from my thesis. This reduction in time, resources, and costs fosters sustainable screening processes.

Finally, CFPS can be directly scaled up without the need for tedious cell-line development and up-scaling, resulting in up to a 100-fold reduction in endotoxin levels compared to conventional bacterial protein production.

STING-driven cancer control upon induced TREX1 KO predicts efficacy of pharmacologic TREX1 inhibition Emilija Marinkovic (1*), Minyi Chen (1*), Nadja Schubert (2,3), Elif Dogan Dar (1), Janet Y. Leung (4), Jack Lohre (4), Jennifer M. Sahni (4), Christine Tun (4), Pavithra Rajeswaran (4), Tanja Mehlo-Jensen (4), Olivia Perng (4), C. Mark Hill (4), Pallavur Sivakumar (4), Michael Barnes (4), Rohit Malik (4), Rayk Behrendt (5), Axel Roers (1) 1 University Hospital Heidelberg, Germany 2 DKMS Life Science Lab gGmbH, Germany 3 TU Dresden, Germany 4 Bristol Myers Squibb, United States 5 University Hospital Bonn, Germany Abstract: Activation of local innate immunity is a key concept in cancer immunotherapy. The CGAS/STING pathway triggers type I IFN responses upon detection of cytosolic DNA and can enhance anti-tumor immunity. Optimal strategies for therapeutic CGAS/STING activation remain unclear. The DNase TREXI is essential for cellular DNA disposal and prevents spontaneous activation of CGAS-driven inflammation. We induced knock-out of Trex1 in mice ('TREX1 iKO') bearing syngeneic tumors. This resulted in immune cell infiltration of the tumor, but not other tissues, and improved type | IFN- and CD8+ T cell-dependent tumor control. TREXI iKO triggered an inflammatory switch of intra-tumoral myeloid cells with enhanced antigen presentation. The CD8+ T cell response was massively invigorated. Combination of TREXI iKO with checkpoint inhibition synergistically enabled complete rejection of most tumors. Low-level STING activation by pharmacological TREX1 inhibition is a promising therapeutic approach.

Counteracting microvascular dysfunction for healthy aging. M. Grunewald, S. Kumar, H. Sharife, E. Volinsky, A. Gileles-Hillel, T. Licht, A. Permyakova, L. Hinden, S. Azar, Y. Friedmann, P. Kupetz, R. Tzuberi, A. Anisimov, K. Alitalo, M. Horwitz, S. Leebhoff, O. Z. Khoma, R. Hlushchuk, V. Djonov, R. Abramovitch, J. Tam, E. Keshet, The Hebrew University, Jerusalem, Israel Abstract: As men and women live longer, they are also spending more years with chronic conditions to which older people are more vulnerable. Effective ways for delaying or slowing-down development of aging-associated frailties and diseases are urgently needed. Current "anti-aging" interventions target specific age-associated processes recognized as "hallmarks of aging" at the molecular and cellular level. Our approach stems from the fact that aging manifests differently in various organs and tissues that deteriorate at different pace, depending on their regenerative capacity and their functional reserve raising the interesting possibility that there might be a hierarchy during the aging process. Here we examine the proposition that vascular aging is a critical upstream process driving the deterioration of many other organs. We provide evidence that counteracting microvascular dysfunction is, on its own, sufficient to confer a holistic geroprotection, increasing both lifespan and healthspan.

Introducing Cell Organization and Disturbance-mapping using Image Activated Cell-profiling (CODIAC) as a new tool for high-throughput characterization of complex cellular phenotypes. Melanie Krause (1), Shengdi Li (1), Dan Lou (2,3), Krisztina Gubicza (2,3), Daniel Kirrmaier (2,3), Fabian Mulder (1), Michael Knop (2,3), Lars Steinmetz (1,4) 1 European Molecular Biology Laboratory (EMBL), Germany 2 German Cancer Research Center (DKFZ), Germany. 3 Zentrum für Molekulare Biologie der Universität Heidelberg (ZMBH), Germany 4 Stanford University School of Medicine, Stanford, USA & Stanford Genome Technology Center, USA Abstract: High-content screening utilizing flow cytometry and genomic sequencing are the state-of-the-art technologies in cell biological research. Recently, image-enabled cell sorting (ICS) has been demonstrated to add spatial and morphological information to classical cell sorting technology. Here we aim to utilize this technology to develop new ways to characterize protein phenotypes within complex cellular pools. Using ICS of fluorescently tagged protein libraries and machine learning, we have devised a novel way to conduct cell organization and disturbance-mapping using image activated cell-profiling (CODIAC). In ICS this method is able to reliably group cells with fluorescently-tagged proteins with similar visual phenotypes into sorting bins used for sequencing. Sorting a baseline distribution as well as cells with chemical perturbance, we are able to identify changes in protein localization and abundance in a pooled fashion at a much faster pace than previously established. We expect this method to have broad applications in the field of high-content screening for the identification of novel drug targets and various medical uses for disease treatment.

MyxoTech: offering a hyper-organized library of microbial dark matter chemistry to pharma companies. Peter Sullivan (1,2), Thomas Hesterkamp (2,3), Ronald Garcia (1,2,3), Jennifer Herrmann (1,2,3), Olga Kalinina (1,2,4), Rolf Müller (1,2,3,4) 1 Helmholtz Institute for Pharmaceutical Research Saarland (HIPS), Germany 2 Helmholtz Centre for Infection Research (HZ!), Germany 3 German Center for Infection Research (DZIF), Germany 4 Saarland University, Germany Abstract: Since the 1960s, researchers at the Helmholtz Center for Infection Research (HZI) and its predecessor have built an unprecedented library of microbial dark matter myxobacteria (phylum Myxococcota). MyxoTech, an HZI spin-off, is building a commercial offshoot of the library that will be made available to the pharmaceutical industry. Ixabepilone, a cancer drug derived from myxobacterial metabolite epothilone B, exemplifies the value of the phylum for drug discovery. Employing omics tools, MyxoTech is comprehensively mapping the distribution of specialized metabolism within the phylum. This will serve as an machine learning model's foundation to (1) strategically expand the library's unmatched chemical diversity and (2) systematically track each strain's impact on the library's pool of chemistry. Novel chemical space is essential for the next generation of therapeutics. With that, MyxoTech is seeking industry partners interested in probing a hyper-organized collection of unprospected natural products.

Your medical Co-Pilot. Heinz Wiendl and Vera Rödel Co-Founders of Prof. Valmed GmbH, Germany Abstract: I am Vera Rödel, CEO of Prof. Valmed®, and together with my Co-Founder Prof. Heinz Wiendl, I am excited to present our work and vision for the future of healthcare. At Prof. Valmed®, we are dedicated to digital innovation and committed to improving patients' lives through holistic, data-driven, and individualized healthcare solutions. Our flagship product, Prof. Valmed®, is an AI-based tool that delivers validated medical information through a user-friendly interface, empowering healthcare professionals to make informed decisions quickly and accurately. Additionally, our Valmed A(I)cademy® offers comprehensive knowledge and skills training to equip healthcare professionals for the digital age, ensuring they are well-prepared to navigate the complexities of modern medicine and deliver superior patient care. We believe that our expertise in AI-driven healthcare solutions can push the boundaries of digital healthcare and create innovative solutions that benefit both humans and animals globally. Together, we can drive the future of healthcare, making it more efficient, accurate, and accessible for all.

Renewable Energy Artificial Production of Hydrogen from rocks and slags: the RockyH2 solution. Eric C. Gaucher (1,3), Samuel Barbier (2,3), Muriel Andreani (2,3), Jean L. Gaucher (1,3) 1 Lavoisier H2 Geoconsult, France 2 Université Claude Bernard Lyon, France 1 3 RockyH2, France Abstract: Can we imitate the natural production of hydrogen from rocks? Can we improve this production? Can we produce industrial volumes? How can we limit energy consumption for this production? What are the most favorable rock types? All those questions are the basics of the challenge of the Artificial Production of Hydrogen from Rocks and Slags. Thus, we need to identify the reduced elements constitutive of minerals that will be able to scavenge oxygen from the water and then release H2. Among these, ferrous iron is a very good candidate, as it is one of the five major elements in the Earth's composition. Other metals (Cu, Cr, Zn, and Pb) in their reduced forms could be very interesting but will be restricted to abnormal geological accumulations. A relative abundant literature exists for one of the processes: the serpentinization. Temperature is a well-controlled parameters but the role of chemistry is unclear. The quest for low-cost efficient catalysts favoring Olivine and Pyroxene dissolution is launched with some first results with aluminum and nickel. Nickel appears to be promising, but excess of catalysts can also be a H2-killer. RockyH2 is a start-up dedicated to the production of hydrogen from rock and slag, developing its catalysts in its own laboratories.

It's alive! In vivo implementation of synthetic CO_2 fixation pathways in the model bacterium E. coli. Alberto Sánchez-Pascuala (1), Nicole Paczia (1), Tobias J. Erb (1,2) 1 Max Planck Institute for Terrestrial Microbiology, Germany. 3 Philipps-University Marburg, Germany Abstract: Synthetic biology approaches make possible the design of synthetic CO_2 fixation pathways never exhibited in nature. One example is the CETCH cycle, a synthetic CO_2 fixation pathway designed to operate faster and more efficient than the naturally evolved Calvin-Benson-Bassham cycle. The CETCH cycle has been extensively validated through in vitro approaches, being its in vivo implementation still under active development. Here, we present our challenging approach by employing Escherichia coli as model microorganism. Accommodating an intricate pathway in a complex metabolic background requires the use of genetic standardization and modularization. The whole enzymatic segment of the CETCH cycle was captured and rewired following the genetic standards of SEVA and MoClo systems. The materialized genetic/metabolic tool, in combination with a set of CETCH-auxotrophic *E. coli* reporter strains, allowed the stepwise implementation and fine-tune of the CETCH cycle genuinely running in vivo.

WE GO BEYOND with sustainable biosurfactants. Lisa Maus, Stefan Liebig Evonik Operations GmbH I Nutrition & Care, Germany Abstract: In today's life, there is a strong demand for natural and mild ingredients that are made from responsibly sourced materials. As a pioneer in glycolipids technologies Evonik has developed and built the world's first rhamnolipid manufacturing plant to produce biosurfactants with the highest standard of sustainability, gentleness, and functionality. These eco-friendly glycolipids are commercially derived through a bio- fermentation process using a responsibly sourced and 100% renewable feedstock. Alongside the strong sustainability profile, Evonik's glycolipids have excellent foaming, cleansing, and solubilization properties making them highly suited for various applications. With Evonik's portfolio of biosurfactants the cleaning revolution is only beginning. For home care and cosmetic applications, products are already developed. Evonik strives to create a platform of commercially available sustainable biosurfactants that embrace a future of sustainable ingredients in a broad range of applications.

Selective Lithium-Ion Conductive Ceramic Membranes for Efficient Lithium Recovery. Elisabeth Gorman, Christian Hying Evonik Operations GmbH, Germany Abstract: Evonik developed a selective lithium-ion conductive ceramic membrane that enables superior separation of lithium ions in an electrochemical membrane process with highest efficiency. The technology allows the direct isolation of pure lithium hydroxide, which is a valuable resource for battery production. This process has the potential to improve the sustainability of the lithium-ion battery industry by reducing waste and conserving resources. The membrane is designed for long-term durability, and the process can be operated continuously. The technology is currently in the incubation phase, with plans for further testing and scale-up. Overall, Evonik's innovative approach to lithium recovery has the potential to close the lithium cycle and improve the sustainability of electromobility.

Enabling Additives for Polymer-Based Electrolytes in Solid-State Batteries Jan Blankenburg, Christos Sarigiannidis Evonik Operations GmbH—Research, Development & Innovation—Creavis, Germany Abstract: Solid polymer electrolytes (SPE) are essential for enabling the next generation of batteries with higher energy density and safety. However, they still face challenges, particularly in ionic conductivity at low temperatures, due to the crystalline nature of PEO-based electrolytes. Addressing these issues, Evonik Creavis has developed a methacrylate with an amorphous, non-crystalline PEG-copolymer side chain. This monomer, being liquid at room temperature, simplifies processing and serves as a versatile precursor for SPE formulations. When utilized in a cured solid-state electrolyte, this novel monomer enhances ionic conductivity and lithium-ion transfer number, especially at ambient temperatures. Additionally, Evonik's surface-modified metal oxides, integrated into the electrolyte matrix, improve mechanical and thermal stability, safeguarding against short circuits and thermal runaway. These advancements promise to enable batteries with increased performance and safety.

Smart Materials New Metal Oxides to Improve Performance and Safety of Lithium-Ion Batteries. Franz Schmidt, Evonik Operations GmbH, Germany
Abstract: Lithium-ion batteries (LIBs) are widely used in portable electronic devices and electric vehicles due to their high-energy density and long cycle life. However, the performance and safety of LIBs is limited by the capacity and stability of the electrode and separator materials. In recent years, Evonik has investigated and successfully developed several metal oxides to enhance the safety, electrochemical performance, and lifetime of LIBs. These metal oxides can be used as thin and protective separator coating and as dry coating on electrode materials on the cathode and anode side. The poster focuses on the next-generation coating of cathode active materials with fumed metal oxides. It summarizes the synthesis and the effects of these coatings on capacity and cycle life. Also new development trends for lithium-ion conductive oxides will be presented.

Transient infrared nanoscopy resolves the millisecond photoswitching dynamics of single lipid vesicles in water. T. Gölz (1,*), E. Baù (1), J. Zhang (1), K. Kaltenecker (2) S. A. Maier (1,3,4), F. Keilmann (1), T. Lohmueller (1), 1 Ludwig-Maximilians-Universität München, 80539 Munich, GER 2 Attocube Systems AG, Germany 3 Monash University, Australia 4 Imperial College London, UK Abstract: S-SNOM is a state-of-the-art super-resolution infrared (IR) microscopy and spectroscopy technique with a resolution of about 20 nm. It is applied to a wide range of materials from 2D materials to biomolecules. However, the studies are performed in a dried state, making it impossible to investigate dynamic systems in their native environment. Recently, we introduced a new in situ s-SNOM method for studying complex dynamical phenomena, such as living biological cells in water over many hours. The method is based on an ultra-thin silicon nitride membrane (10 nm), enabling robust s-SNOM over hours. Here, we show that the method enables the investigation of actively triggered dynamic process in the form of photoswitchable lipid vesicles of submicrometer size in their aqueous environment. Vesicles of such a dimension are challenging to study with conventional fluorescence or phase contrast microscopy, which may further interfere with the photoswitching process of the vesicle. We demonstrate that we can actively photoswitch vesicles between two morphologies and simultaneously image them.

In addition, we report that it is possible to discriminate two photoisomeric states of the photoswitchable lipid vesicle based on intensity differences in their nano-FTIR spectrum. Finally, we implement a new transient SSNOM method to monitor fast dynamic processes by tracking the IR near-field signal traces of a single wavelength at a defined tip position on a vesicle down to 50-ms resolution. We believe our findings will inspire researchers in the field of near-field microscopy to use the in situ s-SNOM technique and the transient near-field signal trace to study complex time dynamics in their native environment on the nanoscale.

Nanodoped Electrospunned Biofunctionalized Scaffold: Advancements for Bone Defect Repair in WNT10A Rare Disease. Elise Perennes (1,2), Nicolas Anton (1,2), Thierry Vandamme (1,2), Ysia Idoux-Gillet (1), Rana Smaida (3), Morgane Meyer (3), Jacky Guoqiang Hua (1), Nadia Benkirane-Jessel (1,3), François Clauss (1,2), Guillaume Conzatti (1,2 *) 1 INSERM/ University of Strasbourg, France 2 University of Strasbourg, France 2 Lamina Therapeutics, France Abstract: In the context of the rare WNT10A disease, the severe dental phenotype associated with bone hypotrophy presents significant challenges for rehabilitation, due to both bone lesions and osteogenesis difficulties. Tissue engineering provides a promising solution through the development of advanced supports. While synthetic polymeric scaffolds offer customizable functions, they often lack inherent biological activity. To address this, we have developed a two-step electrospun composite process that incorporates protein-loaded nanoparticles into polymeric fibers with an encapsulation efficiency of nearly 98%. Our method maintains the structural integrity of the nanoparticles and proteins, enabling controlled kinetic release of proteins over a 14-day period. This scaffold supports the proliferation of human mesenchymal stem cells and promising biological environments. The developed device has the universal potential to incorporate a variety of biomolecules. This innovative bioactive support shows significant promise for treating WNT10A-associated bone hypotrophy and offers new avenues for clinical application in various pathological conditions.

Engineering of Megasynthetases for the Production of Novel Proteasome Inhibitors Leonard Präve (1,2), Wolfgang Kuttenlochner (3), Michael Groll (3) and Helge B. Bode (1,2,4,5) 1 Max-Planck-Institute for Terrestrial Microbiology, Germany 2 Goethe-University, Germany 3 Technical University of Munich, Germany 4 Phillips University Marburg, Germany 5 LOEWE Centre for Translational Biodiversity Genomics (LOEWE-TBG) & Senckenberg Gesellschaft für Naturforschung, Germany Abstract: Many clinically used drugs are derived from bacterial natural products (NPs) that often are produced through nonribosomal peptide synthetases (NRPS), megasynthetases that activate and join individual amino acids in an assembly line fashion. In our lab, these enzymes are engineered by merging different NPRS at specific positions. This results in hybrid enzymes producing novel compounds carrying desired modifications of the target NPs. In this work, we focused on the engineering of megasynthetases producing novel proteasome inhibitors, a pharmaceutical interesting compound class with currently three clinical drugs used for the treatment of myeloma. Derivatives of fellutamides and syrbactins, two NP classes with high proteasome inhibitory activity, were produced fully biosynthetically, isolated and biochemically characterized in their inhibitory potential. Thus, we did not only engineer NRPS but even generated functional artificial hybrids of NRPS and polyketide synthases (PKS).

Engineering metal coordination and characterization of conformational dynamics in a two-domain de novo protein scaffold. Paula Wagner-Egea (1), Florent Delhommel (1,2), Ghulam Mustafa(1), Florian Leiß-Maier (1), Lisa Klimper (1), Thomas Badmann (1), Michael Groll (1), Michael Sattler (1,2), Cathleen Zeymer (1) 1 Technical University of Munich, Germany 2 Helmholtz Munich, Germany Abstract: Incorporating metal cofactors into de novo-designed protein scaffolds is a promising strategy for generating biocatalysts that promote new-to-nature reactions [1]. However, the initial designs are often poorly active, frequently resulting from protein dynamics being suboptimal for catalysis and requiring intensive characterization and engineering for optimization. While the functional dynamics of numerous natural enzymes have been characterized in detail [2], we lack this knowledge for more complex de novo-designed systems. Here, we show an in-depth biochemical and biophysical characterization of a recently designed de novo protein that consists of two domains tethered by flexible linkers: the TIM-Ferredoxin Dimer (TFD) [3]. We identified the effects of temperature and metal binding on conformational distributions through metal binding studies, protein nuclear magnetic resonance spectroscopy, molecular dynamics simulations, and X-ray crystallography. We rationally engineered metal binding sites for lanthanides and various transition metal ions into the scaffold, modifying its conformational dynamics. Our de novo metalloprotein serves as a model to study the interplay between protein structure, function, and dynamics. The results will provide protein engineers with valuable insights, aiding in the development of tailored biocatalysts based on fully de novo- designed proteins. References: [1] A. S. Klein, C. Zeymer, PEDS, 2021, 34, gzab003. [2] N. Tokuriki, D.S. Tawfik, Science, 2009, 324(5924), p. 203–207. [3] S.J. Caldwell, et al., PNAS, 2020, 117(48), p. 30362–30369.

The Rheticus Project. Simon Beck, Thomas Bülter, Ummet Catmak, Martin Demler, Moritz Engemann, Jan Martin Grunwald, Thomas Haas, Juliane Merz, Florian Nattkamp, Julia Niewalda, Georg Schick, Nardana Sivendran Evonik Operations GmbH, Germany Abstract: Climate change is one of the most discussed topics in our society at the moment. It is a common understanding that CO_2 emissions contribute significantly to global warming. Therefore many different approaches are needed to limit or even reduce these emissions. In order to meet this demand, we have to find solutions, efficiently using our resources and at the same time reducing CO_2 emissions. In our present world, the natural photosynthesis is a crucial factor, generating compounds from CO_2 and sunlight and therefore reducing CO_2 in our atmosphere. The Rheticus project mimics the natural photosynthesis designing an artificial photosynthesis system. The energy efficiency, water utilization, and land use of the natural photosynthesis are limiting factors. Artificial photosynthesis improves these factors. Our concept of an artificial photosynthesis is based on the combination of electrolysis and biocatalysis. By this means, bulk chemicals as well as specialties are accessible. Therefore, this concept represents an additional interesting power to chemicals approach to preserve our current living standard in a sustainable manner.

Photorhabdus and Xenorhabdus—killing insects and making drugs. **Alexander Rill (1,2,*), and Helge B. Bode (1,2,3,4) 1 Max-Planck-Institute for Terrestrial Microbiology, Germany 2 Goethe-University Frankfurt, Germany 3 Phillips University Marburg, Germany 4 LOEWE Centre for Translational Biodiversity Genomics (LOEWE-TBG) & Senckenberg Gesellschaft für Naturforschung, Germany** Abstract: Bacteria of the genus Photorhabdus and Xenorhabdus are motile, Gram-negative bacteria that live in symbiosis with entomopathogenic nematodes. Due to their complex life cycle, they have a large number of natural products encoded in biosynthetic gene clusters (BGC). Genetic tools for this genus have been rare and applicable to only a few strains. Therefore, we developed a genetic toolbox based on versatile expression vectors and single plasmid based CRISPR-Cpfl genome editing vector for genetic manipulations of Photorhabdus and Xenorhabdus within a few days. Supported by this toolbox, our research aims to elucidate the complex role of natural products in the life cycle of Photorhabdus and Xenorhabdus. Recent pangenomic analyzes revealed that Photorhabdus and Xenorhabdus share a large fraction of their BGCs, which we consider as essential for the basic life cycle [1]. In addition, there are BGCs that are shared by only a few strains or such exclusively found in a single strain. These singletons represent an important focus of our activation efforts, as we believe they can provide important ecological benefits. This assumption is supported by the fact that most of these singletons are silent under laboratory conditions and that activation of some of them is a challenge that goes beyond "simple" activation. Combinatorial tests with other known biologically active natural products from the same strain showed potentiating effects, suggesting a synergistic role of these singletons derived compounds. [1] Shi, YM., [...] Bode, HB. Global analysis of biosynthetic gene clusters reveals conserved and unique natural products in entomopathogenic mematode-symbiotic bacteria. Nat. Chem. 14, 701–712 (2022).

Multifaceted Engineered Living Materials (ELMs): On-Demand Drug delivery, Biosensors, and Beyond. **Varun Sai Tadimarri (1,2,*), Marc Blanch Asensio (1,2), Dr. Priyanka Dhakane (1), Shrikrishnan Sankaran (1) 1 Leibniz Institute for New Materials, Germany 2 University of Saarland, Germany** Abstract: For decades, bacteria have been used as protein production hosts and are upscaled by employing advanced bioprocessing strategies. There is a growing interest in engineering lactic acid bacteria as live biotherapeutics that produce and deliver recombinant proteins at these sites in the body. The application of the engineered bacteria could be as a production host in the food industry or as a live biotherapeutic product (LBP) or as a engineered living material (ELM) wherein bacteria are being encapsulated within porous materials like hydrogels to improve their biosafety and survival within the body and act as a diagnostic biosensor or as a drug releasing device. In our work, we aim to expand the genetic toolkit and programmability of probiotic Lactobacillus and combine it with biocompatible material systems to develop engineered living materials (ELMs). Testing and characterizing different promoters, signal peptides, and inducible systems to develop smart and responsive microbes has been the primary goal apart from testing the behavior of these microbes within encapsulated material systems. The findings and derived insights will help in characterizing non-model bacterial species such as L. plantarum in terms of protein expression capabilities, apart from providing a guideline to improving their performance when developed as ELMs.

Mycelium-Based Biocomposites: Pioneering Sustainable Alternatives in the Materials Industry. Nazanin Saeidi, Alireza Javadian, Rebekka Volk (2), Carolina Metz, Sofiia Vinnik, Dirk E. Hebel, Karlsruhe Institute of Technology (KIT), Germany Abstract: The rising urgency to transition toward sustainable and circular material systems has led to significant advancements in the development of mycelium-based biocomposites. This research explores the potential of mycelium, the vegetative root network of fungi, as an innovative, bio-based, regenerative, and circular material platform for a wide range of applications. By blending chemistry and bioengineering, we utilize mycelium's natural properties to develop high-performance, bio-based materials. These can be used for insulation, replacing non-renewable materials like polyurethane foam and fiberglass, and in packaging as alternatives to plastics and Styrofoam. In furniture and construction, mycelium-based composite can serve as a sustainable substitute for wood-based panels such as MDF and particle board. Our approach involves cultivating mycelium on various organic waste substrates, including wood, agricultural byproducts like miscanthus and hemp, and furniture waste. The resulting mycelium- based biocomposites exhibit good mechanical properties, excellent thermal and acoustic insulation capabilities, and impressive biodegradability. These characteristics make them highly suitable for replacing conventional materials that are often resource-intensive and environmentally detrimental. The integration of mycelium-based materials into a circular economy framework highlights their potential to reduce carbon footprints, minimize waste, and support regenerative practices. By converting waste into valuable resources, this technology aligns with the principles of bio-based biotechnology and sustainable development. Moreover, mycelium's natural ability to self-assemble and form strong networks offers a unique advantage in producing scalable and customizable material solutions. This research presents a comprehensive analysis of the material properties, production processes, and potential applications of mycelium-based biocomposites. We also discuss the economic and environmental impacts of adopting these materials, highlighting their role in advancing a sustainable and ethical approach to material science. This innovative research not only addresses current environmental challenges but also paves the way for future advancements in sustainable materials technology for diverse range of industrial applications.

Benefits of using artificial intelligence for space operations. Gabriele De Canio, Evridiki Ntagiou, Peter Collins, James Eggleston, European Space Operations Centre, European Space Agency, Germany Abstract: Artificial intelligence (AI) has proven its value in multiple applications, especially in those related to automation, in fields ranging from finance to healthcare, from automotive to space. In 2021, we performed a detailed analysis to investigate areas where Al could be used to support automation of space operations and be scalable. Since then, we engineered a suite of Al-based solutions that encompass various domains, including satellite health monitoring, mission planning, and ground station operations. These solutions are adept at performing tasks such as anomaly detection, time-series forecasting, root-cause investigation, knowledge representation. In this poster, we delineate the achievements realized over the past 3 years, which substantiate the advantages of integrating Al in the automation of space operations. Furthermore, we discuss the hurdles encountered in the development and implementation of Al technologies within the traditionally cautious realm of space operations, which involves intricate technical and human elements.

E-Poster: SkQ1 Treatment Preserves Ovarian Reserve in Mice. Nikoleta Nikou, Maria López Panadés, and Ignasi Roig (*) Universitat Autonoma de Barcelona, Spain Abstract: Women are born with a defined number of follicles, the ovarian reserve, which decreases during their reproductive lifespan, and, finally, when it is depleted, menopause begins. Previously reported data from our lab revealed that the age-associated decrease in primordial follicles is linked to the activation of the DDR. Furthermore, these studies highlighted the importance of oxidative stress in regulating natural menopause in women. Considering all of the above, we attempted to evaluate the effect of an antioxidant treatment that could reverse the effects of ROS in the ovarian reserve preservation. To address this hypothesis, we administered the mitochondria-targeted antioxidant, SkQ1, to young C57BL/6 female mice. To evaluate SKQ1 efficacy, we administered ethanol, an inductor of ROS linked to a diminished ovarian reserve in humans and co-administered SkQ1 and ethanol. Our preliminary findings reveal a significant reduction in primordial follicles in mice exposed to ethanol, suggesting that ethanol consumption diminishes the ovarian reserve. In addition, the SkQ1 treatment could revert these effects on the ovarian reserve. To study the effects of the SkQ1 treatment in natural aging, we administered SkQ1 to young C57BL/6 female mice for 14 weeks. Untreated mice lost around 45% of their primordial follicles, contrary to the SkQ1-treated animals, in which the loss was just 22%. So, SkQ1 consumption rescued approximately 40% of the primordial follicles loss during natural aging. Based on these findings, we propose that the daily consumption of ethanol, even in small doses, could significantly affect the fertility status of mammalian females. However, an SkQ1 treatment could counter balance the ethanol effects and, also, reverts the natural aging effects and highly preserve the mammalian ovarian reserve.

E-Poster: Attacking therapy-evading cancer cells through a new molecular mechanism. Carina Arps-Forker, Nikos Kassapakis, Stephanos Androutsellis-Theotokis, Andreas Androutsellis-Theotokis, Innate Repair LTD Abstract: Aggressive brain tumors are some of the most deadly and hard-to-treat types of cancer. Life expectancy is ~1 year with devastating side effects by current therapies. A population of specialized "cancer stem cells" (CSCs) are at the root of this, as they evade available treatments and regenerate the tumor. To date they remain untargeted. Innate Repair's groundbreaking science effectively and specifically targets CSCs via a new mechanism. We discovered a molecular mechanism that is used by many tissues to regenerate after an injury but is also exploited by CSCs to regenerate the tumor. We are developing new-to-man, advanced and small molecule therapeutics that target this new mechanism as add-ons to current therapies.

E-Poster: The nature of consciousness: the Fermionic Mind Hypothesis. Eva Deli, Institute for Consciousness Studies (ICS), USA Abstract: The nature of consciousness, long-regarded as a philosophical problem, is becoming a fascinating scientific challenge inspired by the expected emergence of artificial consciousness. At its most fundamental level, consciousness, like large language models, is an endothermic information processor and intellect generator. We sought to define the central nature of consciousness by turning to physics. For example, sentiment contagion, or social hysteresis, can elucidate challenges in psychology and social science. We found that perception, formed by a reversible and probabilistic thermodynamic cycle, lends quantum characteristics, such as wave-particle duality, to cognition. Consciousness particle-like nature is a unified identity and a particle-like separation, which strives for constant comparison with the environment. This bias is a spin-like competitive effect, leading to hierarchical social structures. These and other findings indicate that consciousness represents a fermionic system. According to FMH, the brain adopts the environment's operational principles into a fermionic organization, turning consciousness into the fundamental particle of intellect. The fermionic mind hypothesis (FMH) recognizes the fermionic characteristics of consciousness. We use category theory to explore these analogies in more detail. Category theory allows us to confirm that two objects, A and B, in a category, are equivalent if and only if all of the relationships that A has with others in the category are the same as those of B. Based on our interpretation of the Yoneda lemma, consciousness is homomorphic to fermions. The FMH presents a paradigm shift in neuroscience, promising valuable tools for mental health, education, social policy, and artificial intelligence research.

E-Poster: Asclepios IV Mission—Lunar South Pole Analogue Astronaut Missions. Evandros Theodosiou (1), Clara Nogué (2), Federica Torre (3) 1 Imperial College London, UK 2 Bellvitge Biomedical Research Institute—IDIBELL 3 Scuola universitaria professionale della Svizzera italiana (SUPSI) Abstract: Asclepios is an international student association supervised by Space Laboratory, Switzerland. What makes the association so unique is that it is the only analog mission entirely organized and run by students. In order to make the mission as realistic as possible, however, we are mentored by a team of experts in the space field including Claude Nicollier, former ESA astronaut. Asclepios project was started 4 years ago and in July 2020 we ran our first mission. Six analog astronauts spent isolated inside the Grimsel test site tunnels, simulating life inside lunar lava tubes. Analog space missions are tests of space missions that take place in places on Earth with similarities to the space environment. The aim of these missions is to prepare astronauts for the exploration of extraterrestrial environments by allowing them to experience various situations such as isolation or exposure to a hostile environment. Selected analog astronauts from around the world meet regularly throughout the year to prepare for the mission in July. The training includes the development of several skills in the fields of survival, technology, communication and science. We strive to provide solid foundations for our selected analog astronauts benefiting their future careers and building the future generation of the Space industry. Asclepios IV mission is set to take place in July 2024 in Sasso San Gottardo, Switzerland.

E-Poster: Bioplastic production from mycelium-algae co- culture with omega-3 fatty acid at oil/water interface. Sohail Mahmood, Erika Young, Living in Silico Inc., Canada Abstract: Introduction: Breakthrough in biotechnology engineering of bioplastics, notably derived from algae and fungi, have revolutionized the development of a more sustainable and eco-friendly environment through recycling industrial refuse. The use of micro-algae-derived polymers in bioplastic production reduces the threat to marine health and greenhouse gas emissions. A major industrial refuse within nutriceuticals is expired omega-3 fatty acid oil. Thus, we attempted to develop a polymer from these sources along with phytoplankton debris from a co-culture of mycelium forming fungi-algae as a proof-of-concept for hydrogel production. Method: We cultured 25 mL of 80% confluent phytoplankton culture in distilled water with 10 mL of mycelium-forming fungi culture and uncharacterized algae with omega-3 fatty acid substrate in a 75-mL culture flask for 14 days at standard ambient temperature and pressure (SATP) with natural ambient light. Results: Our findings indicate production of an omega-3 fatty acid hydrogel at the water–oil interface from solidification of liquid oil and properties similar to hydrogel through mechanical testing after 14 days. Conclusion: Our study highlights a promising approach to bioplastic production as the algae–fungi co-culture is not as well-known when compared to their individual counterparts. The impact of these findings emphasizes the immense potential of bio-blending algae with fungi to form a sustainable and biodegradable bioplastic. Further research is required to enhance the efficiency and performance of the algae-fungi bioplastics to be applied in real-world settings. Better characterization of the fungi–algae is required to further bioengineer for productivity.

E-Poster: VirCAD—an HPC Platform for Viral Engineering. Abijith Augustine, Sang Lee, Yu Ning Huang, Stefan N. Lukianov, Salve Therapeutics, USA Abstract: Over 5% of newborns suffer from incurable, inherited genetic diseases. Many other debilitating non-congenital diseases with genetic components are activated by environmental triggers (autoimmune, cancer, tissue injury). Current gene therapies are mostly limited to plasmid-based and Adeno-associated virus variants with inefficient response rates and limited use. Better viral delivery methods would expand the available pharmacopeia to produce precision medicines for intractable and incurable genetic diseases. We are building VirCAD© (Virus Computer-Aided Design), an HPC-powered, cloud-accessible, AI/ML-capable bioCAD platform to mine, design, model, and test new viral drugs for improved cell and gene therapies, vaccines, oncolytics, antibiotics, and other categories. These modalities will treat the many inherited and acquired genetic disorders afflicting patients due to their therapeutic specificity, efficiency, and customizability.

E-Poster: Angiosperm Phenological Responsiveness in Relation to Climate Change: Unveiling Novel Insights and Implications for Climate Resilience. Mariam Elmais,Columbia University in the City of New York, USA Abstract Phenology, which refers to the timing of biological events, is of utmost importance for the survival and reproductive success of organisms and serves as an indicator of climate change effects. This study employs long-term temperature records and herbarium specimen data to examine the phenological responses to temperature changes in Ohio, USA. Flowering time is a pivotal event in the plant life cycle, exerting a strong influence on reproductive success and survival. Phenological shifts, such as earlier flowering, have been observed in response to rising temperatures, and our dataset, spanning 115 years, offers valuable insights into the broader reproductive phenology of plants. Understanding the effects of phenological changes on species interactions is crucial for predicting the consequences of climate change. The data analysis reveals a consistent pattern of phenological advancement in response to temperature changes, albeit with variations among species. Ultimately, our results indicate that temperatures in Ohio are rising and that phenological responses to anthropogenic climate change are heterogeneous within communities and across regions, with significant regional variability driven by phenotypic plasticity, local adaptations, and differences in species assemblages. This study represents a crucial step toward comprehending the impacts of changing flowering phenology on species diversity and ecosystem dynamics.

E-Poster: Exploring Global Geological Hydrogen Resources Using Satellite Technology. Yuan Zi, University of Houston, USA Abstract Utilizing public satellite imagery spanning the globe, this research endeavors to identify potential geological hydrogen reservoirs through the detection of distinctive features such as fairy circles and damp areas. By harnessing the power of satellite technology, we aim to map these geological formations on a global scale, unveiling previously untapped resources for renewable energy production. This innovative approach not only promises to expand our understanding of hydrogen-rich geological formations but also lays the groundwork for the development of a comprehensive global renewable energy map. Through the integration of satellite data analysis and geological exploration techniques, this study seeks to pave the way toward a more sustainable and energy-abundant future.

Acknowledgments We thank the abstracts co-chairs Sonja Hatz and Siegfried Schindler, the Poster Session Team: Caroline Lohoff, Claus-Peter Niesert, Nicole Eischer, and the Start-Up Fair Team: Anna Wolfram, Reem Assal, Dragana Savic, Markus Rückels, Julia Schueler, Jan Uphoff, Yulia Kargapolova. Contact .

The manufacturer's authorised representative in the EU is Springer Nature Customer Service Centre GmbH, Europaplatz 3, 69115 Heidelberg, Germany. If you have any concerns regarding our products, please contact ProductSafety@springernature.com

Printed and bound by CPI Group (UK) Ltd, Croydon, CR0 4YY
30/09/2026
02225571-0008